Air-Sea Interaction

Air-Sea Interaction: Laws and Mechanisms provides a comprehensive account of how the atmosphere and the ocean interact to control the global climate, what physical laws govern this interaction, and what are its prominent mechanisms. In recent years, air-sea interaction has emerged as a subject in its own right, encompassing small- and large-scale processes in both air and sea.

A novel feature of the book is the treatment of empirical laws of momentum, heat, and mass transfer, across the air-sea interface as well as across thermoclines, as laws of nonequilibrium thermodynamics, with focus on entropy production. Thermodynamics also underlies the treatment of the overturning circulations of the atmosphere and the ocean. Highlights are thermodynamic cycles, the important function of "hot towers" in drying out of moist air, and oceanic heat transport from the tropics to polar regions. By developing its subject from basic physical (thermodynamic) principles, the book is broadly accessible to a wide audience.

The book is mainly directed toward graduate students and research scientists in meteorology, oceanography, and environmental engineering. The book also will be of value on entry level courses in meteorology and oceanography, and to the broader physics community interested in the treatment of transfer laws, and thermodynamics of the atmosphere and ocean.

Gabriel Csanady is Professor Emeritus and former holder of an endowed Slover Chair of Oceanography in Old Dominion University, Norfolk, VA. He also served as a senior scientist at Woods Hole Oceanographic Institution and as chairman of the Department of Mechanical Engineering at the University of Waterloo. He has been an editor of the *Journal of Geophysical Research* and founder-editor of *Reidel Monographs on Environmental Fluid Mechanics*. He is author of three books: *Theory of Turbomachines* (1964, McGraw-Hill), *Turbulent Diffusion in the Environment* (1973, D. Reidel Publishing Company), and *Circulation in the Coastal Ocean* (1982, D. Reidel Publishing Company).

Air–Sea Interaction

Laws and Mechanisms

G. T. Csanady
Old Dominion University

Illustrations prepared by
Mary Gibson, Toronto

CAMBRIDGE
UNIVERSITY PRESS

32 Avenue of the Americas, New York NY 10013-2473, USA

Cambridge University Press is part of the University of Cambridge.

It furthers the University's mission by disseminating knowledge in the pursuit of education, learning and research at the highest international levels of excellence.

www.cambridge.org
Information on this title: www.cambridge.org/9780521796804

First published 2001

A catalogue record for this publication is available from the British Library

ISBN 978-0-521-79259-2 Hardback
ISBN 978-0-521-79680-4 Paperback

Contents

Chapter 1 **The Transfer Laws of the Air-Sea Interface** 1

1.1 Introduction 1
1.2 Flux and Resistance 3
1.2.1 *Momentum Transfer in Laminar Flow* 4
1.3 Turbulent Flow Over the Sea 7
1.3.1 *Turbulence, Eddies and Their Statistics* 7
1.3.2 *The Air-side Surface Layer* 9
1.3.3 *Properties of the Windsea* 11
1.4 Flux and Force in Air-Sea Momentum Transfer 13
1.4.1 *Charnock's Law* 14
1.4.2 *Sea Surface Roughness* 14
1.4.3 *Energy Dissipation* 15
1.4.4 *Buoyancy and Turbulence* 17
1.5 The Evidence on Momentum Transfer 21
1.5.1 *Methods and Problems of Observation* 21
1.5.2 *The Verdict of the Evidence* 22
1.5.3 *Other Influences* 25
1.6 Sensible and Latent Heat Transfer 28
1.6.1 *Transfer of "Sensible" Heat by Conduction* 29
1.6.2 *Transfer of Water Substance by Diffusion* 31
1.6.3 *Heat and Vapor Transfer in Turbulent Flow* 32
1.6.4 *Buoyancy Flux Correction* 35

1.6.5 *Observed Heat and Vapor Transfer Laws* 36
1.6.6 *Matrix of Transfer Laws* 40
1.6.7 *Entropy Production* 41
1.7 Air-Sea Gas Transfer 44
1.7.1 *Gas Transfer in Turbulent Flow* 45
1.7.2 *Methods and Problems of Observation* 46
1.7.3 *The Evidence on Gas Transfer* 48

Chapter 2 **Wind Waves and the Mechanisms of Air-Sea Transfer** 51

2.1 The Origin of Wind Waves 51
2.1.1 *Instability Theory* 54
2.1.2 *Properties of Instability Waves* 56
2.2 The Wind Wave Phenomenon 59
2.2.1 *Wave Measures* 62
2.2.2 *Wave Growth* 66
2.2.3 *The Tail of the Characteristic Wave* 71
2.2.4 *Short Wind Waves* 74
2.2.5 *Laboratory Studies of Short Waves* 76
2.3 The Breaking of Waves 81
2.3.1 *Momentum Transfer in a Breaking Wave* 82
2.4 Mechanisms of Scalar Property Transfer 86
2.4.1 *Water-side Resistance* 87
2.4.2 *Air-side Resistance* 90
2.5 Pathways of Air-Sea Momentum Transfer 92

Chapter 3 **Mixed Layers in Contact** 97

3.1 Mixed Layers, Thermoclines, and Hot Towers 97
3.2 Mixed Layer Turbulence 100
3.3 Laws of Entrainment 104
3.3.1 *Entrainment in a Mixed Layer Heated from Below* 105
3.3.2 *Mixed Layer Cooled from Above* 108
3.3.3 *Shear and Breaker Induced Entrainment* 110
3.4 A Tour of Mixed Layers 115
3.4.1 *The Atmospheric Mixed Layer Under the Trade Inversion* 116
3.4.2 *Stratocumulus-topped Mixed Layers* 120
3.4.3 *Oceanic Mixed Layers* 124
3.4.4 *Equatorial Upwelling* 129
3.5 Mixed Layer Interplay 132
3.5.1 *Mixed Layer Budgets* 133

3.5.2 *Atmospheric Temperature and Humidity Budgets* 136
3.5.3 *Oceanic Temperature Budget* 136
3.5.4 *Combined Budgets* 137
3.5.5 *Bunker's Air-Sea Interaction Cycles* 140

Chapter 4 **Hot Towers** 146

4.1 Thermodynamics of Atmospheric Hot Towers 147
4.1.1 *The Drying-out Process in Hot Towers* 148
4.1.2 *The Thermodynamic Cycle of the Overturning Circulation* 152
4.2 Ascent of Moist Air in Hot Towers 158
4.2.1 *Hot Tower Clusters* 160
4.2.2 *Squall Lines* 164
4.3 Hurricanes 167
4.3.1 *Entropy Sources in Hurricanes* 172
4.3.2 *Thermodynamic Cycle of Hurricanes* 175
4.4 Oceanic Deep Convection 178
4.4.1 *Observations of Oceanic Deep Convection* 181

Chapter 5 **The Ocean's WarmWaterSphere** 187

5.1 Oceanic Heat Gain and Loss 189
5.1.1 *Mechanisms of Heat Gain* 194
5.2 Oceanic Heat Transports 197
5.2.1 *Direct Estimates of Heat Transports* 198
5.2.2 *Syntheses of Meteorological Data* 199
5.3 Warm to Cold Water Conversion in the North Atlantic 204
5.3.1 *Cold to Warm Water Conversion* 205
5.4 The Ocean's Overturning Circulation 208
5.4.1 *The Role of the Tropical Atlantic* 211
5.4.2 *Heat Export from the Equatorial Atlantic* 213
5.5 What Drives the Overturning Circulation? 216
5.5.1 *CAPE Produced by Deep Convection* 217
5.5.2 *Density Flux and Pycnostads in the North Atlantic* 219

References 225
Index 237

Chapter 1

The Transfer Laws of the Air-Sea Interface

1.1 Introduction

Hurricane Edouard had just passed by Cape Cod when I wrote these lines, after giving us a good scare, and keeping meteorologists of local TV stations out of bed all night. Approaching on a track along the East Coast, Edouard remained a category 3 hurricane, with 180 km/h winds, from the tropics to latitude 38°N. This is where it left the warm waters of the Gulf Stream behind, quickly to lose its punch over the much cooler Mid Atlantic Bight, and to be degraded to category 1, with 130 km/h winds, still enough to uproot a few trees on the Cape.

Edouard's fury came from water vapor, as it ascends the "eye-walls" (Figure 1.1) that surround a hurricane's core, condensing and releasing its latent heat of evaporation. The heat makes the moist air buoyant, turning the eyewalls into a giant chimney with an incredibly strong draft. The draft sucks in sea-level air, causing it to spiral toward the core in destructive winds and to drive waters against nearby coasts in storm surges. The fast air flow over warm water also ensures intense heat and vapor transfer to the air, sustaining the hurricane's strength. Over colder water, where not enough water evaporates, the hurricane dies: The lifeblood of a hurricane is intense sea to air transfer of heat and water vapor. On the other hand, as hurricane winds whip the waters along, they transfer some of their momentum downward. The loss of momentum acts as a brake on the hurricane circulation, keeping the winds from completely getting out of hand.

A hurricane also mimics on a small-scale the global atmospheric circulation, which is similarly "fueled" by latent heat released from condensing water vapor. This happens in "hot towers," concentrated updrafts of the InterTropical Convergence Zone (ITCZ), and also in somewhat less vigorous updrafts within extratropical storms. Many of the

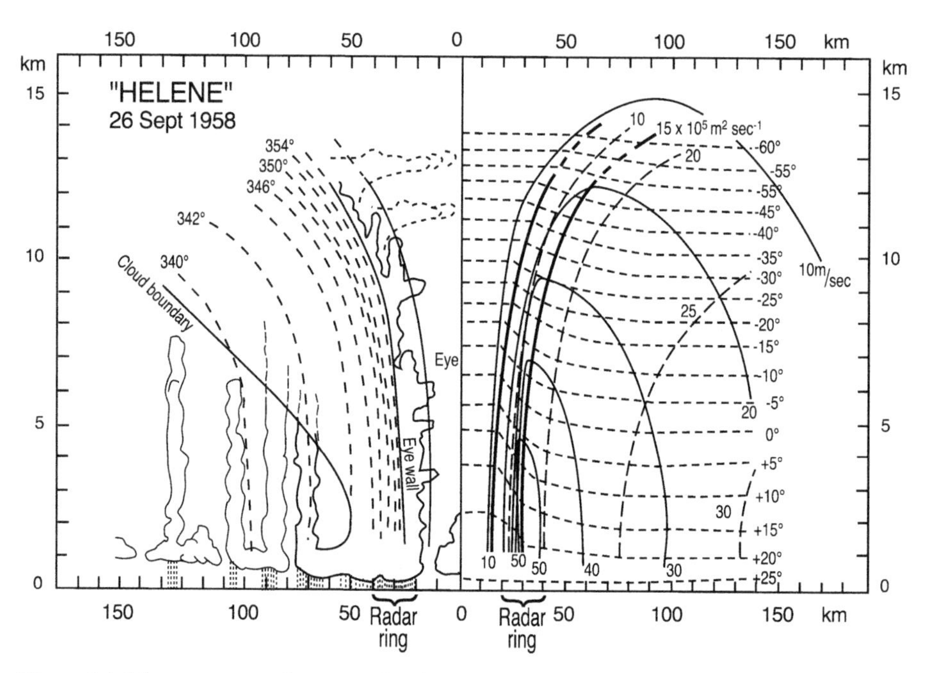

Figure 1.1 Mean structure of a mature hurricane ("Helene," 26 Sept. 1958) in cross section, supposing axial symmetry. The left-hand half shows the boundaries of the eye-wall (solid lines, bending outward with height) and illustrates the cloud structure. The broken lines are contours of constant "equivalent potential temperature," the absolute temperature in degrees Kelvin that the air would have with all of the latent heat in its vapor content released, and the pressure brought down to sea level pressure. In the right-hand half section, thin full lines are contours of constant wind speed in m s^{-1}(the thick lines repeat the eye wall boundaries), the broken lines are angular momentum contours, the dotted lines contours of temperature in °C. The maximum wind speed is in excess of 180 km/h. Note the stratiform cloud (dashed lines in the left half) extending to 13 km height, to the top of the troposphere, where the temperature is -55°C, $= 218$ K. Satellites see this "cloud-top" temperature. From Palmén and Newton (1969).

latter draw their vapor supply from the warm Gulf Stream and its Pacific counterpart, the Kuroshio, ocean currents transporting massive amounts of heat from warm to cold regions. Hot towers make their presence known to travelers crossing the equator, and wake them from their slumber when updrafts toss around their jetliner, as high as 10 or 12 km above sea level. Heat release in the updrafts, and compensating cooling and subsidence, are part of a thermodynamic cycle that energizes various atmospheric circulation systems, including the easterly winds of the tropics and subtropics, and the westerlies of mid-latitudes. The winds in turn sustain sea to air heat and vapor transfer, supplying the fuel, moist air, for the updrafts. The associated air to sea transfer of momentum from the winds is again the control on the strength of the atmospheric circulation.

Important to the operation of hurricanes and to large-scale atmospheric and ocean circulation systems is therefore in what amount, and by what mechanisms, momentum, heat and vapor pass from one medium to the other. The rates of transfer, per unit time and unit surface area, depend on a variety of conditions and processes; relationships between the rates and the variables influencing them are the "transfer laws" of the air-sea interface that we seek in this chapter. As all laws of physics, these too are distilled from observation, and, as most such laws, they are more or less accurate approximations. Their establishment requires painstaking work, hampered by difficulties of observation at sea. After nearly a century of research by many scientists from a variety of nations, there are still many uncertainties affecting the transfer laws.

1.2 Flux and Resistance

Transfers of momentum, heat and mass, are all *irreversible* processes. A number of texts deal with irreversible molecular processes of transfer, viscosity, heat conduction or diffusion, but their common thermodynamic characteristics have only engaged the interests of scientists relatively recently. De Groot's seminal synthesis (1963) bears the title "Thermodynamics of Irreversible Processes," while a later development (De Groot and Mazur 1984) is called "Non-equilibrium thermodynamics." These monographs develop the subject for molecular transfer processes, and show that their laws have the general form:

$$Flux = Force/Resistance \tag{1.1}$$

where the "Force" has the character of a potential gradient, the "Resistance" of inverse conductivity.

Irreversible processes change the entropy of the system in which they occur. Entropy changes because it flows in and out of the system, and also because internal irreversible processes generate it. The rate of entropy generation, the internal entropy "source" term in the entropy balance, is always positive, according to the Second Law of thermodynamics. "To relate the entropy source explicitly to the various irreversible processes that occur in the system" is the main preoccupation of nonequilibrium thermodynamics (De Groot and Mazur, 1984). When only one Force is acting, the entropy production rate is proportional to the product of Flux and Force. Absorbing the proportionality factor in the Force, entropy production can be made equal to the Flux-Force product. With several Fluxes and their conjugate Forces present, a similar standardization of the Forces yields the entropy production rate as the sum of the Flux-Force products, a result known as "Onsager's theorem."

The transfer laws of the air-sea interface are also relationships between Fluxes and Forces in the sense of nonequilibrium thermodynamics. They are, however, the result of an interaction between turbulent flows in air and water, and wind waves on the sea surface, and are more complex than linear relationships between a Flux and a Force with a constant Resistance. They are empirical laws of physics depending on material properties, properties of the turbulent flows in air and water, and of wind waves. Their usual form is an implicit Flux-Force relationship:

$$func\left(Flux, Force, \sum_{i=1}^{n} X_i\right) = 0 \tag{1.2}$$

where the X_i are n variables having measurable influence on the transfer law.

An important requirement of a physical law is that it must be independent of units of measurement. This dictates the use of a consistent system of units, and leads to Buckingham's theorem, according to which all physical laws are expressible as relationships between nondimensional combinations of variables, in appropriate products and quotients. Therefore, yet another way to state the transfer law of Equation 1.2 is:

$$func\left(\sum_{i=1}^{m} N_i\right) = 0 \tag{1.3}$$

where N_i are nondimensional combinations of the variables, including the Flux and the Force. Their number, m, is less than the $n + 2$ of Equation 1.2, usually by the number of measurement units in the dimensional relationship 1.2. Such formulations of the transfer laws are most useful if either the Flux or the Force appears in only one of the N_i; that variable can then be treated as the dependent one, the others deemed independent.

De Groot and Mazur (1984) discuss Flux-Force relationships valid locally, between heat flux and temperature gradient, and analogous quantities in other irreversible processes, while the most useful formulation of the air-sea transfer laws is between a property *difference* across a layer of air above the interface, and the flux across the interface, in what we might call a "bulk" relationship. To illustrate the difference between local and bulk relationships, and also to give a taste of the classical results of nonequilibrium thermodynamics, next we discuss viscous momentum transfer in a simple situation.

1.2.1 Momentum Transfer in Laminar Flow

Suppose that air and water are two semi-infinite viscous fluids in contact at the $z = 0$ plane, with the upper fluid impulsively accelerated to a velocity $u = U = const.$ at time $t = 0$. In the absence of other forces, and as long as the flow remains laminar and unidirectional, shear stress between layers accelerates the lower fluid while retarding the upper one. The shear stress τ, force per unit area, equals viscosity times velocity gradient (see e.g., Schlichting 1960):

$$\tau = \rho\nu\frac{\partial u}{\partial z} \tag{1.4}$$

where ρ is density and ν is kinematic viscosity. A layer of fluid between two levels δz apart experiences a net force equal to the difference in shear, which then accelerates the fluid:

$$\frac{\partial(\rho u)}{\partial t} = \frac{\partial \tau}{\partial z} \tag{1.5}$$

where the left-hand side is mass times acceleration or rate of change of horizontal momentum ρu. A legitimate interpretation of this relationship is that the shear stress is equivalent to vertical flux of horizontal momentum, the difference of which across the layer increases the local momentum.

In this light, the previous relationship, Equation 1.4, is now seen as one between a Flux (of momentum) and a Force, the gradient of the velocity $\partial u/\partial z$, a local law, valid at any level z. The dynamics is contained in Equation 1.5. Multiplying that equation by u, we arrive at the energy balance:

$$\frac{\partial(\rho u^2/2)}{\partial t} = u\frac{\partial \tau}{\partial z} \tag{1.6}$$

which, after rearrangement and substitution from Equation 1.4, transforms into:

$$\frac{\partial(\rho u^2/2)}{\partial t} = \frac{\partial}{\partial z}\left(\nu\frac{\partial}{\partial z}(\rho u^2/2)\right) - \rho\nu\left(\frac{\partial u}{\partial z}\right)^2. \tag{1.7}$$

The first term on the right is the divergence of viscosity times the gradient of kinetic energy, legitimately interpreted as energy flux. The divergence of this quantity signifies vertical energy transfer from one location to another, leaving the total energy unchanged. The second term, however, is always negative, and signifies loss of mechanical energy, its transformation into heat through viscosity. The heat added to the air or water increases its entropy at the rate of heat generation divided by absolute temperature. This then is the entropy source term, locally, level by level, equal to the product of the Force $\partial u/\partial z$ and (by Equation 1.5) the Flux τ, conforming to Onsager's theorem.

The fluid properties, viscosity and density, are constant in either medium, but change at the interface: They will bear indices a, w, for air above, water below. Writing down Equations 1.4 and 1.5 separately for air and water, and eliminating τ, we have two second order differential equations for u to solve. The boundary conditions are as follows: Far above the interface the velocity is the undisturbed U, far below it is zero. At the interface, the velocity and the shear stress are continuous. The solution follows the standard approach to such problems, see e.g., Carslaw and Jaeger (1959). The results are:

$$u_a = u_0 erfc\left(\frac{z}{2\sqrt{\nu_a t}}\right) + U erf\left(\frac{z}{2\sqrt{\nu_a t}}\right) \tag{1.8}$$

$$u_w = u_0 erfc\left(\frac{-z}{2\sqrt{\nu_w t}}\right)$$

where u_0 is the common interface velocity. The boundary condition of continuous interface stress yields a relationship for u_0:

$$\frac{u_0}{U} = \left(1 + \frac{\rho_w\sqrt{\nu_w}}{\rho_a\sqrt{\nu_a}}\right)^{-1}. \tag{1.9}$$

The solution represented by Equations 1.8 and 1.9 reveals the velocity distributions to be error functions and complementary error functions of the distance from the interface, portraying air-side and water-side boundary layers of thickness $2\sqrt{\nu t}$, which grow with the square root of time. The water-side velocities are much slower than the air-side ones: The typical value of u_0/U is $1/200$. This can be anticipated from Equation 1.5, which shows accelerations to be inversely proportional to density. The density of water is about 800 times greater than the density of air, balanced somewhat in Equation 1.9 by the kinematic viscosity of the air being some 16 times greater than that of water.

From the solution we find the value of the interface stress, alias momentum flux from air to water:

$$\tau_i = \frac{U}{R} \tag{1.10}$$

with

$$R = \frac{1}{\rho_a}\sqrt{\frac{\pi t}{\nu_a}} + \frac{1}{\rho_w}\sqrt{\frac{\pi t}{\nu_w}}.$$

The result is clearly of the form of Equation 1.1, constituting a bulk relationship between the interface momentum flux and the undisturbed velocity difference between air and water, which plays the role of the conjugate Force. The Resistance R consists of two additive components, identifiable as air-side and water-side resistance, respectively. Each component is proportional to the boundary layer thickness on that side, and inversely proportional to dynamic viscosity $\rho\nu$. With the values of material properties substituted, the air-side resistance turns out to be some 200 times greater than water-side resistance, so that the latter is for all practical purposes negligible.

The momentum transfer law must be reducible to a nondimensional form, containing fewer variables. One such form is:

$$C_D = \frac{2}{\sqrt{\pi}}\mathrm{Re}^{-1}\left(1 + [\rho_a/\rho_w\sqrt{\nu_a/\nu_w}]\right)^{-1} \tag{1.11}$$

with $C_D = \tau_i/\rho_a U^2$ a drag coefficient or nondimensional interface momentum flux, and $\mathrm{Re} = 2U\sqrt{\nu_a t}/\nu_a$ a Reynolds number based on air-side boundary layer thickness. Counting the density ratio and the viscosity ratio as two separate parameters, the nondimensional version of the transfer law contains four variables, versus seven in the dimensional formulation. The reduction by three corresponds to the three units of measurement – mass, time and length – quantifying the dimensional variables.

In Equation 1.11, the density-viscosity ratio term is small compared to unity, so that a sufficiently accurate form of the transfer law is the much simpler: $C_D = \frac{2}{\sqrt{\pi}}\mathrm{Re}^{-1}$. A lesson to be learned here is that not all variables playing a role in momentum transfer necessarily have a significant impact on the interface transfer law: Nobody could argue that the density or viscosity of water is irrelevant to momentum flux, yet neither significantly affects it in this example.

Does the bulk relationship, Equation 1.10, conform to Onsager's theorem? The total energy dissipation is the integral of the local value $\rho\nu(du/dz)^2$. Using the approximate formula taking into account air-side resistance only, neglecting u_0, and integrating on the air side from zero to infinity, we find the total dissipation to be $U^2/(\sqrt{2}R)$, or momentum Flux U/R times $U/(\sqrt{2})$. The latter is then the conjugate Force in the bulk version of the viscous transfer law.

The laminar flow example treated here is an overidealization of conditions near the sea surface, but its overriding weakness is that hydrodynamic instability causes laminar shear flow to break down into the chaotic motions of turbulence in a very short time. In turbulent flow, different and more complex laws govern momentum transfer. The one important feature of the laminar momentum transfer law that carries over into turbulent flow is that the air-side Resistance still dominates. Perhaps paradoxically, this is because, whichever way momentum gets across the interface, the light air still has a hard time moving the much heavier water around.

1.3 Turbulent Flow Over the Sea

1.3.1 Turbulence, Eddies and Their Statistics

Turbulence consists of a continuous succession of chaotic movements by parcels of fluid, analogous perhaps to molecular agitation, but occurring on a much larger than molecular scale. Moving parcels of fluid displace other fluid that eventually has to fill in the space vacated. This is known as continuity. Irregular and ephemeral closed flow structures arise in this manner, loosely called eddies. The details of eddy motion are complex, yet "stochastic" average properties of the flow (averages over many "realizations" in statistical theory, time-mean properties in practice) obey ascertainable laws, not unlike laws that quantify the macroscopic effects of molecular agitation.

The chaotic motions of turbulence are three-dimensional, so that at a fixed point there are velocity fluctuations along all three coordinate axes, u', v', w', even if the mean velocity has the same "alongwind" direction, $\overline{u} > 0, \overline{v} = 0, \overline{w} = 0$ (primes distinguish fluctuations from mean quantities carrying overbars). The mean square velocity fluctuations are then nonzero and their square roots provide measures of eddy velocity, a velocity "scale," such as $u_m = \sqrt{\overline{u'^2}}$. They also define the important Turbulent Kinetic Energy, TKE per unit mass in J kg^{-1}:

$$E_t = \frac{1}{2}(\overline{u'^2} + \overline{v'^2} + \overline{w'^2}) \tag{1.12}$$

Eddies also stir up the fluid, and if some fluid property is unevenly distributed, they try to equalize it. Thus, when mean flow momentum $\rho\overline{u}$ varies in the vertical, fluctuating vertical eddy motions of velocity w' bring faster fluid from the momentum-rich region, which locally appears as excess velocity, positive u'. Averaged, the effects of these eddy motions add up to eddy transport of momentum, $\rho\overline{u'w'}$, also known as Reynolds flux of momentum or Reynolds stress, after Osborne Reynolds who first formulated equations of motion for a turbulent fluid with Reynolds stresses included.

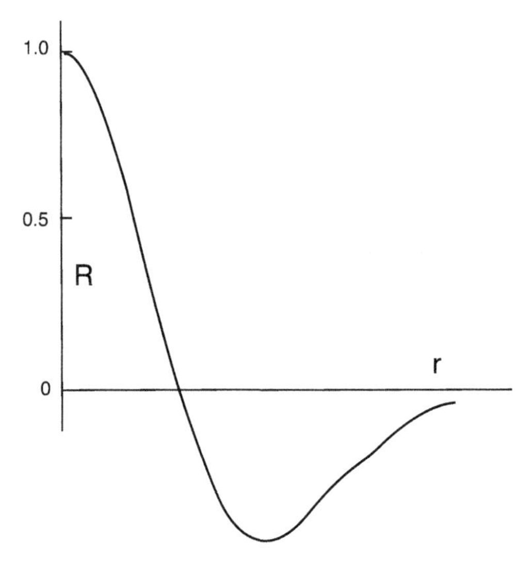

Figure 1.2 Typical correlation function of a turbulent velocity component, $\overline{w'(x)w'(x+r)}$, against distance r in the perpendicular direction. Adapted from Townsend (1956).

Another important turbulence property is characteristic eddy size. This can again be assigned only from statistical properties of the flow, traditionally from a two-point correlation function, such as $\overline{w'(x)w'(x+r)}$, the mean product of the vertical velocity component at two along-wind locations a distance r apart. See Townsend (1956) for a fuller discussion. Figure 1.2 shows the typical shape of such a correlation function, with a negative correlation loop required by continuity, on the principle of "what goes up must come down." The distance where the correlation function drops to zero is a measure of eddy size, or a "length-scale" of turbulence, say ℓ.

Correlation functions contain more information. According to a well-known theorem of statistics, a two-point velocity correlation function is the Fourier transform of an energy spectrum that assigns portions of kinetic energy to wavenumbers k (radians per unit length), and vice versa, the spectrum function $\phi(k)$ is the Fourier transform of the correlation function. The most useful correlation function in this context is $\overline{u'(x)u'(x+r)}$, between alongwind velocity fluctuations at downwind distances. The corresponding energy spectrum of turbulence peaks at a wavenumber k_p, which is close to ℓ^{-1}derived from the $\overline{w'(x)w'(x+r)}$ correlation. An alternative choice for eddy length scale is then $\ell = k_p^{-1}$. A physical interpretation of the spectrum is that reciprocal wavenumbers are characteristic dimensions of smaller and larger eddies, the values of the spectrum function a measure of their energy.

Apart from length and velocity scales, an important property of turbulence is the rate at which it dissipates energy, conventionally denoted by ε, in W kg^{-1}. Energy dissipation is the work of the sharpest instantaneous velocity gradients that occur in the eddying motion; viscous shear stress times the velocity gradient being the rate at which mechanical energy is converted into heat. Laboratory observations of many different types of turbulent shear flow revealed the general "similarity" principle that the dissipation rate is proportional to u_m^3/ℓ, varying from one part of the flow to another with the velocity and length scales as this product does. The proportionality constant changes, however, with the boundary conditions on the shear flow, as well as with the different possible choices for velocity and length scales of the turbulence.

The same similarity principle applies to other properties or effects of turbulence and constitutes the great simplifying factor in an otherwise almost untreatably complex phenomenon: once we have information on the variation of the velocity and length scales of turbulence in space or time, we are often able to quantify other properties of the flow. This was first recognized by Ludwig Prandtl, who introduced the concept of a "mixing length" for the eddy length scale, and used it to considerable advantage in constructing theories for different species of turbulent shear flow subject to simple boundary conditions, such as the flow in boundary layers, jets and wakes. The empirical finding, that two independent variables characterizing a turbulent shear flow are sufficient to describe other flow properties, is analogous to the thermodynamic principle that two state variables are all that is needed to determine different properties of a pure substance.

Energy dissipation occurs in the sharpest velocity gradients and therefore at the smallest scales (i.e., at the highest wavenumbers). Kolmogorov (1941) hypothesized that the spectrum function well beyond the peak of the spectrum toward the dissipative range (in the "inertial subrange") depends only on the wavenumber k and the energy dissipation ε (instead of separately on ℓ and u_m). This implies by dimensional reasoning:

$$\phi(k) = a\varepsilon^{2/3}k^{-2/3} \tag{1.13}$$

with a a constant, equal to 0.47 according to Lumley and Panofsky (1964). Observations of the spectrum in the intermediate range thus yield the rate of energy dissipation. Recalling that ε is proportional to u_m^3/ℓ, known ℓ allows the velocity scale to be determined: This is the basis of the so-called "dissipation method" of determining wind stress (see below).

According to the similarity principle of turbulence, the Reynolds stresses should be proportional to density times the square of the velocity scale, $-\rho\overline{u'w'} = const.\rho\overline{u'^2}$, choosing u_m for the velocity scale, as suggested above. An alternative legitimate choice for the velocity scale is therefore the "friction velocity" $u^* = \sqrt{-\overline{u'w'}}$, particularly useful where the Reynolds stress is constant in a region of the flow. This is (nearly) true of the airflow at low levels over the sea, where the Reynolds stress differs little from τ_i, the effective shear force on the interface (that includes any pressure forces acting on wind waves), alias momentum flux from air to water.

1.3.2 The Air-side Surface Layer

Air flow above the sea is variable, but changes in atmospheric conditions take place slowly enough to regard the mean wind speed steady at a few tens of meters above the interface, in what we will call a surface layer. Nor does the mean wind direction vary noticeably with height in this layer, only the wind speed: $\overline{u} = U(z)$, $\overline{v} = 0$, $\overline{w} = 0$. The mean velocity is thus a function only of the distance z above the mean position of the

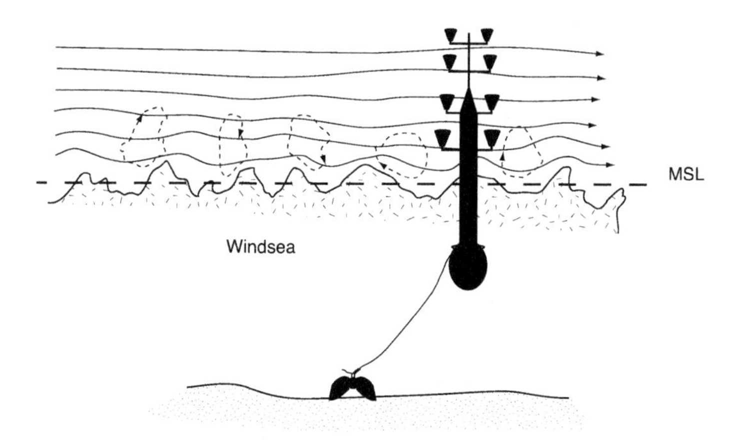

Figure 1.3 The windsea with air flow and eddies over it, and a spar buoy carrying anemometers recording the wind speed at different levels.

interface (known as the Mean Sea Level, MSL). What makes the problem of observing any property in the surface layer very difficult, is that, under wind, the interface is a highly irregular surface that also rapidly changes its shape. The visible structures on the wind-blown interface are wind waves in common parlance, but to avoid even a suggestion of regular parallel-crested water waves we will call them collectively the "windsea." A U.S. Navy Hydrographic publication (Bigelow and Edmondson, 1947) distinguishes between "sea" and "swell," two different wave-like phenomena, "sea" under storms, "swell" what is left over from a storm, more or less regular parallel-crested waves propagating away from the region where the storm generated the "sea." Windsea is a less confusing name than sea, and is certainly descriptive.

Figure 1.3 illustrates the surface layer above the windsea, indicating the air flow, eddies possibly tied to individual irregular waves, and a spar buoy with instruments to observe the mean wind at different levels. Smith (1978) gave details of such a "stable platform"; it was designed to withstand waves of 18 m height crest to trough, albeit protruding only 12.5 m above MSL. In moderate winds, waves are only 2 m height or less, and instruments on platforms similar to Smith's (e.g., fixed towers or ship masts) are able to determine the mean wind at several levels.

Such observations have revealed that, some distance above the windsea, the turbulent air flow has the same character as over a solid boundary, in what is known as a "wall layer." The mean velocity increases with distance above the sea surface, while the Reynolds flux of momentum, $\rho\overline{u'w'}$, that dwarfs viscous stress, is approximately constant with height from just above the waves to 10 m or more, and equal to the effective interface stress τ_i. The latter, the net horizontal force on the interface, includes pressure forces on the inclined surfaces of wind waves, as well as shear stress. The effective interface shear stress defines the friction velocity, $u^* = \sqrt{\tau_i/\rho}$, which then serves as the velocity scale of the turbulent flow in the entire surface layer. Above the waves, where the flow has the character of a wall layer, also described as the constant stress layer, observations have shown the eddy length scale ℓ to be proportional to the distance above the smoothed air-sea interface, z. Other properties of turbulent flow in this region are then expressible in terms of these two scales.

One effect of the eddies is to smooth out mean velocity variations, acting much as viscosity. According to the similarity principle, the vertical gradient of the mean velocity should depend only on the velocity and length scales of the eddies, i.e., by dimensional reasoning:

$$\frac{z}{u^*}\frac{dU}{dz} = \kappa^{-1} = const. \tag{1.14}$$

where the constant κ has the empirically determined value of about 0.4, and is known as Kármán's constant, after one of the great fluid dynamicists of the early twentieth century. Integration from some reference level z_r now results in:

$$\frac{U(z)}{u^*} = \frac{U(z_r)}{u^*} + \kappa^{-1}\ln\left(\frac{z}{z_r}\right). \tag{1.15}$$

The reference level is arbitrary, except that it has to be in the constant stress layer, where the velocity and length scales of the eddies are u^* and $const.\ z$. The velocity at a given level $U(z)$ must be independent of the choice of z_r, however, implying a relationship between reference level height and velocity:

$$U(z_r) - u^*\kappa^{-1}\ln(z_r) = const. \tag{1.16}$$

For the right-hand side constant not to depend on the unit of length or time, it must have the dimension of a velocity, and contain a constant times the logarithm of a length. Writing r for that length, $Cu^* - u^*\kappa^{-1}\ln(r)$ for the right-hand side with C a dimensionless constant, we arrive at the following form of the velocity distribution:

$$\frac{U(z)}{u^*} = \kappa^{-1}\ln\left(\frac{z}{r}\right) + C. \tag{1.17}$$

We anticipate the length r and the constant C to depend on the interplay of the windsea and the air-side turbulence. For the present, they are two empirical parameters of the velocity distribution over the windsea.

Countless observations support this "logarithmic law" in the atmospheric surface layer over the sea. Roll (1965) lists fourteen sets of field observations that do so over various natural water surfaces. In semi-logarithmic representation, at constant u^*, the velocity distributions, $U(\ln z)$ are straight lines, displaced upward or downward according to how much velocity change occurs between the interface and the top of the waves. That displacement depends on just how vigorously the wave-bound eddies stir up the air: the more stirring, the less velocity change. The stirring is the work of the windsea.

1.3.3 Properties of the Windsea

The waves of the windsea are just as chaotic as turbulence, and under simple conditions their stochastic average properties also obey simple laws. Surface elevation is the windsea analog of velocity in turbulence, a random function of time at a fixed location or of location at a fixed time, that defines a frequency or wavenumber spectrum. Chapter 2 discusses windsea properties in detail; here we only catalog the wave-related variables that might influence air-sea momentum flux, with a view to connecting the two empirical parameters in Equation 1.17 to properties of the windsea.

Under a steady wind, and in the absence of waves originating from a distant storm ("swell"), the phase velocity $C_p = \sqrt{g/k_p}$ of a gravity wave (g is the acceleration of gravity, k_p the wavenumber at the peak of the spectrum) defines the "characteristic wave." To a casual observer, larger waves appear to progress with phase velocity C_p, and to have a dominant wavelength of $2\pi/k_p$, amidst much other complexity.

Everyday observation shows that the height of the characteristic wave grows with distance from an upwind shore. Far enough from such a shore, the wave field becomes saturated, and the wave height stops growing. Here the height of the characteristic wave, $H_{1/3}$, defined as the average height of the 1/3 highest waves, depends only on friction velocity $u^* = \sqrt{\tau_i/\rho}$ and g. Thus, $gH_{1/3}/u^{*2} = const.$, with similar relationships for other wave properties. Notice that u^{*2}/g is a waveheight-scale, u^* a common velocity scale of the windsea and the surface layer turbulent shear flow.

Closer to an upwind shore, under a steady and horizontally uniform wind, and again in the absence of swell, waves grow from small to large waveheight with distance from shore (with "fetch" F), the characteristic wave's phase velocity increasing, wavenumber decreasing in the process. Under these idealized conditions (in "local equilibrium" with the wind), properties of wind waves depend on fetch F as well as on friction velocity u^*, and gravitational acceleration g. The phase velocity of the characteristic wave, C_p, or its nondimensional version C_p/u^* (known as "wave-age") serves as a surrogate variable for fetch, F. Nondimensional long-wave properties, such as $gH_{1/3}/u^{*2}$, only depend on one nondimensional parameter, conveniently wave-age.

The properties of the shortest surface structures, not always wave-like, also depend on surface tension σ. The kinematic version of this variable, $\gamma = \sigma/\rho$, is convenient in dimensional argument.

Wind waves facilitate momentum transfer, because horizontal pressure forces may act on their inclined faces, and contribute to τ_i, the net force of the air on the water surface per unit horizontal area. Pressure and shear forces on the interface are also what cause waves to grow with fetch. When waves decay, they hand over momentum to the water-side shear flow, adding to the momentum transferred from the air via viscous shear stress. Even while wind waves grow, they also continuously lose momentum to the water-side shear flow, to a small extent through viscous and turbulent drag on orbital motions, but mostly through "breaking," a complex turbulent overturning motion.

From our point of view in this chapter, wind waves may be thought to open another pathway of air-sea momentum transfer, on top of viscous shear. Somewhat surprisingly, while the long waves of the spectrum carry most of the horizontal momentum transport of the wind wave field, they neither gain nor lose momentum very fast, except on beaches, or perhaps in very strong winds. Short waves of the spectrum, on the other hand, are steep, efficient at extracting momentum from the air flow, prone to breaking, and thus short-lived. Circumstantial evidence suggests that they are responsible for a considerable fraction of the total air-sea momentum transfer. Viscous shear stress meanwhile remains active in momentum transfer: It is difficult to imagine circumstances under which fast air flow in contact with short or long waves would not exert viscous stress.

Returning to Equation 1.17, this summary of wave effects shows that the parameters r and C, quantifying wave influence on the velocity distribution in the surface layer, could depend on the force of gravity g, kinematic surface tension γ, friction velocity u^*, and the nondimensional parameter of wave age C_p/u^*. Viscosity and density of air and water still influence the shear-stress pathway of momentum transfer and should not be forgotten.

1.4 Flux and Force in Air-sea Momentum Transfer

The flux of momentum from air to sea, alias effective interface shear stress $\tau_i = \rho u^{*2}$, is the Flux we wish to relate to a conjugate Force. In the bulk version of the viscous momentum transfer law, Equation 1.10, we found the velocity $U/\sqrt{2}$, realized in the upper portion of the growing air-side boundary layer, to be the conjugate Force. Something similar should prove a suitable choice again, the wind speed at a level well above the waves, say at $z = h$. The standard practical choice is $h = 10$ m. The Force $U(h)$ is then supposed to drive the Flux $\tau_i = \rho u^{*2}$. Putting $z = h$ in Equation 1.17 converts it into a complex implicit relationship between Flux and Force:

$$\frac{U(h)}{u^*} = \kappa^{-1} \ln\left(\frac{h}{r}\right) + C \tag{1.18}$$

where r and C also depend on u^*, as well as on other wave parameters, as just discussed. A nondimensional form of their functional relationship is:

$$C, \frac{gr}{u^{*2}} = func.\left(\frac{C_p}{u^*}, \frac{\gamma g}{u^{*4}}\right). \tag{1.19}$$

If viscosity has a significant effect on interface processes, also ν_a/ν_w and $u^{*3}/g\nu_w$ should be considered. Dividing the nondimensional Force $U(h)/u^*$ by the square root of gh/u^{*2} results in $U/\sqrt{gh}$, a more appropriate nondimensional Force, not containing the conjugate momentum Flux. The similar $u^*/\sqrt{gh}$ is a convenient nondimensional Flux variable. It should also be remembered here that the height h is a proxy for eddy size in the surface layer, an important physical factor in momentum transfer, not the incidental location of a recording instrument.

Equation 1.19, with possibly the viscous variables added, suggests a fairly complex momentum transfer law. The example of the laminar flow transfer law suggests, however, that some of the possible influences may not be noticeable. A drastic simplification would be if instead of Equation 1.18 we had just $U(h)/u^* = const.$ A hypothesis to this effect in fact guided early years of research on momentum transfer, when the focus was on the wind-speed dependence of the momentum flux. Constant

$u^*/U(h)$ means constant drag coefficient $C_D = u^{*2}/U^2$. Within the limited wind speed range explored, and in light of considerable scatter in the observed value of the drag coefficient, a constant value seemed then a reasonable conclusion. Most data on momentum transfer are still presented today in the form: drag coefficient versus (dimensional) wind speed. The latter may be taken to be a proxy for nondimensional $U(h)/\sqrt{gh}$, with the denominator a constant scale velocity of about 10 m/s, for the usual reference height of $h = 10$ m.

1.4.1 Charnock's Law

Later work revealed that the drag coefficient increases with wind speed. Almost half a century ago, Charnock (1955) reported the distribution of wind speed with height over a reservoir, and expressed the results in the form:

$$\frac{U(z)}{u^*} = \kappa^{-1} \ln\left(\frac{gz}{u^{*2}}\right) + C \tag{1.20}$$

where C is a constant, not for just one velocity profile but at all observed wind speeds and directions, and according to Charnock approximately equal to 12.5.

Putting $z = h$ in Equation 1.20 brings it to the form of Equation 1.18, with $r = u^{*2}/g$, the waveheight scale, C a universal constant. We shall refer to it as Charnock's law. Another way to write it is:

$$\frac{U(h)}{\sqrt{gh}} = \frac{u^*}{\sqrt{gh}}\left[C - 2\kappa^{-1} \ln\left(\frac{u^*}{\sqrt{gh}}\right)\right]. \tag{1.21}$$

Because $u^*/\sqrt{gh}$ is always much less than 1.0, its logarithm is negative, so that the square-bracketed expression is positive, with a value typically around 30, and slowly decreasing with increasing u^*. Alternative statements of Charnock's law are:

$$C_D = (C - 2\kappa^{-1} \ln[u^*/\sqrt{gh}])^{-2}$$

$$R = (C - 2\kappa^{-1} \ln[u^*/\sqrt{gh}])/u^*$$

with C_D the drag coefficient, R the Resistance to momentum transfer. The most convenient graphical representation of the law is friction velocity against wind speed, $u^* = func\,[U(h)]$, or the inverse of Equation 1.21, as Amorocho and DeVries (1980) pointed out some years ago. This minimizes the scatter of observed values.

1.4.2 Sea Surface Roughness

Over a "rough" solid surface, the experiments of Nikuradse (1933), using walls roughened by glued-on sand-grains of mean diameter r, showed the velocity distribution in the wall layer to be:

$$\frac{U(z)}{u^*} = \kappa^{-1} \ln\left(\frac{z}{r}\right) + 8.5. \tag{1.22}$$

A comparison with Charnock's law leads to the result that the sea surface behaves as a solid surface of "sand-grain roughness" r, where:

$$r = 3.064 \frac{u^{*2}}{g}. \tag{1.23}$$

To take a typical situation, an 8 m s^{-1} wind calls forth a friction velocity of $u^* = 0.3$ m s^{-1}, and a sand-grain roughness of the sea surface of $r = 3$ cm or so. This contrasts with a characteristic waveheight of some 1.2 m in this wind at long fetch. Sand-grains 0.03 m in diameter closely packed on a smooth surface would mimic short waves in their effects on the velocity distribution over the sea surface in an 8 m s^{-1} wind. The comparison suggests that the sand-grain roughness length according to Equation 1.23 reflects the height of the surface disturbances mainly responsible for the drag of the air on the sea surface. We should add the caveat that the analogy with solid roughness is imperfect because wind waves are mobile, solid roughness elements are not, so that the mechanisms of momentum transfer may differ between them. To the extent that the analogy holds, it singles out short waves as the primary transferrers of momentum.

Meteorologists have fallen into the habit of reporting data on air-sea momentum transfer in terms of a "roughness parameter" z_0 (a length) that combines r and C, defined by the following alternative statement of the logarithmic law:

$$\frac{U(z)}{u^*} = \kappa^{-1} \ln \left(\frac{z}{z_0} \right).$$

The parameter z_0 according to Charnock's law, with constant $C = 12.5$, is:

$$z_0 = 0.011 \frac{u^{*2}}{g} \tag{1.24}$$

some 300 times smaller than the sand-grain roughness, with no relationship at all to any observable structures on a wind-blown sea surface.

One important point about the "roughness" of the sea surface, whichever way it is quantified, is that it is not an externally imposed parameter of the dimension of length. It arises from wind action on the water surface and could in principle depend on any or all of the wave parameters as well as viscosity. To the accuracy, and within the range of validity, of Charnock's law, it depends only on the two parameters u^* and g. To use a sea-surface roughness length in dimensional analysis as an external variable, side by side with u^* and g, is a serious conceptual error (unfortunately not uncommon, e.g., Maat et al., 1991).

1.4.3 Energy Dissipation

Does Charnock's law pass muster in nonequilibrium thermodynamics by conforming to Onsager's theorem? With $U(h)$ the Force, ρu^{*2} the Flux, their product equals by Charnock's law:

$$\rho u^{*2} U(h) = \rho u^{*3} \left(\kappa^{-1} \ln \left(\frac{gh}{u^{*2}} \right) + C \right). \tag{1.25}$$

The left-hand side of this equation is the work done by the wind stress on the air layer underneath the level h, energy transfer downward. As this is unquestionably dissipated in some manner by the underlying shear flow in air and water, and by the windsea, Onsager's theorem is satisfied by the bulk relationship that we call Charnock's law.

To examine the details of energy dissipation, we need the Turbulent Kinetic Energy (TKE) equation, derived from Reynolds' equations of motion for a turbulent fluid (see e.g., Businger 1982), in a simplified form, valid for the constant stress layer with unidirectional flow:

$$\frac{\partial \overline{E_t}}{\partial t} = -\overline{u'w'}\frac{dU}{dz} - \frac{\partial(\overline{w'p'}/\rho + \overline{w'E_t'})}{\partial z} - \varepsilon \tag{1.26}$$

where E_t is TKE defined in Equation 1.12, as energy per unit mass. Multiplied by density, the first term on the right contains the Reynolds stress $-\rho\overline{u'w'} = \tau_i = \rho u^{*2}$ multiplied by the mean velocity gradient, clearly the local Flux-Force product analogous to what we have seen in viscous momentum transfer. In this equation, the Flux-Force product plays the role of TKE production rate. The second term on the right is a divergence, of "pressure work," the velocity-pressure correlation, plus vertical flux of E_t. The divergence represents transfer of energy from one level to another; integrated from the interface up to some level, it yields pressure work transferring energy to wind waves. In the (nearly) constant stress layer above the waves, E_t is constant by the similarity principle, so that its flux, and its time-derivative on the left, both vanish. The same similarity principle also yields the velocity gradient (see Equation 1.14 on page 11), and the energy dissipation rate $\varepsilon = \rho u^{*3}/\kappa z$, that turns out to equal the TKE production rate (i.e., the local Flux-Force product). The flux-divergence term is then also insignificant above the waves. All these relationships are approximate and valid only from some level above the waves to levels where the Reynolds stress remains close to τ_i.

Integrating local energy dissipation from the lowest conceivable level where the constant stress layer formula holds, $z = u^{*2}/g$, to $z = h$, we find:

$$\int_{u^{*2}/g}^{h} \frac{\rho u^{*3}}{\kappa z} dz = \frac{\rho u^{*3}}{\kappa} \ln\left(\frac{gh}{u^{*2}}\right) = \rho u^{*2}(U(h) - Cu^*) \tag{1.27}$$

where the second equality comes from Charnock's law, showing the integrated dissipation to equal the downward energy transfer at level h, minus the downward energy transfer at level u^{*2}/g. With C about 12, $U(h)$ some $30u^*$, only 60% of the downward energy transfer is dissipated between the integration limits, the rest handed down to lower layers. Because u^{*2}/g is typically only 1 cm, most of the remaining dissipation must take place on the water side. It is indeed already stretching a point to suppose constant stress layer formulae valid so close to the sea surface, so that the energy transfer to the water side may be even greater than $C\rho u^{*3}$.

How does this compare with viscous energy transfer across the interface? If all of the effective interface stress $\tau_i = \rho u^{*2}$ were viscous stress, the energy transfer would equal $\tau_i u_0$, with u_0 the velocity at the interface. Typically, the interface velocity is $u^*/3$, some 36 times smaller than Cu^*, so that downward energy transfer at the bottom of the constant stress layer dwarfs what viscous shear stress alone could conceivably accomplish. We return to this point in Chapter 3, where Figure 3.8 shows that, within one waveheight below the interface, energy dissipation is some 30 times greater than what turbulent shear flow on the water side is responsible for.

1.4.4 Buoyancy and Turbulence

The premises underlying Charnock's law are that in the constant stress layer over the sea, properties of turbulent flow depend only on the velocity and length scales, u^*, z, and that wave influences are expressible through the same velocity scale u^* plus the acceleration of gravity, g. In physical terms, the two chaotic processes arising from hydrodynamic instability, shear flow turbulence and windsea, between them govern air-sea momentum transfer, both behaving as, say, a perfect gas, their properties depending on just two variables.

The simple scaling of the turbulent flow in the constant stress layer no longer holds, however, when air density varies owing to heating, cooling, or evaporation. Upward sensible heat transfer or evaporation from the sea surface makes lower layers of air lighter than layers above, an unstable arrangement that results in chaotic gravitational convection, a species of turbulence different from the shear flow variety. Downward heat flux at the sea surface, on the other hand, generates air heavier than above, a stable arrangement. Whatever the source of density variations, the differential gravity force on a parcel lighter than its environment, known as buoyancy, tends to speed up upward eddy motions, while their greater than average density propels heavier parcels downward. Upward heat flux or evaporation thus intensifies vertical eddy motion, while downward heat flux does the opposite, retarding vertical motions arising from shear flow turbulence. Vertical eddy momentum transport, the Reynolds stress $-\rho\overline{u'w'}$, then also depends on heat and vapor fluxes at the air-sea interface.

The proximate cause of enhanced or depressed eddy motion in the presence of heat or vapor flux is the buoyancy or net upward gravitational force per unit mass, acting on a fluid parcel that is slightly lighter or heavier, as the case may be, than its average environment. If the density anomaly is ρ', the buoyancy is $b' = -g\rho'/\overline{\rho}$. The density *defect* $-\rho'$ is in turn proportional to the excess temperature θ' and excess vapor concentration χ' of an air parcel, $-\rho'/\overline{\rho} = \theta'/T + 0.61q'$, where T is absolute temperature and $q' = \chi'/\overline{\rho}$, specific humidity excess, q the standard variable in meteorology representing vapor concentration. The factor 0.61 comes from the different molecular weights of water vapor and air (see e.g., Garratt, 1992).

Eddy motions transport heat and humidity just as they transport momentum. The Reynolds fluxes of temperature and humidity are $\overline{w'\theta'}$ and $\overline{w'q'}$; the corresponding Reynolds flux of buoyancy is:

$$\overline{w'b'} = \frac{g}{T}\overline{w'\theta'} + 0.61g\overline{w'q'}. \tag{1.28}$$

Positive (upward) flux of heat or vapor implies positive buoyancy flux $\overline{w'b'}$.

From another point of view, the product of upward velocity and positive buoyancy force represents work done on the air parcel by the force of gravity, tending to increase its kinetic energy. By the same token, downward buoyancy flux implied by downward Reynolds flux of heat and vapor means work done against gravity, a sink for the kinetic energy of moving air parcels. The parcels must then somehow gain energy to sustain their motions. This interplay of turbulence and buoyancy is portrayed by the turbulent kinetic energy (TKE) equation, expanded from its form in Equation 1.26 to include buoyancy work:

$$\frac{\partial \overline{E_t}}{\partial t} = \overline{w'b'} - \overline{u'w'}\frac{dU}{dz} - \frac{\partial(\overline{w'p'}/\rho + \overline{w'E_t'})}{\partial z} - \varepsilon. \tag{1.29}$$

The first term on the right represents energy gain or loss on account of buoyancy, the other terms are as discussed above.

As we have seen, in the absence of buoyancy flux, dependence of turbulence properties on only two scales implies that both the production and the dissipation terms are proportional to u^{*3}/z. The buoyancy flux in the constant stress layer depends, however, on heat and vapor fluxes imposed at the boundary, that is on other independent variables, and it cannot vary with just u^* and z. This then implies that some or all other terms in the TKE balance must vary with the buoyancy flux. The key additional external variable affecting the TKE balance is the interface buoyancy flux $B_0 = \overline{w'b'}(0)$. The properties of the mean flow as well as of the turbulence then depend on B_0 as well as on u^* and z. A modified similarity principle for the buoyancy-affected shear flow is that its properties depend on these three parameters only (Monin and Yaglom, 1971).

One way to take interface buoyancy flux into account is by means of a length scale, L, introduced into the literature by Obukhov (1946):

$$\frac{1}{L} = -\frac{\kappa B_0}{u^{*3}} \tag{1.30}$$

which serves as a proxy for B_0 in dimensional argument. Apart from the constant κ and the negative sign, both retained here to conform to historical custom, the Obukhov length contains only the two interface fluxes, of momentum (represented by u^*) and buoyancy. Negative B_0 or positive L signifies energy drain on the turbulence, positive B_0 or negative L extra energy supply. The meteorological literature refers to these as stable and unstable conditions, respectively.

Velocity gradients in the shear layer above the waves now depend on B_0, represented by L, as well as on u^* and z. Dimensional analysis leads to the following expanded version of Equation 1.14:

$$\frac{dU}{dz} = \frac{u^*}{\kappa z}\phi\left(\frac{z}{L}\right) \tag{1.31}$$

with $\phi(z/L)$ an unspecified function. Under "neutral" conditions, when the air is neither stable nor unstable, i.e., at vanishing B_0, hence $z/L \to 0$, $\phi(z/L)$ must tend to unity. Large positive B_0 generates vigorous convection and reduces surface stress-induced mechanical turbulence to insignificance. At moderately high positive B_0, or z/L of

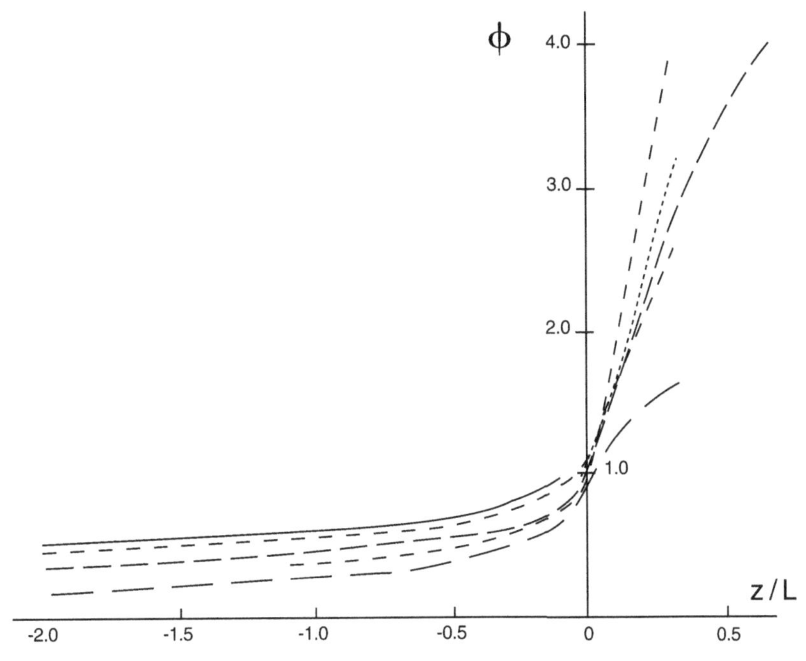

Figure 1.4 The empirical stability function $\phi(z/L)$ as recommended by different authors. Even at z/L close to zero, the uncertainty is seen to be high. From Yaglom (1977).

order -1, compound mechanical-convective turbulence prevails and Equation 1.31 is useful. At the other extreme, large negative buoyancy flux overwhelms mechanical turbulence to the point of completely eliminating it. At moderately high negative B_0 (i.e., positive and suitably small z/L), Equation 1.31 is again valid. The negative buoyancy flux in the TKE equation signifies work against gravity, that is, increase of potential energy as eddies bring lighter fluid down from higher levels. The production term must balance this loss of TKE, resulting in less vigorous shear flow turbulence, and sharper mean velocity gradients.

Boundary layer meteorologists have explored buoyancy effects on the atmospheric surface layer in detail and proposed several different empirical formulae for the function $\phi(z/L)$, separately for stable and unstable conditions. Figure 1.4 after Yaglom (1977) shows some of these. We may conclude from the differences between the formulae that the corrections are known only within a factor of two, and that only at small $|z/L|$. Most widely used are the formulae summarized by Deardorff (1968); they are, in the stable case, $L > 0$:

$$\phi\left(\frac{z}{L}\right) = 1 + \beta\frac{z}{L} \tag{1.32}$$

and in the unstable case, $L < 0$:

$$\phi\left(\frac{z}{L}\right) = \frac{1}{(1 - \alpha z/L)^{1/4}} \tag{1.33}$$

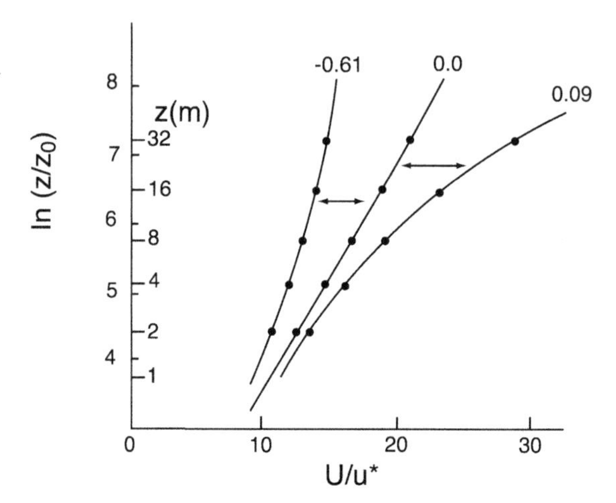

Figure 1.5 Velocity profiles over land in stable, neutral, and unstable conditions, marked by a parameter ("gradient Richardson number"), of the same sign as, but inversely proportional to, the Obukhov length L. Arrows mark typical departures from the logarithmic neutral profile, of some $\Delta U = 5u^*$. Over the ocean, typical departures are generally somewhat less. From Garratt (1992).

with α, β constants. Integration now recovers the logarithmic law plus correction terms depending on z/L:

$$\frac{U(z)}{u^*} = \kappa^{-1} \ln\left(\frac{z}{r}\right) + C + \kappa^{-1}\psi\left(\frac{z}{L}\right) \tag{1.34}$$

with:

$$\psi\left(\frac{z}{L}\right) = \beta\frac{z}{L}$$

in the stable case; while in the unstable case we have:

$$\psi\left(\frac{z}{L}\right) = -\left[\ln\left(\frac{1+x^2}{2}\right) + 2\ln\left(\frac{1+x}{2}\right) - 2\tan^{-1}(x) + \pi/2\right]$$

where $x = (1 - \alpha z/L)^{1/4}$. In the unstable case, at constant u^*, the more vigorous turbulence reduces the wind speed at a fixed level, compared to the undisturbed wall layer, while less vigorous stirring under stable conditions increases it, (Figure 1.5).

Putting $z = h$ and $r = u^{*2}/g$ in Equation 1.34 yields a corrected form of Charnock's law that connects the Force $U(h)$ to the three interface fluxes, of momentum, heat and vapor, the latter two through B_0. Taking the buoyancy-related term in Equation 1.34 to the left-hand side, we are back at Charnock's law, but for a "corrected," or "neutral," nondimensional velocity, $U(h)/u^* - \kappa^{-1}\psi(h/L)$:

$$\frac{U(h)}{u^*} - \kappa^{-1}\psi\left(\frac{h}{L}\right) = \kappa^{-1}\ln\left(\frac{gh}{u^{*2}}\right) + C \tag{1.35}$$

with the same constant C as before, and with correction terms as given following Equation 1.34. This is how observations on momentum flux are usually reported, corrected for buoyancy flux to a neutral value of the wind speed.

1.5 The Evidence on Momentum Transfer

Meteorologists have expended a great deal of effort on the empirical determination of a momentum transfer law for the air-sea interface. Hundreds of contributions to the subject have appeared in the literature in the past few decades. Some excellent review articles (Donelan, 1990; Garratt, 1977; Smith, 1988) summarize the evidence, and some recent papers describing major observational projects paint a vivid picture of the difficult problems of observation at sea (DeCosmo et al., 1996; Smith et al., 1992; Yelland and Taylor, 1996). Here we can only give a glimpse of this vast effort, prior to examining the evidence on the momentum transfer law.

Dimensional arguments above revealed the potential complexity of such a law, while Charnock's formula for it, distilled from observation, held out the hope of great simplification. We ask now: Does the much larger body of evidence available today, within the accuracy of the observations, contradict Charnock's law? If that law holds only within a range of parameter space, what is that range? What other law should replace it outside that range?

1.5.1 Methods and Problems of Observation

Progress toward a consensus on the momentum transfer law was slow, because observation at sea of any variable is difficult. The earliest estimates of momentum flux over the sea came from the "profile method." This exploited the logarithmic law, by observing wind speed at several levels, and if a substantial logarithmic range could be identified, inferred not only u^*, but also the displacement of the logarithmic line from where it would have been over a smooth solid surface. The displacement defined a roughness length z_0 or alternatively a drag coefficient C_D. The process of fitting a straight line in a semi-log plot to a sparse set of points, provided by observation at a few levels on a spar-buoy, tower, or a ship-mast, is not a particularly accurate process, however. Errors were fairly high, both in the friction velocity and in the roughness length, the latter an exponential function of the observed log-line displacement.

"Direct" observation of the Reynolds flux $\overline{u'w'}$, meaning via two separate instruments at a single level, recording time series of u' and w', runs into other difficulties. It is necessary to average the velocity product over a fairly long time, one hour is common, to make sure that all of the surface layer eddies have been fairly sampled. However, the wind speed itself changes in such a long period: Choice of an averaging period in effect defines what changes are deemed fluctuations, and what slower ones contribute to a "trend" in the average wind speed and Reynolds flux. This problem is common to the determination of any "mean" quantity in turbulent flow, which means extracting it from a long but finite record. Additional problems are the alignment of instruments with the true vertical on a fixed tower, or allowing for ship-mast motions on board ship (in order to ensure that the vertical velocity fluctuations are not contaminated by the horizontal wind). Tower or mast structures may also interfere with the air flow. Upto 20% errors in individual observations of Reynolds fluxes have been common. Yelland et al. (1998) carried out a comprehensive investigation of flow distortion effects on a ship, reporting: "Originally, the four anemometers [on the foremast of the research vessel Charles Darwin] gave drag coefficient values that differed by up to 20% from one to another," and were up to 60% too high. They also found flow distortion to depend on wind direction relative to the ship.

The "dissipation method" of determining wind stress is perhaps most convenient: It consists of extracting from a time series a velocity spectrum, and determining energy dissipation from the observed spectral amplitudes in the inertial subrange of wavenumbers, with the aid of Equation 1.13. With ε known at some height h above Mean Sea Level, and its distribution supposed governed by the surface layer law, $\varepsilon = u^{*3}/\kappa h$, u^* follows. Because only relatively high frequencies need to be observed, the dissipation method requires shorter averaging periods than the direct observation of $\overline{u'w'}$, therefore allowing more frequent observations on a cruise. Its weakness is, however, that it relies on allowances for stability effects on the energy balance of turbulence, and on the values of some empirical constants in the underlying theory. As Yelland and Taylor's (1996) careful discussion points out, the method "works" nevertheless, yielding estimates of the Reynolds flux in good agreement with direct correlation measurements, or more accurately, with similar scatter and a not too different mean value.

1.5.2 The Verdict of the Evidence

In light of the difficulties of observation, what is surprising is that a reasonably well-defined momentum transfer law emerges, not that the scatter of individual observations is large. As an example of what a well-conducted major experiment yields, consider the results reported by Yelland and Taylor (1996), and discussed further by Yelland et al. (1998). The observations come from large open stretches of the Southern Ocean at fairly high latitudes, as well as from the subtropical Atlantic near the Azores. The authors employed the dissipation method, and presented drag coefficient and friction velocity against ten-meter wind speed, corrected for buoyancy effects to "neutral" values, observed on thousands of occasions, apparently the largest single data set available on momentum transfer today. Figures 1.6 to 1.9 portray their results, over the moderate to high wind speed range encountered, and separately with higher resolution in low winds. As pointed out before, the friction velocity and the wind speed may be taken to be proxies for the nondimensional parameters $u^*/\sqrt{gh}$ and $U/\sqrt{gh}$.

The friction velocity-wind speed relationship is remarkably tight, as such observations go, (Figure 1.6) defining a momentum transfer law between wind speeds of about 3 and 25 m s^{-1}. At 10 m s^{-1} wind speed, the range of the scatter expressed as U/u^* is about 8, or ± 4 (Figure 1.7). Figure 1.8 shows the averaged drag coefficient-wind speed relationship, the focus of many prior summaries of observations, with Charnock's law as dotted lines, the lower one with the constant C equal to 11.3, the upper one with 10.2. Data above 6 m s^{-1} fit with few exceptions between these two lines, error bars considered.

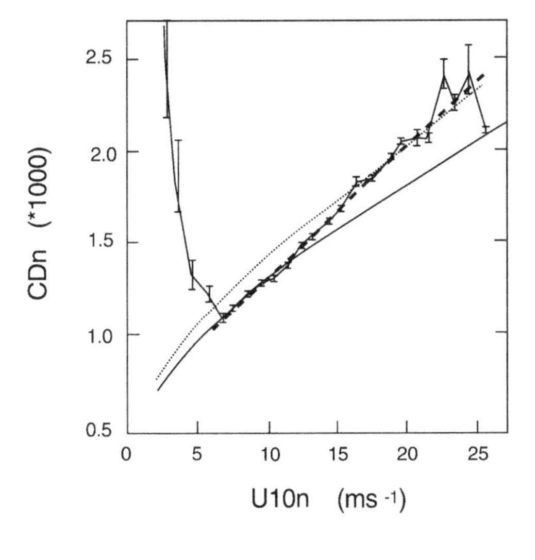

Figure 1.6 Drag coefficient averaged over many observations in the open ocean, at wind speeds above 6 m s^{-1}, with error bars indicating standard deviations. Charnock's law is shown by dotted lines, with constants $C = 11.3$ (lower line) and $C = 10.2$ (upper line). From Yelland and Taylor (1996).

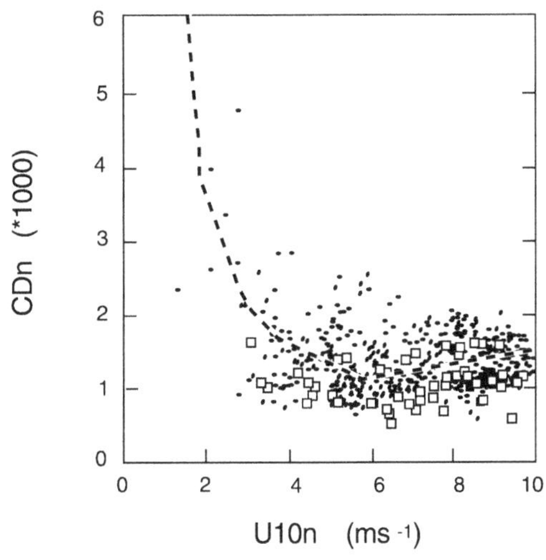

Figure 1.7 Drag coefficient at low wind speeds, individual observations. From Yelland and Taylor (1996).

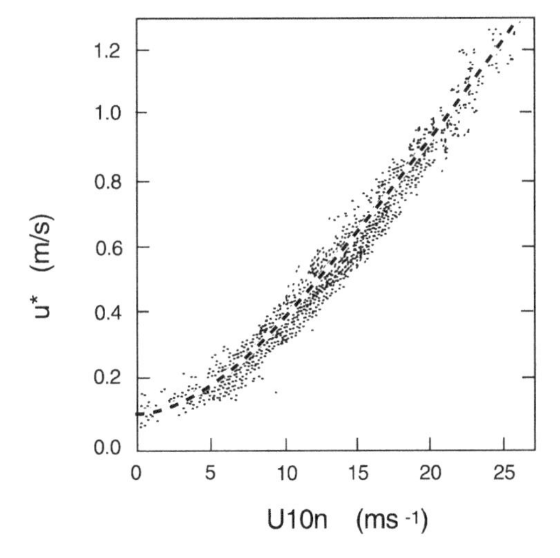

Figure 1.8 Individual observations of friction velocity at moderate winds and higher. From Yelland and Taylor (1996).

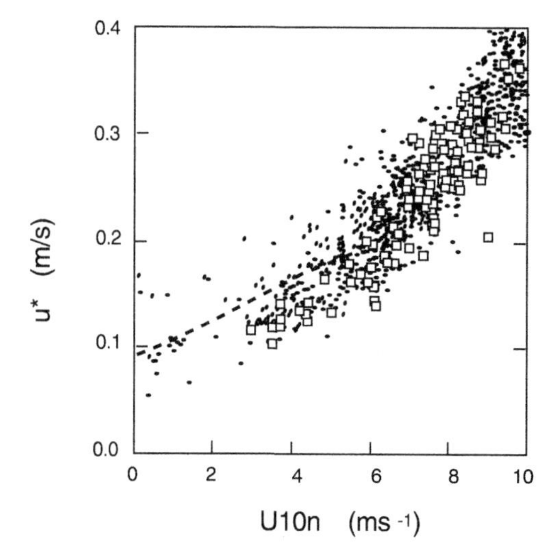

Figure 1.9 Friction velocities at low wind speeds. From Yelland and Taylor (1996).

In a remarkably rigorous further analysis of these results, Yelland et al. (1998) concluded that the drag coefficients reported earlier were on the average 6% too high, owing to airflow distortion by the research vessel. Correcting the data brings them very close to what a prior authoritative summary of momentum flux data found: Smith (1988) confirmed Charnock's law, and yielded the best estimate for C of 11.3. Thus, Yelland and Taylor's averaged data agree with the prior estimate within an error in the U/u^* ratio of less then 1.0, or less than 3%. The averaged data also firmly fix the constant in Charnock's law at $C = 11.3$. Yelland et al.'s (1998) analysis furthermore ties the much larger scatter of individual observations to airflow distortions varying with wind direction. All in all, Smith's (1988) synthesis of earlier information and Yelland et al.'s report on the sources of errors ten years later constitute as good support for Charnock's law as one could possibly get.

Another way to express the verdict of the evidence is that Charnock's law yields the ten-meter wind speed $U(10)$, as a multiple of the friction velocity, with an error of less than one u^*, which is the order of magnitude of the neglected common surface velocity of air and water. The conclusion holds for winds between 6 and 25 m/s, in the open ocean, with corrections applied for buoyancy flux to reduce observed U to the equivalent wind speed under neutral conditions. The corrections for buoyancy flux could not be directly verified, but a fairly wide range of z/L having been covered, the relatively low scatter of the observed U/u^* ratio argues that the corrections "worked."

Charnock's law with buoyancy flux corrections applied thus falls into the category of many other laws of physics valid under certain idealized conditions (steady wind, no swell, etc.), similarly to Hooke's law or the "perfect" gas law, and within a limited range of the Force, the wind speed. Within those limitations, it is accurate to a few percent, in the U/u^* ratio.

The situation today may be contrasted with the confusing state of the subject just a decade ago. A number of different formulae claimed to represent the observations, as individual investigators packaged their conclusions into one coded form or another. Most common were formulae for the drag coefficient $C_D = u^{*2}/U^2$, although many investigators offered recipes for the "roughness" of the sea surface as expressed by z_0. Much of the difficulty in arriving at a consensus was indeed due to this unfortunate focus on z_0, which amplifies errors owing to the exponential dependence of z_0 on the constant in the logarithmic law. According to a study of Blanc (1985), C_D according to ten different formulae then in use varied within wide limits, by about ±20%.

1.5.3 Other Influences

At least some of the scatter in observed $U(h)/u^*$ values is bound to be due to genuine influences of other external variables, quantified by the nondimensional parameters in Equation 1.19, but not appearing in Charnock's law. One independent nondimensional variable of potential importance is the Keulegan number, $Ke = u^{*3}/g\nu_a$. This may be regarded as the ratio of the waveheight scale u^{*2}/g that appears in Charnock's law, and the viscous length scale, ν_a/u^*. Keulegan (1949) showed that above a "critical" value of this number instability waves appear on a two-fluid interface. On a wind blown water surface, as several authors reported, short, sharp-crested waves become visible above a wind speed of about 6 m s^{-1}, when $Ke = 100$, or so. This is at the lower limit of validity of Charnock's law, suggesting that in weaker winds the sea surface may behave similarly to a smooth solid wall, above which a viscous sublayer separates the interface from the wall layer.

Outside the viscous sublayer, in the wall layer above a smooth surface, the velocity distribution is:

$$\frac{U(z)}{u^*} = \kappa^{-1} \ln\left(\frac{u^* z}{\nu_a}\right) + 5.7. \tag{1.36}$$

The independent variable here is z divided by the viscous length scale, yielding a Reynolds number. At suitably low wind speeds, when waveheights are small, the smooth law may well apply over the sea and provide a transfer law with $z = h$. This is far from certain, however, horizontal motions are unhindered in a fluid surface, unlike in a solid wall, and eddies in contact with the free surface may behave differently, (e.g., surface tension variations may affect them) modifying the mechanism of viscous momentum transfer.

Experimental evidence on momentum transfer in light winds has been sparse and conflicting until recently. Smith (1988) gave the smooth law as the best representation of the limited data then available in the low wind speed range. Later studies of Geernaert et al. (1987), Bradley et al. (1991), and Yelland and Taylor (1996) have confirmed the validity of the smooth law around a wind speed of 4 m s^{-1}, but showed a much more definite increase with reducing U, see Figure 1.7 above and Figure 1.10 here, from Bradley et al. (1991). In a 1 m s^{-1} mean wind, Bradley et al.'s data cluster around $C_D = 2.5 \times 10^{-3}$, or u^* of 0.05 m s^{-1}, instead of $C_D = 1.56 \times 10^{-3}$, according to the smooth law. Yelland and Taylor's results depart from the smooth law already

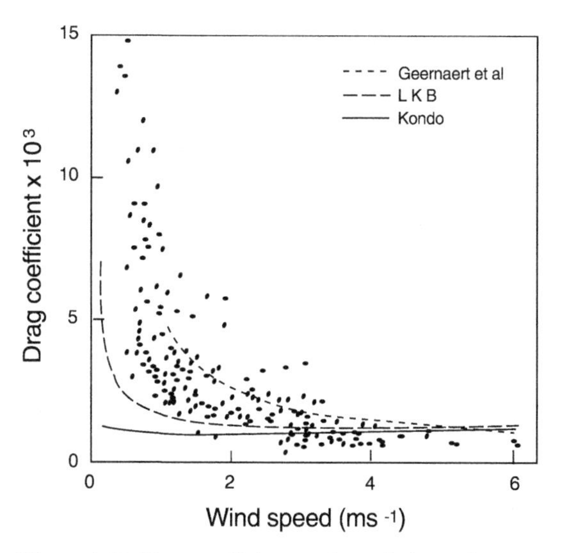

Figure 1.10 Drag coefficients at low wind speeds, from Bradley et al. (1991). The lines show formulae recommended by different authors.

at 3 m s^{-1}, although the observations taken in the subtropical Atlantic remain closer to it than the Southern Ocean ones. The very high drag coefficients at very low wind speeds, reported in all three recent studies, were obtained by the dissipation method, the validity of which under such challenging conditions remains to be confirmed. Another point is that, as the language of meteorological forecasts teaches us, light winds are also variable, so that $U + u'$ may even vary between positive and negative values. The mean wind stress then falls between extremes of the product $\rho|U + u'|(U + u')C_D$, at a higher value than $\rho U^2 C_D$. For $U \ll \sqrt{u'^2}$ it could be much higher.

Even more puzzling than the high drag coefficients at low wind speeds, are some very low ones, found by Sheppard et al. (1972) and Portman (1960), over inland lakes in light winds. The velocities they observed at each level were higher than the smooth law predicts (Figure 1.11), showing the interface to be "supersmooth," at friction velocities less than about 0.18 m s^{-1}, the drag coefficients very low. Similarly, low drag coefficients were reported by Barger et al. (1970) over artificially produced sea slicks in Buzzard's Bay in Massachusetts, that depressed the surface tension of the water surface. The "supersmooth" character of the water surface in these observations may have been the result of energy drain from air-side turbulence to the free surface, similar to the effect of drag-reducing chemicals on boundary layers over solid surfaces. Exactly how, and under precisely what conditions, such drag reduction over water surfaces occurs, is not clear. Much thus remains to be learned about momentum transfer in very light winds.

One other factor that may cause significant departures from Charnock's law is "wave age," quantified by C_p/u^*. Field studies of wave age effects are of recent date, however, and the results have remained controversial for some time. A major field study off the Dutch coast code-named HEXMAX (the main experiment of the Humidity Exchange

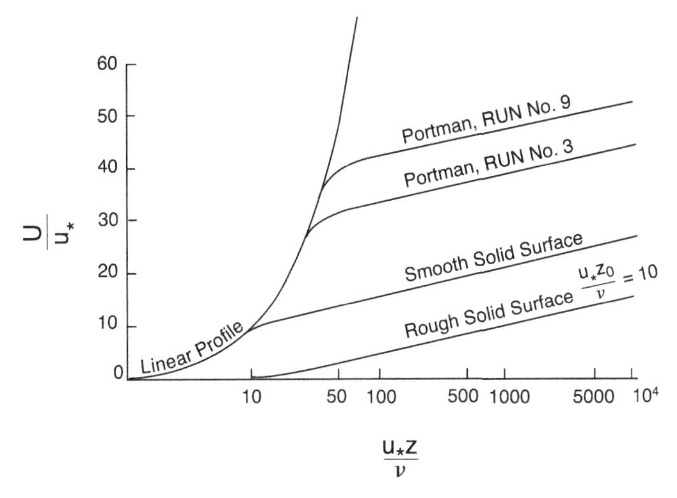

Figure 1.11 Logarithmic velocity distributions in light winds over "supersmooth" surfaces in Lake Michigan, observed by Portman (1960), compared with typical distributions over a smooth and rough solid surface. From Csanady (1974).

over the Sea, HEXOS, program, see Maat et al., 1991 and Smith et al., 1992) led to the conclusion that the constant C in Charnock's law is constant only at long fetch (i.e., C_p/u^* around 30). At younger age it varies according to the HEXOS results as:

$$C = 1.83 + \kappa^{-1} \ln(C_p/u^*) \tag{1.37}$$

supposedly valid within the range of C_p/u^* from 4 to 14 or so. Smith et al. state this result in terms of the nondimensional roughness length as: $gz_0/u^{*2} = 0.48/(C_p/u^*)$.

Figure 1.12 from Donelan et al. (1993), shows the data underlying this law, in a log-log plot. The ordinate is proportional to $-C$, if the logarithmic scale is replaced by a linear one ($C = 11.5$ at $gz_0/u^{*2} = 10^{-2}$, decreasing by 5.76 with each decade, to a minimum of about $C = 5$ at the highest observed points). On the abscissa, the logarithmic scale of C_p/u^* unduly compresses the explored range of this variable. At the high C_p/u^* end at the left, the large cluster of badly scattered C-values comes from mature waves, apparently subject to many unexplained influences. At rather lower wave age, the HEXOS law, Equation 1.37, supposedly applies as far as the field data reach, to about $C_p/u^* = 4$. The scatter is considerable, however. Only laboratory data are available at still lower C_p/u^*, seen on the right of Figure 1.12: they seem to represent a different population of randomly scattering values. This is how Donelan et al. (1993) interpret the data, contending that laboratory waves differ qualitatively from field waves. Toba et al. (1990) argue, on the other hand, that laboratory waves correctly portray the roughness of sea surface at very low C_p/u^*. In that case, according to Figure 1.12, the "constant" C in Charnock's law first increases then decreases with increasing C_p/u^*.

The solution of this conundrum seems to be that the wave age dependence of wind stress found in the HEXOS project was in reality an effect of long waves shoaling and

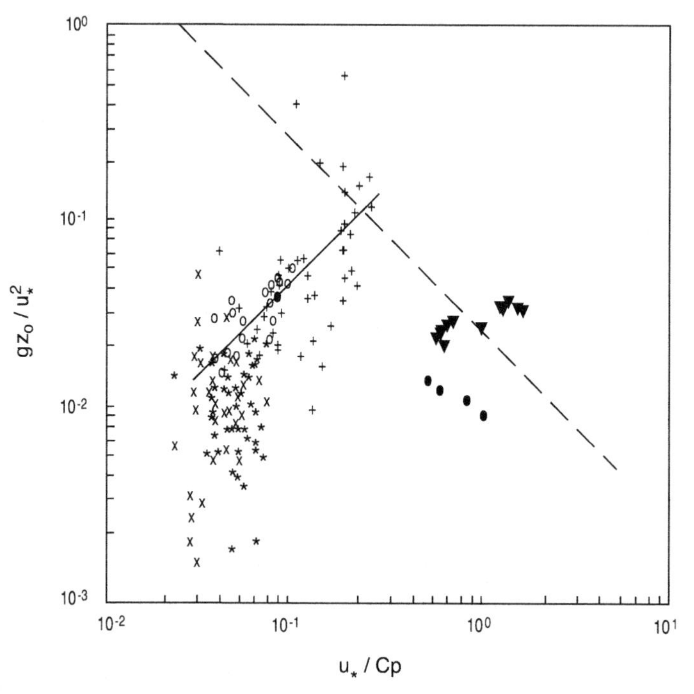

Figure 1.12 Nondimensional roughness gz_0/u^{*2} versus inverse wave age u^*/C_p, representing velocity distributions observed over the windsea. The lines show supposed functional relationships deduced from these data by two different groups of investigators. From Donelan et al. (1993).

steepening in shallow water (Oost 1997), and that it does not apply in the open ocean. As Yelland et al. (1998) stated in their report on flow distortion effects on Yelland and Taylor's (1996) data: "This study examined open ocean drag coefficient measurements for evidence of significant anomalies that can be related to the sea state or wave age. None were found." Dobson et al. (1994) reached a similar conclusion. Yelland and Taylor (1996) also point out that the large scatter of data on wind stress is likely due to flow distortion varying with wind direction relative to the ship or other platform of observation.

This completes our discussion of the momentum transfer law of the air-sea interface. Charnock's law corrected for buoyancy flux emerges as an empirical formula valid within close error bars in moderate to fairly strong winds. We turn next to transfer laws for scalar properties.

1.6 Sensible and Latent Heat Transfer

The air-sea interface is the seat of a whole array of complex thermodynamic interactions: it receives energetic short-wave solar radiation but reflects a large fraction of it, allowing the rest to pass through and warm the top ten meters or so of the oceanic surface layer. The interface also sends long-wave radiation back into the atmosphere and space. Side by side with radiant transfers, the interface allows, and even facilitates through its complex topography, the passage of heat via the molecular processes of conduction and diffusion. Usually heat transfer proceeds from the ocean to the atmosphere (only under rare conditions in reverse) along two pathways, as "sensible" and latent heat transfer. Sensible heat raises or lowers air temperature, but is the junior partner: The bulk of the heat transfer from the world ocean to the atmosphere occurs via evaporation and the attendant transfer of latent heat. The radiant heat transfers indirectly influence interface molecular transfers through the heat balance of the air-side and water-side "mixed layers," but that is another story (see Chapter 3). Here our object is to establish the dependence of sensible and latent heat Fluxes on appropriate Forces.

1.6.1 Transfer of "Sensible" Heat by Conduction

Sensible heat transfer is the simpler problem. Again, as in dealing with momentum transfer, consider first a highly idealized case, that of two stagnant fluids in contact, with the temperature of the upper fluid impulsively raised to θ_∞, the lower fluid remaining at $\theta = 0$. The subsequent heat exchange via conduction is subject to the following simple heat balance (see e.g., Carslaw and Jaeger, 1959):

$$\rho c_p \frac{\partial \theta}{\partial t} = -\frac{\partial Q}{\partial z} \tag{1.38}$$

where Q is heat flux in W m^{-2}, c_p specific heat at constant pressure. The heat flux is subject to the Force-Flux relationship:

$$Q = -k_t \frac{\partial \theta}{\partial z} \tag{1.39}$$

where k_t is conductivity. Elimination of the heat flux from the last two equations yields the differential equation for temperature anomaly θ:

$$\frac{\partial \theta}{\partial t} = K \frac{\partial^2 \theta}{\partial z^2} \tag{1.40}$$

where $K = k_t/\rho c_p$ is now "thermometric conductivity" of air or water. Similarly to the momentum balance, 1.5, this equation states that the rate of change of heat content $\rho c_p \theta$ equals the divergence of heat flux, $-k_t(\partial\theta/\partial z)$. Boundary conditions at the interface are continuity of temperature θ, as well as of heat flux:

$$Q_i = -k_{ta} \frac{\partial \theta_a}{\partial z} = -k_{tw} \frac{\partial \theta_w}{\partial z} \quad \ldots . (z = 0) \tag{1.41}$$

where Q_i is the heat flux across the interface, counted positive upward, a and w subscripts designating the air side or the water side of the interface. The mathematical problem of solving Equation 1.40 with these boundary conditions is the same as we dealt with in momentum transfer, and the resulting transfer law can be written down at once:

$$-Q_i = \frac{\theta_\infty}{R} \tag{1.42}$$

where θ_∞ plays the role of a Force, and the Resistance R is:

$$R = \frac{1}{\rho_a c_{pa}}\sqrt{\frac{\pi t}{K_a}} + \frac{1}{\rho_w c_{pw}}\sqrt{\frac{\pi t}{K_w}} \tag{1.43}$$

where the thermometric conductivity bears indices a and w for air and water. The Resistance again increases as "thermal" boundary layers of thickness $2\sqrt{Kt}$ grow on the two sides of the interface. The product of density and specific heat is much higher for water than for air, however, so that the water-side Resistance again makes a negligible contribution to the total, even more so than in momentum transfer. Dropping this part of the Resistance, the heat transfer law becomes the simple:

$$-\frac{Q_i}{\rho_a c_{pa}\theta_\infty} = \sqrt{\frac{K_a}{\pi t}} \tag{1.44}$$

The right-hand side has the dimension of velocity, and is the rate at which the air-side boundary layer thickens. It may also be thought of as a heat transfer velocity: multiplied by the initial air-side heat content per unit area, $\rho_a c_{pa}\theta_\infty$, it yields the heat flux. Scalar-property transfer laws are often expressed in terms of such a transfer velocity. The nondimensional form of the heat transfer law, Equation 1.44, is:

$$\frac{H_i}{v_t\theta_\infty} = -\sqrt{\frac{1}{\pi}} \tag{1.45}$$

where $H_i = Q_i/\rho_a c_{pa}$ is "kinematic" heat flux, or temperature flux, $v_t = \sqrt{K_a/t}$ heat transfer velocity.

Heat conduction changes the entropy of the fluid, at the rate of heat supply divided by absolute temperature:

$$\frac{dS}{dt} = \frac{\rho c_p}{T}\frac{\partial\theta}{\partial t} = \frac{k_t}{T}\frac{\partial^2\theta}{\partial z^2}. \tag{1.46}$$

Rearrangement of the right-hand side turns this equation into:

$$\frac{dS}{dt} = \frac{\partial}{\partial z}\left(\frac{k_t}{T}\frac{\partial\theta}{\partial z}\right) - \frac{k_t}{T^2}\left(\frac{\partial\theta}{\partial z}\right)^2 \tag{1.47}$$

where the first term on the right is the divergence of a quantity legitimately labeled entropy flux; the second term always negative, and equal to heat flux times temperature gradient, the local Flux-Force product, divided by T^2. This is the local entropy source term in molecular heat conduction, discussed in greater detail by De Groot and Mazur (1984). Onsager's theorem is clearly satisfied. The bulk relationship in Equation 1.44 also conforms to this theorem if we take the Flux to be the interface heat flux, the force $\theta_\infty/\sqrt{2}T^2$: the Flux-Force product is then the integral of the local entropy source on the air side.

1.6.2 Transfer of Water Substance by Diffusion

Sea to air latent heat transfer is accomplished by the water substance, as water molecules cross the interface by phase change. The phase change, evaporation, requires a supply of latent heat from the water side to the surface, and a gradient of vapor concentration in the air to drive upward diffusive vapor flux. The Resistance to transfer then comes from limitations on the ability of water molecules to diffuse in the air, and to conduct heat in the water. Thermodynamic equilibrium dictates that the air in contact with the interface be saturated with vapor at the surface temperature. Energy conservation dictates that the heat flux in the water be equal to the vapor flux in the air times the latent heat of evaporation.

We again consider first the highly idealized case analogous to the previous transfer problems, dry air suddenly replacing moister air above a stagnant water surface. Establishment of thermodynamic equilibrium at once induces evaporation and upward vapor flux via molecular diffusion on the air side. The attendant latent heat flux cools the water surface, calling forth upward conduction of heat on the water side.

The diffusion equation in the air is:

$$\frac{\partial \chi}{\partial t} = D_v \frac{\partial^2 \chi}{\partial z^2} \tag{1.48}$$

where χ is vapor concentration in kg/m^3 and D_v diffusivity of water vapor in air. In the meteorological literature, the standard variable for vapor concentration is specific humidity $q = \chi/\rho_a$, a nondimensional quantity, with ρ_a the density of moist air. In later chapters, we will work with q, as well as with ρ_v, the density of vapor in the air, which is the same as vapor concentration χ. Here we treat vapor as any other constituent of the atmosphere, the conventional measure of which is concentration, its diffusion described by Equation 1.48.

This equation is again of the same form as we solved in the viscous momentum transfer problem. The boundary conditions are here: High above the interface vapor concentration is the relatively dry χ_d. At the interface, thermodynamic equilibrium prescribes the saturation concentration $\chi^*(\theta_s)$, a function of the surface temperature θ_s alone. Solving Equation 1.48 with these boundary conditions yields the interface vapor Flux, or evaporation rate E in kg m^{-2} s^{-1}:

$$E = -D_v \frac{\partial \chi}{\partial z} = (\chi^*[\theta_s] - \chi_d)\sqrt{\frac{D_v}{\pi t}} \tag{1.49}$$

This is undetermined until we find the surface temperature, and the corresponding saturation concentration. Temperatures on the water side obey the heat conduction in Equation 1.40. Boundary conditions are as follows: Well below the interface the undisturbed water temperature is θ_0, while at the interface the heat flux is $Q_i = LE$, with L the latent heat of evaporation, the temperature θ_s. The appropriate solution of the heat conduction equation is:

$$\frac{LE}{\rho_w c_{pw}} = (\theta_0 - \theta_s)\sqrt{\frac{K_w}{\pi t}} \tag{1.50}$$

The solution shows that the surface temperature θ_s is less than the undisturbed temperature below, on top of a "cool skin" or thermal boundary layer owing its existence to upward heat flux.

To connect the air-side results to the water-side results, we need the relationship of water temperatures θ to equilibrium vapor concentration, $\chi^*(\theta)$. The perfect gas law connects the latter to equilibrium vapor pressure, $e^*(\theta) = R_v T \chi^*(\theta)$, with T absolute temperature, R_v gas constant of water vapor. The Clausius-Clapeyron equation (see e.g., Emanuel, 1994) relates the derivative $de^*(\theta)/d\theta$ to temperature; for the expected small temperature differences on the water side of the interface this relationship may be written as:

$$e^*(\theta_s) = e^*(\theta_0) - (\theta_0 - \theta_s)e^*(\theta_0)\frac{L}{R_v T^2} \tag{1.51}$$

For a nearly constant $R_v T$, the perfect gas law allows replacement of e^* by χ^*, so that the last equation is equivalent to:

$$\frac{L\chi^*(\theta_0)}{R_v T^2}(\theta_0 - \theta_s) = \chi^*(\theta_0) - \chi^*(\theta_s) \tag{1.52}$$

Substituting the temperature difference from Equation 1.50, and adding the result to Equation 1.49, we find the transfer law for water vapor:

$$E = \frac{\chi^*(\theta_0) - \chi_d}{R} \tag{1.53}$$

where E is the Flux of interest, while the Force is the vapor concentration difference, between saturation at the undisturbed water temperature, and the dry air. R is the following Resistance to vapor transfer:

$$R = \sqrt{\frac{\pi t}{D_v}} + \frac{L^2\chi^*(\theta_0)}{R_v T^2 \rho_w c_{pw}}\sqrt{\frac{\pi t}{K_w}} \tag{1.54}$$

As in our previous examples, this again consists of an air-side and a water-side resistance. The thermodynamic properties R_v, L affect the water-side resistance only, but that resistance is typically again some 100 times smaller than the air side. The evaporation law is then, to a good approximation, the simpler $E = v_t(\chi^*(\theta_0) - \chi_d)$, with $v_t = \sqrt{D_v/\pi t}$, a vapor transfer velocity. In form this is identical with the sensible-heat transfer law via molecular conduction, *mutatis mutandis*, diffusivity replacing conductivity, etc.

1.6.3 Heat and Vapor Transfer in Turbulent Flow

In turbulent flow, Reynolds fluxes of temperature $\overline{w'\theta'}$ and of humidity $\overline{w'q'} = \overline{w'\chi'}/\rho$ are again the main vehicles of heat and vapor transport to or from the interface on the air side, above the waves, just as Reynolds stress is for momentum. A major difference, however, is that the final step at the interface has to be transfer by molecular conduction or diffusion, there being no alternative pathway analogous to pressure forces in momentum transfer. This means that conductive or diffusive boundary layers develop on both sides of the interface that impede scalar property transfer as in the stagnant fluid case we analyzed. Eddy motions, however, confine the boundary layers to the immediate vicinity of the interface, counteracting the tendency of the boundary layers to grow, and exchanging fluid with the boundary layers. The irregular activity of the windsea enhances these exchanges. Our analysis above showed that the air-side boundary layers constitute the main impediment to property transfer.

The transfer laws for heat and vapor are thus an outcome of a complex interplay among molecular conduction and diffusion, wind waves, and the eddies of the turbulent flow on the air side of the interface. Short, steep wind waves tend to disrupt viscous or diffusive sublayers, reducing their thickness and the temperature or humidity change across them. Along-interface eddy motions are divergent in some locations, convergent in others. Divergence brings air from above down to the interface, helping to keep boundary layers thin. We discuss these mechanisms of scalar transfer at the end of the next chapter after dealing with wind waves; here we merely note that molecular properties are likely to influence scalar property transfer to a greater degree than they do momentum transfer.

Well above thin viscous or diffusive boundary layers, the influence of molecular properties should become imperceptible. Above the waves, in the constant stress layer, Reynolds fluxes of heat and humidity, $\overline{w'\theta'}$ and $\overline{w'q'}$, are (very nearly) equal to the mean interface fluxes $Q_i/\rho_a c_{pa}$ and E/ρ_a: this is also a constant flux layer. Upward temperature flux, $\overline{w'\theta'}$ positive, results from a surface temperature θ_s higher than temperatures in the constant flux layer. As long as the corresponding buoyancy flux is small enough, the stability is nearly "neutral," the properties of the turbulent flow remain as before, eddy size proportional to distance above the interface, friction velocity characterizing eddy motions. Gradients of mean temperature then depend only on the temperature flux and the two scales of turbulence:

$$\frac{d\theta}{dz} = func\,(\overline{w'\theta'}, u^*, z). \tag{1.55}$$

There are four variables in this equation and three units of length, time, and temperature. Hence, they can be combined into a single nondimensional variable that should be constant. Introduction of a temperature scale containing the temperature flux simplifies the dimensional argument: let $\theta^* = -\overline{w'\theta'}/u^*$, the negative sign chosen so that θ^* has the same sign as $\theta(z) - \theta_s$. Equation 1.55 in a nondimensional form is then $d\theta/dz = const.\ \theta^*/z$, a local relationship between Force $d\theta/dz$ and Flux $u^*\theta^*$ that integrates to a logarithmic law:

$$\frac{\theta(z)}{\theta^*} = \frac{\theta(z_{rt})}{\theta^*} + \kappa^{-1} \ln\left(\frac{z}{z_{rt}}\right) \tag{1.56}$$

with z_{rt} a reference level in the constant flux layer and κ a constant. Observations have verified the logarithmic distribution of temperature in the constant flux layer.

Remarkably, the constant κ turns out to be approximately the same as the constant in the logarithmic law for velocity, $\kappa = 0.4$.

Putting $z = h$ in Equation 1.56, and subtracting θ_s/θ^* on both sides, turns the equation into a heat transfer law, connecting interface flux (through θ^*) to the temperature excess or deficiency $\theta(h) - \theta_s$. Because the temperature distribution must be independent of the choice of a reference level, we may replace $\theta(z_{rt})$ and z_{rt} by a constant and a length as we did in the momentum transfer law and write:

$$\frac{\theta(h) - \theta_s}{\theta^*} = \kappa^{-1} \ln\left(\frac{h}{r_t}\right) + C_t \tag{1.57}$$

where the parameters r_t and C_t depend on processes in the conductive boundary layer as well as on the influence of the windsea.

As in the case of the momentum transfer law, this relationship can be cast in the form common in the meteorological literature, after absorbing the constant C_t in r_t:

$$\frac{\theta(h) - \theta_s}{\theta^*} = \kappa^{-1} \ln\left(\frac{h}{z_t}\right) \tag{1.58}$$

with z_t playing the same role as z_0 in the velocity distribution, so that it is referred to as a temperature roughness length.

Analogous results follow from similar arguments for mean vapor concentration, expressed by specific humidity:

$$\frac{q(h) - q_s}{q^*} = \kappa^{-1} \ln\left(\frac{h}{r_q}\right) + C_q \tag{1.59}$$

where $q^* = -\overline{w'q'}/u^*$, r_q, and C_q are a length and a dimensionless constant characterizing the humidity profile, and q_s is the saturation humidity at the sea surface temperature. The meteorological version of this humidity transfer law is:

$$\frac{q(h) - q_s}{q^*} = \kappa^{-1} \ln\left(\frac{h}{z_q}\right) \tag{1.60}$$

where z_q is a humidity roughness length. The value of the empirical constant κ is again approximately the same as in the velocity distribution. Observations have verified the logarithmic distribution also for humidity.

Are these results on scalar property transfer in turbulent flow compatible with Onsager's theorem? To answer this question we have to determine entropy production associated with Reynolds fluxes of heat and vapor, $\overline{w'\theta'}$ and $\overline{w'q'}$. Such fluxes can only exist if there are fluctuations in temperature and humidity, that is if the variances $\overline{\theta'^2}$ and $\overline{q'^2}$ are nonzero. We will refer to these key properties of nonhomogeneous turbulent flow as TTV, Turbulent Temperature Variance and THV, Turbulent Humidity Variance. Balance equations for either derive from the equation of conduction or diffusion, that is Equations 1.40 or 1.48 expanded to include temperature or humidity advection by eddies, see e.g. Businger (1982). The TTV equation valid in the constant flux layer is:

$$\frac{\partial}{\partial t}(\overline{\theta'^2}/2) = -\overline{w'\theta'}\frac{\partial\bar{\theta}}{\partial z} - \frac{\partial}{\partial z}\left[\overline{w'\theta'^2}/2 - k_t\frac{\partial}{\partial z}(\overline{\theta'^2}/2)\right] - \varepsilon_t \tag{1.61}$$

where $\varepsilon_t = k_t\overline{(\nabla\theta')^2}$ is a positive definite quantity containing the magnitude of the fluctuating temperature gradient vector $\nabla\theta'$, a generalized version of the local entropy source term in the simple molecular diffusion example (Equation 1.47). The terms of the TTV equation resemble those in the TKE equation, being in order the production term of TTV, the divergence of TTV flux, and TTV dissipation. The production term is the Force-Flux product in turbulent heat transfer. In steady state, and integrated over the available depth so that the flux divergence term vanishes, the production equals the dissipation, the latter proportional to an entropy source term of magnitude $dS/dt = c_p\varepsilon_t/T^2$. The Force-Flux product in turbulent heat transfer thus generates TTV, as a first step in entropy production through TTV dissipation, complying in this manner with Onsager's theorem.

An almost identical set of calculations yields the result that the humidity Force-Flux product, $-\overline{w'q'}\frac{dq}{dz}$, integrated over depth, equals the THV dissipation rate ε_q, proportional to an entropy source of magnitude $c_p\varepsilon_q$.

1.6.4 Buoyancy Flux Correction

The above "neutral" scalar property distributions hold only as long as the implied heat or vapor fluxes are small enough for buoyancy effects to be negligible. Strictly speaking, a neutral temperature of vapor concentration distribution is a contradiction in terms. It is nevertheless a useful idealization, a standard of comparison that separates wind and buoyancy effects on scalar property transfer.

The interface buoyancy flux B_0, defined by Equation 1.28, influences eddy motions as we have seen in connection with momentum transfer. Therefore, it also affects the temperature and humidity distributions, just as it influences the velocity distribution. In dimensional argument, the Obukhov length L conventionally represents interface buoyancy flux. The temperature and humidity gradients in the constant flux layer then depend on L, as well as on $\theta^*$and u^*, and are of the form:

$$\begin{aligned}\frac{d\theta}{dz} &= \frac{\theta^*}{\kappa z}\phi_t\left(\frac{z}{L}\right)\\ \frac{dq}{dz} &= \frac{q^*}{\kappa z}\phi_q\left(\frac{z}{L}\right)\end{aligned} \tag{1.62}$$

with ϕ_t, and ϕ_q functions to be determined by observation.

A number of different empirical formulations for these functions have appeared in the literature. By general agreement they are the same for temperature and humidity, $\phi_t = \phi_q$.

In the stable case, with buoyancy negative, Obukhov length positive, the simple consensus formula, $\phi_t\,(z/L) = 1 + \beta z/L$, applies, the same as the correction for the velocity gradient, although with some variation in the value of the constant β.

Under unstable conditions, $\phi_t\,(z/L) = (1 - \alpha z/L)^{-1/2}$, a different power, and a different constant α, than in the velocity gradient correction, the constant α again with some variations.

Garratt (1992) gives the following ranges of the constants, recommended by different investigators: $\beta = 4.0$ to 6.9 for momentum, 4.7 to 9.2 for temperature or humidity; $\alpha = 15$ to 28 for momentum, 9 to 14 for temperature or humidity. One concludes that the constants are only known within ± 5, so that $\beta = 5$, for either momentum or scalar, $\alpha = 20$ for momentum, and $\alpha = 10$ for scalars, are in the middle of the ranges recommended.

Integration of Equations 1.62 yields corrections for buoyancy to be added to the right-hand sides of Equations 1.57 and 1.59. These are, in the stable case:

$$\psi_t \left(\frac{h}{L} \right) = \beta z / L \tag{1.63}$$

exactly the same as the correction to the logarithmic law for velocity, while in the unstable case:

$$\psi_t \left(\frac{h}{L} \right) = 2\kappa^{-1} \ln[(1 + x^2)/2] \tag{1.64}$$

with again $x = (1 - \alpha h/L)^{1/4}$.

Subtracting the above corrections from both sides of the corrected Equations 1.57 and 1.59, leaves the right-hand sides as they were, describing the neutral temperature of humidity distributions. Subtracting the same from an *observed* temperature or humidity distribution therefore recovers what the distribution would be without buoyancy effects, in the neutral case. From such corrected distributions the unknown parameters r_t and C_t, r_q, and C_q, can then be extracted. They express conductive and diffusive boundary layer and wave influences on the temperature and humidity distributions. It is through observations corrected to the neutral case that we have gained most of such information.

1.6.5 Observed Heat and Vapor Transfer Laws

Measurement of heat flux and temperature, vapor flux, and vapor concentration over the sea runs into even more severe practical problems than corresponding observations of momentum flux and velocity distribution. Temperature and humidity differences in the wall layer are either quite small, or the influence of buoyancy flux on them is relatively large. Therefore their "neutral" values are either not well defined by the observations, or obtained upon substantial correction for buoyancy flux, a source of significant error, given the uncertainties of the constants in the corrections. Another inherently error-prone step is inferring vapor concentration from the difference of two measurements: the wet bulb and dry bulb temperatures. Both can be degraded by salt water spray. The mean velocity-temperature and velocity-humidity fluctuation products, $\overline{w'\theta'}$ and $\overline{w'q'}$, require a suitably long, but not too long, observation period, similar to momentum flux measurement. For a detailed account and assessment of problems bedeviling the field observation of temperature and humidity, see the report of DeCosmo et al. (1996) on the most extensive such project to date, HEXOS. Earlier detailed discussions of difficulties in the observation of heat and vapor fluxes over the sea are those of Smith (1980) and Large and Pond (1982).

Equations 1.57 and 1.59 are *not* how the literature reports most observations on heat and vapor transfer. Standard practice still relies on alternative scalar transfer laws of the form:

$$
\begin{aligned}
-\overline{w'\theta'} &= C_T U(h)(\theta(h) - \theta_s) \\
-\overline{w'q'} &= C_E U(h)(q(h) - q_s)
\end{aligned}
\qquad (1.65)
$$

and reports the coefficients C_T and C_E. These transfer laws do not exploit what we know about eddy transfer processes, and do not conform to Onsager's theorem, relating the scalar Fluxes to two Forces, the wind speed driving momentum flux and the temperature or humidity excess driving scalar fluxes. They are, however, practical and simple to use. Whether or how C_T and C_E vary with wind speed and buoyancy flux is what a number of workers have investigated and presented in the form of equations or graphs.

In a lucid and authoritative paper, Smith (1988) summarized the empirical evidence on heat and vapor transfer available prior to HEXOS, in terms of the coefficients C_T and C_E. He concluded that the neutral value of the heat flux coefficient is a constant 1.0×10^{-3}, the evaporation coefficient is 1.2×10^{-3}, independent of wind speed (or rather that the scatter of observed coefficients is too large to clearly show wind-speed dependence). The simplicity of the results is strongly in favor of using the formulae in Equation 1.65 in most practical situations.

DeCosmo et al. (1996) also concluded on the basis of the HEXOS observations that the scatter of the data does not allow an empirical determination of wind-speed dependence of the coefficients, and gave the constant neutral values $C_T = (1.14 \pm 0.35) \times 10^{-3}$ and $C_E = (1.12 \pm 0.24) \times 10^{-3}$ as best estimates, identical for all practical purposes. The error bars are high, however. Figure 1.13 a,b show the HEXOS data as vapor flux versus wind speed-humidity product and as an evaporation coefficient versus wind speed. As may be seen, these data do not rule out weak wind-speed dependence of the evaporation coefficient: They are equally consistent with an increase

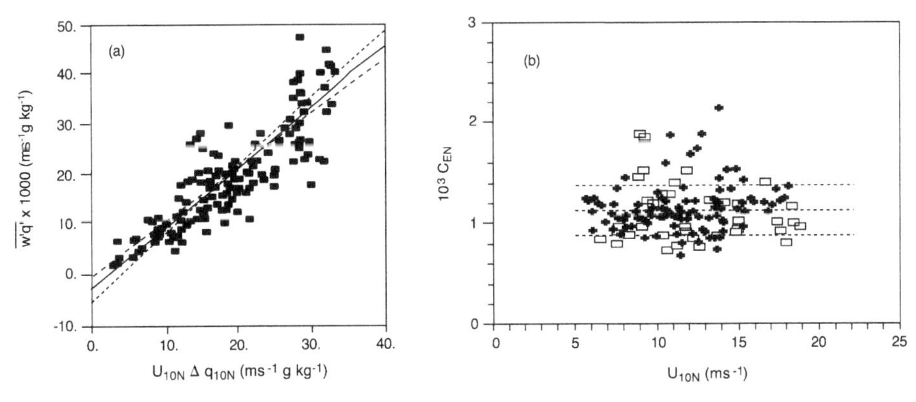

Figure 1.13 HEXOS data on vapor flux: (a) individual flux observations reduced to neutral conditions, versus wind speed-specific humidity difference (between the surface and the 10 m level); (b) exchange coefficients C_E, reduced to neutral conditions, versus wind speed. From DeCosmo et al. (1996).

of the coefficient by 20 or 30% over the range of the data, especially if the high outliers are deleted. The scatter of the HEXOS data on the heat flux coefficient is even greater than on the vapor flux, but again, they do not rule out slow increase over the range of wind speeds encountered.

Observations on the distribution of mean temperature and humidity with height in the surface layer are usually reported as data on the roughness lengths z_t and z_q. Such data have shown that while z_0 of the velocity distribution increases with wind speed, the similar scalar roughnesses z_t, z_q decrease. Also, the latter are much smaller than z_0, by some two orders of magnitude (all three corrected for buoyancy to neutral values). These are indeed striking empirical facts known for some time, but most investigators shied away from an attempt to interpret them, no doubt because of the great scatter of data. The HEXOS data also show much scatter as we have seen, yet DeCosmo et al. (1996) succeeded in showing the relationship of the three roughness lengths, in graphs of the logarithms of z_0, z_t, and z_q, versus wind speed (Figure 1.14). The graphs show that, within the range of the HEXOS data, the logarithm of the median z_0 increases by about as much as median z_t or z_q decrease.

The implied approximate relationship $z_0 z_t = const.$ (with wind speed, or equivalently with u^*) is therefore a reasonable representation of the evidence. Because, according to Charnock's law, z_0 varies as u^{*2}, z_t and z_q should then vary as $1/u^{*2}$. Boundary layer processes affecting z_t or z_q presumably depend on the wave variables u^* and g, plus fluid properties controlling the viscous, conductive, or diffusive boundary layers – viscosity ν_a, thermometric conductivity K_a in the case of heat transfer, diffusivity D_v, in the case of evaporation. The only length scale one can form from these variables under the constraint that it should vary as $1/u^{*2}$ is $K_a^{4/3} g^{1/3}/u^{*2}$, or a similar combination with viscosity or diffusivity replacing thermometric conductivity. Such a combination would only differ from the one containing conductivity by a nearly constant Prandtl number or Schmidt number, both of order one in air. An ansatz similar to Charnock's inspired idea is then to use this length scale $K_a^{4/3} g^{1/3}/u^{*2}$ in Equation 1.57 for r_t:

$$\frac{\theta(h) - \theta_s}{\theta *} = C_s + \kappa^{-1} \ln \left(\frac{u^{*2} h}{K_a^{4/3} g^{1/3}} \right) \tag{1.66}$$

where C_s is the scalar-property equivalent of C in Charnock's law. For the record, we also write down an analogous neutral vapor transfer law:

$$\frac{q(h) - q_s}{q*} = C_s + \kappa^{-1} \ln \left(\frac{u^{*2} h}{D_v^{4/3} g^{1/3}} \right) \tag{1.67}$$

where C_s may differ slightly from the heat transfer law constant, on account of Schmidt number–Prandtl number differences. To the extent that the inverse u^{*2} dependence of the thermal roughness length z_t captures the wind-speed dependence, C_s should be constant over the range of the HEXOS observations. Fitting Equation 1.67 to DeCosmo et al.'s value of z_q at 10 m/s wind speed, we find for the constant $C_s = -1$, not significantly different from zero, and the same constant for Equation 1.66.

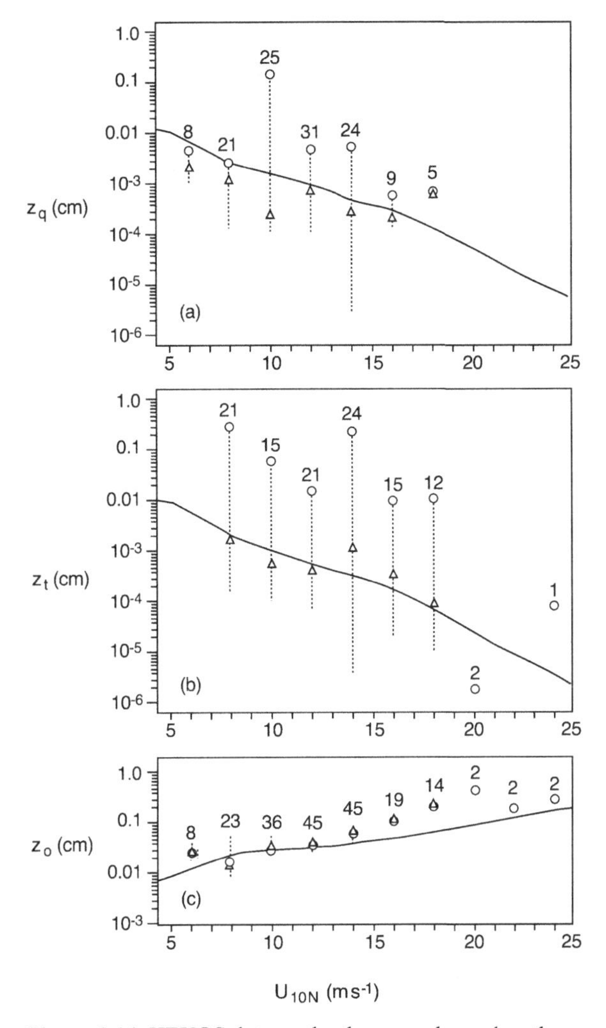

Figure 1.14 HEXOS data on the three roughness lengths z_0 and z_t and z_q, representing observed velocity, temperature, and humidity profiles over the ocean, versus wind speed. The lines are faired to the median of the number of observations noted on the graph. From DeCosmo et al. (1996).

From the last two equations we calculate the neutral heat or vapor coefficients to increase with wind speed from 1.02×10^{-3} to 1.2×10^{-3} within the range of the HEXOS data. This, of course, follows the data in Figure 1.13 remarkably well, because we used the empirical values of the humidity roughness length in Equation 1.67. The slow increase of the neutral heat or vapor flux coefficient with wind speed is in fact a slightly better representation of the data than a constant C_E, or C_T.

While it is gratifying to find a possible logarithmic law version for heat and vapor transfer, the basis for Equation 1.66 or 1.67 is an approximate relationship of roughness parameters (subject to notoriously high scatter) derived from data on a single cooperative experiment, HEXOS. As Yelland et al. (1998) have shown, the shoaling of waves have influenced the HEXOS data on momentum flux – they may also have influenced heat and vapor transfer. Nevertheless, earlier observations already established that the scalar roughness length was much smaller than z_0 (Donelan, 1990; Smith, 1988), and that it at least did not increase with u^*. Until a better formulation evolves, the above relationships are explicit logarithmic laws derived from observation, without any unknown parameters.

Further, in connection with the posited inverse relationship of z_q to z_0, we note that the nondimensional variable in Equation 1.67 splits naturally into:

$$\frac{u^{*2}h}{K_a^{4/3}g^{1/3}} = \left(\frac{u^{*3}}{K_a g}\right)^{4/3}\left(\frac{gh}{u^{*2}}\right) \tag{1.68}$$

a Keulegan number-like variable to the four thirds power times the same nondimensional variable that appears in Charnock's law. This is eminently reasonable on general physical grounds: The Keulegan number-like variable is the ratio of the wave length scale u^{*2}/g to the conductive length scale K_a/u^*, and it may be taken to be a measure of a wave-conductive (or diffusive, if D_v replaces K_a) boundary layer thickness. Substitution into Equation 1.67 yields an expression similar to Charnock's law with a different constant, plus a term representing the wave-conductive boundary layer influence:

$$\frac{q(h) - q_s}{q^*} = C_s + \kappa^{-1}\ln\left(\frac{gh}{u^{*2}}\right) + \kappa^{-1}\ln\left(\frac{u^{*3}}{D_v g}\right)^{4/3}. \tag{1.69}$$

We will return to the question of wave-boundary layer influence at the end of the next chapter.

1.6.6 Matrix of Transfer Laws

Neutral laws of heat and vapor transfer are a contradiction in terms, as several authors have noted, and it is desirable here to add the buoyancy flux corrections to all three momentum, heat and vapor transfer laws, and list them together in a matrix of three Forces in function of the three Fluxes. The Matrix is:

$$\begin{aligned}
\frac{U(h)}{\sqrt{gh}} &= \frac{u^*}{\sqrt{gh}}\left(M\left[\frac{u^*}{\sqrt{gh}}\right] - B\left[\frac{\theta^*}{T}, q^*\right]\right)\\
\frac{\theta(h) - \theta_s}{T} &= \frac{\theta^*}{T}\left(S\left[\frac{u^*}{\sqrt{gh}}, \frac{u^{*3}}{K_a g}\right] - B\left[\frac{\theta^*}{T}, q^*\right]\right)\\
q(h) - q_s &= q^*\left(S\left[\frac{u^*}{\sqrt{gh}}, \frac{u^{*3}}{K_a g}\right] - B\left[\frac{\theta^*}{T}, q^*\right]\right).
\end{aligned} \tag{1.70}$$

On the left-hand side are the nondimensional Forces; on the right the three Fluxes represented by $u^*/\sqrt{gh}$, θ^*/T, q^*. The functions on the right, M, S, and B represent neutral momentum and scalar transfer laws and buoyancy corrections. The neutral laws are:

$$M\left(\frac{u^*}{\sqrt{gh}}\right) = 2.5\ln\left[\frac{gh}{u^{*2}}\right] + 11.3$$

$$S\left[\frac{u^*}{\sqrt{gh}}, \frac{u^{*3}}{K_a g}\right] = 2.5\ln\left[\frac{gh}{u^{*2}}\right] - 1.0 + 2.5\ln\left[\frac{u^{*3}}{K_a g}\right]^{4/3}.$$

The buoyancy corrections in the stable case, for either momentum or scalar properties are:

$$B\left(\frac{\theta^*}{T}, q^*\right) = 8\frac{gh}{u^{*2}}\left[\frac{\theta^*}{T} + 0.61q^*\right]$$

while in the unstable case the momentum transfer correction is:

$$B\left(\frac{\theta^*}{T}, q^*\right) = 2.5X(x)$$

differing from the scalar correction that is:

$$B\left[\frac{\theta^*}{T}, q^*\right] = 5\ln\frac{1+x^2}{2}$$

where x and X are auxiliary variables defined as follows:

$$x = \left(1 + 8\frac{gh}{u^{*2}}\left[\frac{\theta^*}{T} + 0.61q^*\right]\right)^{1/4}$$

and

$$X(x) = \left[\ln\frac{1+x^2}{2} + 2\ln\frac{1+x}{2} - 2\tan^{-1}(x) + \pi/2\right].$$

For constant Force, any of the Equations 1.70 defines a surface in u^*, θ^*, q^* space. A triplet of observed Forces $U(h), \theta(h) - \theta_s, q(h) - q_s$ defines three surfaces, hopefully intersecting in a single point. It would be nice to explore the mathematical properties of this system of rather complex equations.

1.6.7 Entropy Production

In our previous analyses of air-sea transfer processes for momentum, heat and vapor, we have shown that the local Force-Flux products are generators of TKE, TTV, and THV, the dissipation rates of which, multiplied by appropriate proportionality constants, are all entropy sources. The Force-Flux products, multiplied by the same constants, thus represent entropy transfer to turbulence, which sustains three different kinds of entropy sources.

A puzzling fact came to light, however, in the effect of buoyancy on turbulence: the TKE Equation 1.29 acquired an extra term, the local buoyancy flux that, when positive, acts as another TKE generator. The local buoyancy flux derives from interface fluxes of heat and vapor. How does this fit into the framework of nonequilibrium thermodynamics?

In molecular transfer processes, Onsager's theorem for multiple Fluxes acting in tandem states that the rate of entropy production equals the sum of the conjugate Force-Flux products, provided that the Forces are "properly chosen," meaning that they absorb the proportionality constants connecting to entropy generation. In air-sea transfer, Equation 1.70 reveals a much more complex interdependence of the three Fluxes – of momentum, heat, and vapor, and their conjugate Forces – than the linear relationships governing molecular transfer processes. Does Onsager's theorem nevertheless apply to these complex multiple fluxes?

The fundamental equation for calculating entropy production, in a gas such as moist air, is a statement of energy conservation in terms of entropy:

$$T\frac{ds}{dt} = \frac{de}{dt} + p\frac{dv}{dt} \tag{1.71}$$

where T, s, e are in order: absolute temperature, specific entropy in $\mathrm{J\,kg^{-1}\,K^{-1}}$, internal energy in $\mathrm{J\,kg^{-1}}$; p is pressure, $v = \rho^{-1}$ is specific volume; and t is time. The left-hand side is the heat added locally to the system, the right-hand side the change of internal energy and work done on the environment.

One entropy source in turbulent flow is viscous energy dissipation by the eddies, the ε of the TKE Equation 1.29. Viscous stress acting on the mean velocity gradient adds very little, $\nu_a(dU/dz)^2$, and could be added without affecting the argument. The dissipated kinetic energy appears as locally added heat, increasing entropy:

$$\varepsilon = T\frac{ds}{dt} = \overline{w'b'} - \overline{u'w'}\frac{dU}{dz} - \frac{dF_t}{dz} \tag{1.72}$$

where $\overline{w'b'}$ is the buoyancy flux, with $b' = gT'/T + 0.61gq'$, and we have written F_t for the vertical flux of eddy energy in Equation 1.29. The rate of entropy change is now the right-hand side divided by T. We may then identify F_t/T as entropy flux, its divergence as the net outflow of entropy, and the stress-rate of strain product, $-\overline{u'w'}(dU/dz)/T$, as the rate of entropy transfer to turbulence. This leaves the buoyancy flux term, $\overline{w'b'}/T$, work done by gravity, that is neither a flux-divergence nor an admissible contribution to internal entropy production because it can be negative.

The total entropy change, however, is now not only ε/T. Whether buoyancy flux comes from heat or vapor flux, there is extra entropy production on account of at least one other irreversible process, which has to be added to Equation 1.72 to find the total.

To calculate entropy production due to Reynolds heat flux alone, we return to Equation 1.71, and note that local changes in internal energy and specific volume occur because vertical eddy motions advect fluid parcels from a neighboring level: $de/dt = c_v w'(d\overline{T}/dz)$, and $dv/dt = w'(d\overline{v}/dz)$. The perfect gas law, $pv = RT = (c_p - c_v)T$, and the hydrostatic equation $dp/dz = -g/v$ now allow us to calculate the two terms in Equation 1.71:

$$\frac{de}{dt} + p\frac{dv}{dt} = w'\left(c_p\frac{d\overline{T}}{dz} + g\right) \tag{1.73}$$

yielding the rate of change of entropy:

$$\frac{ds}{dt} = \frac{w'}{\overline{T} + T'} \left(c_p \frac{d\overline{T}}{dz} + g \right). \tag{1.74}$$

The temperature fluctuation T' is small compared to the absolute temperature, so that the multiplier of the bracketed quantity is approximately $(w'\overline{T})(1 - T'/\overline{T})$, with a stochastic average value of $-\overline{w'T'}/\overline{T}^2$. The mean entropy production rate is then:

$$\frac{d\overline{s}}{dt} = -c_p \frac{\overline{w'T'}}{T^2} \frac{d\overline{T}}{dz} - \frac{g\overline{w'T'}}{T^2}. \tag{1.75}$$

This contains the heat flux times temperature gradient, i.e., again the positive definite Flux-Force product (for downgradient heat transfer), divided by T^2, and the buoyancy flux $\overline{w'b'} = g\overline{w'T'}/T$ divided by T.

The buoyancy flux term occurs with a negative sign, so that there is less entropy production with upward heat and buoyancy flux than because of downgradient heat flux on its own. More entropy production attends downward heat and buoyancy flux.

The total entropy change resulting from heat flux plus TKE dissipation is the sum of the right-hand sides of Equations 1.72, divided by T, and 1.75. Surprisingly, the two buoyancy flux terms cancel. A calculation of entropy production attendant on eddy vapor flux, following the same recipe as for heat flux, shows that a similar fate befalls the buoyancy flux because of vapor flux. The physical explanation is that the heating of air parcels, or adding lighter water vapor to them, increases their potential energy and reduces the strength of the entropy source associated with heat flux. The gain is short lived, however. The potential energy is converted to kinetic energy of convective turbulence (as the TKE equation shows) and dissipated via ε in the end anyway.

Downward or negative buoyancy flux adds to the entropy source because of heat flux, and diminishes that because of momentum flux, owing to potential energy gain. In the TKE balance, it must be compensated for by local shear production of turbulence, or by TKE import via the TKE flux term from a region where shear flow or convection produces turbulence. The potential energy gain of the fluid is energy regained from the chaotic motions of turbulence, akin to mechanical energy gain from heat, i.e., chaotic molecular motions.

In any case, what is left for the total entropy transfer rate to turbulence is the sum of conjugate Flux-Force products. This also equals the sum of entropy sources associated with the three Fluxes, integrated over the range of the eddy motions. In this indirect manner, the air-sea transfers, with their complex Flux-Force relationships, conform to Onsager's theorem.

An interesting aside to Equation 1.75 is that the right-hand side may be written as:

$$-c_p \frac{\overline{w'T'}}{T^2} \frac{d(T + gz/c_p)}{dz}$$

where the quantity $T + gz/c_p$ is known as potential temperature. When this is constant with height, entropy production because of heat flux vanishes. This can only be true of an irreversible process like heat flux if the flux vanishes, too. The Reynolds flux of heat therefore vanishes in an atmosphere of constant potential temperature. Texts in meteorology explain that a parcel of air moved adiabatically up or down in such an atmosphere changes its temperature owing to expansion or compression just so as to remain at the temperature of its environment. There is then no mechanism for producing temperature fluctuations, and there can be no Reynolds flux of heat.

1.7 Air-Sea Gas Transfer

Atmospheric gases, such as carbon dioxide, oxygen, radon, and a number of others, are also present in the ocean in dilute solution. The air-sea exchange of some of these gases is important to various global balances (e.g., of CO_2 to the carbon cycle or of O_2 to marine photosynthesis). Diffusing gas molecules make slow progress among the tightly packed molecules of liquid water, so that their diffusivity in water, D_w, is much less than in air, D_a, typically by some four orders of magnitude. In consequence, the Resistance to the transfer of these and similar gases across the air-sea interface resides primarily on the water side, in contrast to the other scalar property transfers we discussed earlier.

Another complication of gas transfer is that a given concentration of gas on the air side of the interface has to be in thermodynamic equilibrium with the concentration on the water side. This is called Henry's law:

$$S\chi_a = \chi_w \tag{1.76}$$

where S, a measure of the solubility of the gas in seawater, decreases with increasing temperature. There are large differences in solubility between different gases. Jähne et al. (1985) quote for radon at 23°C $S = 0.25$, for helium at the same temperature, $S = 0.0082$.

A simple molecular diffusion model again reveals some essentials of gas transfer. Suppose that a gas constituent suddenly materializes in the air above the interface, in a constant concentration χ_a, over a water surface containing the same gas in a concentration χ_w, greater or smaller than the equilibrium concentration on the water side, $\chi_w \neq S\chi_a$. Solving the diffusion Equation 1.48 on both sides of the interface, with continuous flux and thermodynamic equilibrium (Henry's law) prescribed at the interface, we find for the interface gas flux in kg m^{-2} s^{-1}:

$$F_i = \frac{\chi_w - S\chi_a}{\sqrt{\frac{\pi t}{D_w}} + S\sqrt{\frac{\pi t}{D_a}}}. \tag{1.77}$$

The structure of this formula is the same as in previous molecular transfer problems: Flux F_i equals Force $(\chi_w - S\chi_a)$ over Resistance. As in the similarly idealized cases of momentum, heat, and vapor transfer, the Resistance naturally splits into air-side and water-side components. In contrast to those other fluxes, however, the water-side resistance dominates gas transfer by a wide margin, on account of low water-side diffusivity, plus low solubility for many gases. Writing $\chi_s = S\chi_a$ for the water-side concentration in equilibrium with the imposed air-side concentration, the effective Force is $(\chi_w - \chi_s)$. The transfer law is then to a good approximation: $F_i/(\chi_w - \chi_s) = k$, with $k = \sqrt{D_w/\pi t}$ a gas transfer velocity (geochemists and chemical engineers call this a "piston velocity"). This transfer law is again of the same form as other simplified laws of scalar transfer via molecular conduction or diffusion, except that the Resistance is now predominantly on the water side. Air-side and water-side boundary layers again characterize the concentration distributions on the two sides of the interface, but the water-side layer is much thinner, the concentration difference across it much greater. That difference for all practical purposes equals the total Force driving the transfer, $\chi_w - S\chi_a$, leaving concentration changes negligible on the entire air side, as well as underneath the diffusion boundary layer on the water side.

1.7.1 Gas Transfer in Turbulent Flow

The last sentence remains true in turbulent flow. This is because, however much the eddies stir up the fluid on the two sides of the interface, the last step in crossing the interface must be via the exchange of molecules (i.e., diffusion). The diffusive boundary layer on the water side remains very thin, and constitutes for practical purposes the entire Resistance to gas transfer in turbulent flow. The thickness of this boundary layer, ranging from 20 to 200 μm (Jähne, 1991), depends on the interplay of molecular diffusion with eddies, and is also affected by short waves.

As may be expected then, a gas transfer law similar to the laminar flow result, Equation 1.77, holds in turbulent flow, with the same Flux and Force, but with an empirically determined Resistance. The literature usually reports the latter as a gas transfer velocity (or piston velocity, equivalent to inverse Resistance). Because practically all of the concentration change in the water occurs across the thin water-side diffusive boundary layer, the Force equals the concentration on the water side of the interface, in equilibrium with the air-side concentration, minus the concentration below the boundary layer. Concentration changes on the air side, attributable to gas transfer, are very small.

Variables influencing the gas transfer velocity, apart from solubility and diffusivity in water, include the friction velocity in water, $u_w^* = \sqrt{\tau_i/\rho_w}$, the common velocity scale of wave and eddy motions exchanging fluid between the diffusive boundary layer, and the turbulent interior of the water side. A key property of such motions is their surface divergence, $\partial u/\partial x + \partial v/dy = u_w^*/\Lambda$, where Λ is their length scale. Positive surface divergence induces upward vertical motion that holds the diffusive boundary layer against the interface, aiding gas transfer. We discuss this interaction further at the end of the next chapter. Here, the important point is that motions of the smallest length scales Λ are most effective in keeping the boundary layer thin. The large eddies of turbulence are much too large to interact with the diffusion sublayer, ruling out any influence from flume- or mixed-layer depth. The same is true of realistic values of the Obukhov length L (i.e., buoyancy flux). The viscous length scale $\Lambda = \nu_w/u_w^*$ characterizes the size of the smallest eddies, with surface divergence of order u_w^{*2}/ν_w. They are likely to play an important role in gas transfer.

The interface gas flux F_i should then depend on the Force $\chi_w - S\chi_a$, and at least on the three variables, diffusivity of the gas in water D_w, friction velocity u^*_w, and viscosity ν_w. The nondimensional form of such a relationship is:

$$\frac{F_i}{u^*_w(\chi_w - \chi_s)} = func\,(Sc)\,. \tag{1.78}$$

where Sc is the Schmidt number, $Sc = \nu_w / D_w$. The left-hand side is nondimensional gas transfer velocity; recent publications denote it by k^+ or k/u^*_w, with k the gas transfer velocity.

Various investigators reported that the appearance of short waves coincides with rapid increase of gas transfer. This points to a possible role of gravity, g, and wave age C_p/u^*. At least in stronger winds when spray and bubbles appear, surface tension may also affect gas transfer; a convenient way to represent it in dimensional argument is surface tension divided by density, $\sigma/\rho_w = \gamma$. Furthermore, as the phrase "oil on troubled waters" suggests, surfactant concentration per unit area affects short waves at low wind speeds, and may influence gas transfer. The nondimensional Flux to Force ratio could therefore be a more complex function of these various possible influences than Equation 1.78 would suggest. What does observation say?

1.7.2 Methods and Problems of Observation

The possible dependence of gas transfer on an array of parameters is only one problem. Very low concentrations (what one would normally call "trace") and correspondingly low fluxes have to be measured or inferred from other evidence. This dictates a long effective duration for such experiments. Although only a considerable handicap in laboratory studies, in the open ocean the need for long period measurements greatly restricts what can be directly observed.

Many careful laboratory studies of the past two decades have nevertheless established the dependence of air-water gas transfer on at least the two key control variables of friction velocity and Schmidt number. The studies were carried out in wind wave tunnels, most of them linear, a few circular, connected to elaborate apparatus for determining the trace gas concentrations both in the air and in the water. In large linear tunnels, the waves are noticeably fetch-dependent, while the observable gas concentrations are averages over the entire contact area of air and water. Samples of air and water have to be collected for long periods, and leakage is a serious problem, degrading the observed Flux-Force relationship.

Owing to the necessarily long duration of an experiment, the usual technique of determining gas transfer velocity in the laboratory involves some unavoidable approximations. Jähne et al. (1985) give the following details: the concentration of a trace gas in water, underneath the infinitesimal diffusion boundary layer, obeys the equation:

$$\frac{\partial \chi_w}{\partial t} = -\frac{\partial(\overline{w'\chi'_w})}{\partial z} - \lambda\chi_w \tag{1.79}$$

where λ is the radioactive decay time scale of a gas such as Radon; for most other gases λ vanishes. The Reynolds flux $\overline{w'\chi_w'}$ immediately below the diffusion sublayer is taken to equal the surface flux $F_i = k\,(S\chi_a - \chi_w)$, written as concentration difference across the diffusion boundary layer times the unknown mass transfer velocity k. Integration over water depth h yields now:

$$k\,(S\chi_a - \chi_w) = h\left(\frac{\partial \chi_w}{\partial t} + \lambda\chi_w\right). \tag{1.80}$$

In an experiment lasting Δt seconds, in practice 0.5 to 20 hours, the water-side concentration changes by $\Delta\chi_w$, so that the average time derivative is $\Delta\chi_w/\Delta t$. The average concentrations in air and water then yield an average transfer velocity k, from Equation 1.80. Its nondimensional version as reported in the literature is $k^+ = k/u_w^*$, with u_w^* the friction velocity in *water*, $u_w^* = u_a^*\sqrt{\rho_a/\rho_w}$.

The laboratory technique of gas collection is not transferable to the open ocean, but a suitably long time-scale phenomenon, the radioactive decay of radon gas, comes to the rescue. Taking advantage of that, the "radon evasion method" is at present our only source of direct evidence on oceanic gas transfer (other, indirect, methods are discussed by Broecker et al., 1996). The radon content of the ocean's surface mixed layer is the daughter product of radioactive decay of radium, present in seawater in dilute solution. Radon decays on a time scale (half life of 3.82 days) that is long enough for its loss to the atmosphere to noticeably affect its concentration in the mixed layer. The observed mixed layer concentration therefore provides a basis for determining radon flux to the atmosphere.

Broecker and Peng (1982) describe the method of calculating the transfer velocity of radon. Figure 1.15, from their monograph, shows a typical distribution of radon in the surface layers of the ocean. The abscissa is radioactive decay rate, $\lambda\chi$, a quantity proportional to concentration. Below the mixed layer, radium and radon are in equilibrium, and their decay rates are equal. The same decay rate and radon concentration would prevail in the mixed layer, were it not for loss to the atmosphere, which is devoid of radon. In the mixed layer, the diffusion Equation 1.79 governs the mass balance of radon, with a production term added, representing the decay of radium that produces the radon:

$$\frac{\partial \chi_n}{\partial t} = -\frac{\partial(\overline{w'\chi_n'})}{\partial z} - \lambda_n\chi_n + \lambda_m\chi_m \tag{1.81}$$

where index m refers to radium, index n to radon. As Figure 1.15 shows, the concentration of radon is constant in the mixed layer. The concentration of radium is the same in the mixed layer as below, and so is therefore the production rate of radon. The atmosphere is devoid of radon, so that the surface flux is $k_n\chi_n$, by the definition of the gas transfer coefficient k_n. If (and this is a big *if*) the mixed layer is in steady state equilibrium with the atmosphere, the time dependent term drops out. With both decay terms constant in the mixed layer, the flux gradient term in Equation 1.81 is also constant, equal to surface flux $k_n\chi_n$ divided by mixed layer depth h. Rearranging the

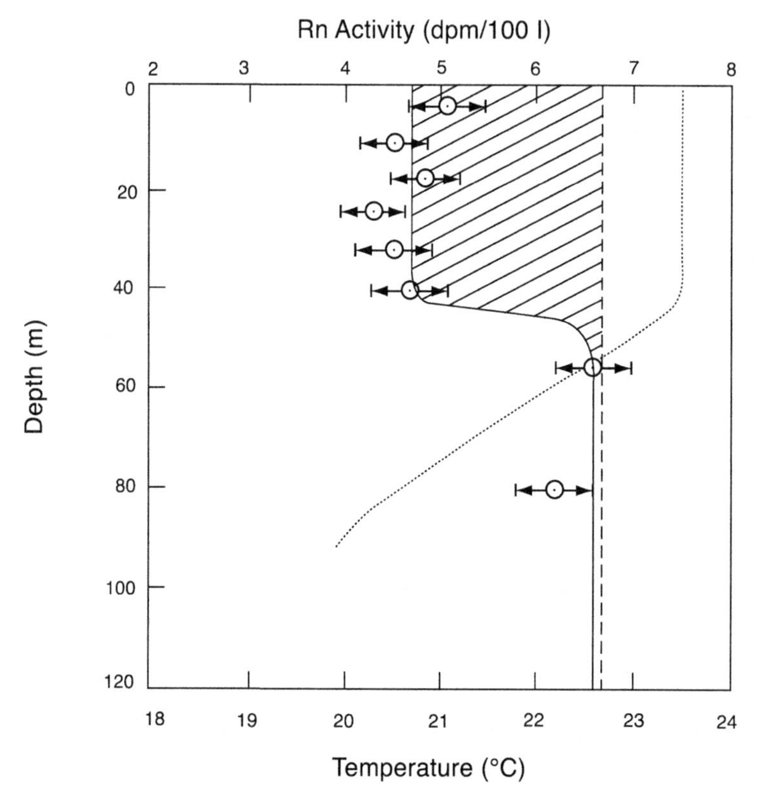

Figure 1.15 Concentration of radon gas versus depth in the Atlantic at 24°S, 35°W, round dots with error bars. The dotted line is the temperature distribution, showing a mixed layer of 40 m depth. The dashed line shows the concentration the radon would have if it were not escaping from the mixed layer to the atmosphere. The radon deficit, shaded, is a measure of the radon's rate of escape. From Broecker and Peng (1982).

variables yields the following formula for the transfer velocity:

$$k_n = h\lambda_n \left(\frac{\lambda_m \chi_m}{\lambda_n \chi_n} - 1 \right). \tag{1.82}$$

The ratio of the disintegration rates in the figure is about $6.7/4.7 = 1.42$, the decay rate of radon is 2.1×10^{-6} s^{-1}, mixed layer depth is about 40 m, yielding $k_n = 3.6 \times 10^{-5}$ m s^{-1}. Corrected to a Schmidt number of 600, this is very nearly 4×10^{-5} m s^{-1}, the middle of the range in laboratory determined transfer velocities.

1.7.3 The Evidence on Gas Transfer

Jähne et al. (1987), and Jähne et al. (1985) summarized a number of laboratory observations of gas transfer in various laboratory facilities, from small circular to large linear tunnels. Figure 1.16 shows the results from large linear tunnels, with data for different gases converted to a common standard by supposing k to vary with the inverse square root of Schmidt number, $Sc^{-1/2}$. In the presence of a surfactant film,

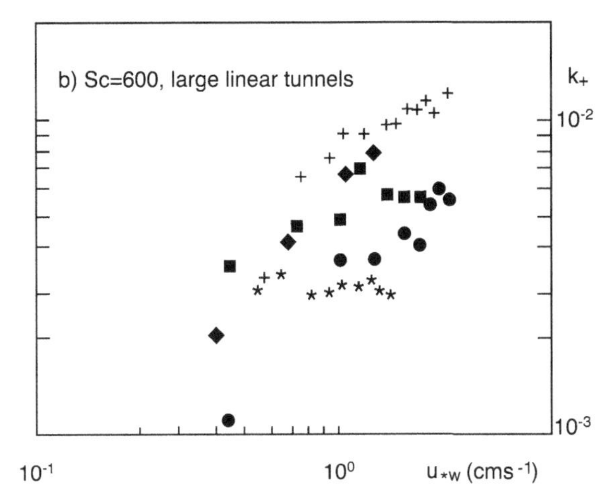

Figure 1.16 Nondimensional gas transfer velocity k^+, observed in large linear laboratory flumes, reduced to a Schmidt number of 600, versus friction velocity in water, u_w^*. Crosses are Jähne et al.'s results over a clean water surface, asterisks the same with a surface film. Solid symbols are data of other investigators. The upper bound of the data defines a nearly linear relationship, presumably characterizing a clean surface. From Jähne et al. (1987).

k^+ remains constant as u^* increases; over a clean surface, however, it grows roughly linearly with the friction velocity, as the figure shows, in a wind speed range of 3 to 15 m s^{-1}. In the middle of the range, near $u_w^* = 0.01$ m s^{-1}, the transfer velocity is about $k = 4 \times 10^{-5}$ m s^{-1}. Observations in smaller linear and circular wind wave tunnels yielded similar results. The scatter of the observations was fairly large, however, and there was a tendency toward much lower coefficients at low wind speeds.

Laboratory data also confirmed the Schmidt number dependence of the transfer coefficient to be $Sc^{-1/2}$ at $u^* > 0.008$ m s^{-1} or so. In lighter winds this changes to $Sc^{-2/3}$, when the surface behaves as a solid wall. The strong temperature dependence of the transfer coefficient is fully accounted for by the corresponding Schmidt number change, so that there is no separate influence of viscosity on the transfer coefficient normalized to a constant Schmidt number. The nearly linear dependence of k^+ on the friction velocity in laboratory observations (all on "young" waves) must then be related to some other variable. An interesting finding was that the mean square surface slope of the laboratory windsea, which increases with wind speed, roughly matches the increase in k^+. We will return to this point in the next chapter, after discussing the physics of the windsea.

Using the radon evasion method, Peng et al. (1979) have reported radon transfer velocities observed on shipboard over the world ocean in the course of the GEOSECS expeditions in 1972–73. The range of k_n that they found is similar to the range of transfer velocities yielded by laboratory studies (Figure 1.16).

As that figure shows, laboratory results unequivocally show significant increases of gas transfer rate with increasing wind speed. Peng et al. (1979) find little correlation between these two variables. They point out, however, that the radon evasion method "is not the way to study the wind velocity dependence of gas exchange." The reason is that the long half life of radon greatly smoothes wind variations. They demonstrate this by time-integrating Equation 1.81 for a 3-week period, using laboratory data for k_n. The results show that the transfer velocities derived from laboratory data do a very decent job of representing the moving average of radon concentration. Peng et al. conclude that there is remarkably good agreement between laboratory and field data, if only the capacitance effect of the mixed layer on radon concentration is taken into account. It is also noteworthy that Peng et al. (1979) quote a global mean radon transfer velocity of about 3.4×10^{-5} m s^{-1}, which is what one would expect from laboratory data for average winds.

No direct method exists for determining the oceanic transfer rate of nonradioactive gases. The accepted indirect method is to correct for Schmidt-number dependence the observed gas transfer coefficient for radon, relying on laboratory data. One source of error in such conversions is that the diffusion coefficients are only known to within 10 or 20% (Jähne et al. 1985). The laboratory values of transfer coefficients for various gases furthermore depend also on imprecisely known solubilities, and are subject to experimental errors of their own, as already mentioned. Waves in laboratory flumes differ from open ocean waves in lacking low wavenumber components, so that laboratory determined gas transfer rates only give a reasonable approximation to open ocean values if only short waves play a significant role in gas transfer. Bubble plumes generated by breaking long waves in strong winds may, for example, open another pathway to gas transfer, not available in laboratory flumes.

The small concentration differences on the air side that arise from gas transfer across the interface make the determination of Reynolds fluxes of trace gases, $\overline{w'\chi'}$, in the air over water inaccurate to impossible, by the standard micrometeorological method (i.e., by simultaneous determination of w' and χ' at some fixed level $z = h$), and finding the correlation. CO_2 measurements in this manner at a coastal ocean site by Smith and Jones (1985) yielded fluxes an order of magnitude higher than the above reviewed field and laboratory studies. Broecker et al. (1996) showed that the high fluxes conflict not only with other observations, but also with much geochemical evidence on the carbon cycle. As Wesely (1986) pointed out, the high concentrations observed by Smith and Jones may have resulted from horizontal advection (i.e., were not the outcome of local mass transfer from the water). Smith and Jones (1986) mention that their fluxes were not correlated with $u^*(\chi_w - S\chi_a)$, confirming their nonlocal origin. Although the editor of the *Journal of Geophysical Research*, in which the last three quoted papers appeared, called the story of the over-water Reynolds flux of CO_2 "an important scientific controversy," the unreliability of its micrometorological measurement simply underlines the well known fact that the principal resistance to gas transfer resides in the water-side diffusive sublayer, not on the air side.

Returning finally to the symbolic result in Equation 1.78, the only certain variation of the nondimensional transfer velocity k^+ at constant Schmidt number is its increase with friction velocity. There is every reason to believe that short wind waves play a role in this behavior. This conveniently ushers in the next chapter.

Chapter 2

Wind Waves and the Mechanisms of Air-Sea Transfer

One outstanding lesson of the previous chapter is that wind waves play a central role in air-sea transfer processes. In this chapter, we examine how wind waves first arise on a wind-blown water surface, what laws govern their growth, what processes are responsible for their decay, and how exactly they facilitate the transfer of momentum and of scalar properties across the air-sea interface.

2.1 The Origin of Wind Waves

In a triumph of early twentieth century fluid mechanics, a combination of theory and laboratory experiment elucidated the origin of turbulence, as Schlichting (1960) has authoritatively chronicled. The theoretical advance was the development of stability theory for laminar shear flow in a boundary layer, with the "Orr-Sommerfeld Equation" (OSE) as its centerpiece and principal tool for analytical investigations. This equation describes the behavior of small-amplitude wave-like disturbances on laminar shear flow that spontaneously grow to large amplitude ("instability waves"). Instability waves in laminar boundary layer flow became known as "Tollmien-Schlichting waves," after the theoreticians who explored their properties.

For some time, such waves eluded efforts at observing them in the laboratory, because laminar flow free of residual turbulence is almost impossible to maintain. Finally, in a wind tunnel made painstakingly clear of other disturbances, the *experimentum crucium* of Schubauer and Skramstad (1947) demonstrated that Tollmien-Schlichting waves of highest theoretical growth rate were the initially observed surface undulations. This verified the theory on the supposition that minute random disturbances of all other wavelengths were present from the beginning, but only the most unstable ones grew to an observable amplitude. The theoretically predicted exponential growth for the instability waves also meant that they would rapidly outgrow the small amplitude assumption. Indeed, laboratory Tollmien-Schlichting waves soon break down into chaotic motions and eventually into full-blown turbulence.

A similar story unfolded in the second half of the century about the origin of wind waves, which we will take to mean the first appearance of small-amplitude waves on an initially smooth water surface under suddenly arising wind. This had remained an enigma in spite of numerous attempts at explanation. Ursell (1956) wrote in a review of the subject: "Wind blowing over a water surface generates waves in the water by physical processes which cannot be regarded as known."

In his book, Phillips (1977) gives a full account of two later theories attempting to explain wave generation: one his own theory that holds pressure fluctuations in the turbulent wind responsible, and two, the theory developed by Miles (1957, 1959, 1962) and Brooke-Benjamin (1959, 1960), that attributes wave growth to coupling between waves and a "critical" layer in the air flow where the wind speed equals wave celerity. The latter theory has attracted a great deal of attention. It concentrates on the air side, ignores shear flow on the water side, and treats viscous effects approximately, if at all taking them into account. In light of later evidence, the most damaging overidealization of this theory was failing to apply the full set of four interface boundary conditions, relying on ad-hoc approximations instead. The Miles-Brooke Benjamin theory nevertheless gave much insight into the dynamics of the air flow over a wavy interface, even if it explains neither the first appearance of waves on an initially smooth surface upon sudden wind, nor the later growth of waves from moderate to larger height.

As in the case of turbulence, the theoretical explanation of wind wave origin finally came from the Orr-Sommerfeld equation supported by laboratory experiment. The theory was more complicated than in the case of turbulence near a solid boundary, because it had to take full account of shear flow both in the air and in the water, as well as of four interface boundary conditions. The latter require continuous normal and tangential velocity and shear stress, and surface pressures in air and water differing on account of surface tension. Wuest (1949) first formulated the stability problem in these terms (originally in a doctoral thesis dated 1941), followed later independently by Lock (1953). Wuest calculated instability waves on some overidealized basic flows and demonstrated the essential importance of satisfying all four interface conditions. Lock took the undisturbed shear flow velocity distribution to be laminar on both sides of the interface, and discovered "air-side" and "water-side" waves with different stability properties. The air-side wave motions were much like Tollmien-Schlichting waves that only induced small amplitude oscillations on the water surface. The water-side wave motions in the shear flow engendered larger surface displacements and traveled at a speed close to the propagation speed of free surface waves on stagnant water. Lock's (1953) study anticipated later work, but lack of observational evidence left it in limbo for years.

The first observations of wave growth under sudden wind, in a laboratory flume (wind tunnel over a water channel), came a quarter of a century later, with the aid of the microwave backscatter technique that enables satellites to observe the ocean surface (Larson and Wright, 1975). After a sudden start, the air flow became fully turbulent in the first second, so that no air-side instability waves could be observed. On the water surface, however, regular short waves appeared spontaneously, their amplitude increasing for some ten seconds before breaking down into irregular motions. The experimental technique yielded the growth rate of individual spectral components of surface elevation. The spectral densities grew exponentially, indicating instability waves. The wavelengths observed were quite short, close to the wavelength of capillary-gravity waves of minimum speed, which is about 1.7 cm. Theoretical growth rates in the same wavelength range were then calculated by Valenzuela (1976) from the Orr-Sommerfeld equation with all four interface conditions. He found encouraging agreement between theory and observation, although the lack of information on the velocity distribution in the Larson and Wright (1975) experiment made the application of the stability theory somewhat uncertain.

The comprehensive laboratory and theoretical investigation of Kawai (1979) fully established the agreement of theory and observation. In his seminal paper, Kawai described how waves develop on an initially smooth surface, upon turning on the air flow in a laboratory flume: "A shear flow first starts and grows in the uppermost thin layer of the water, and then the appearance of waves follows several seconds later. The waves that appear initially are long-crested and regular, and so they are distinguished from those appearing later, which are short-crested, irregular, and accompanied by forced convection." By forced convection, Kawai means turbulence in the water; short-crested, irregular waves are laboratory wind waves. The initial period during which the waves were long-crested and regular lasted some ten seconds.

Kawai also calculated the growth rates of the waves over the range of wavelengths observed, from the OSE with the full set of four boundary conditions. He used the observed laminar shear flow profile in the water, and an empirical turbulent velocity distribution in the air. As Schubauer and Skramstad thirty odd years before, Kawai ended up by concluding that the observed "initial wavelets are the waves whose growth rate by the instability mechanism is maximum."

A crucial aspect of the above scenario, as calculated or observed, is that when the initial regular waves break down into irregular wind waves, the water-side shear flow simultaneously breaks down into turbulence. The theoretical instability waves consist of wave like motions in the air and in the water, as well as on the interface. When one of these interacting flow structures breaks down into chaotic motions, so must the others. Putting it another way, with the water side in mind, the water-side instability waves are the equivalents of Tollmien-Schlichting waves in laminar boundary layer flow, satisfying different boundary conditions, however. As these instability wave motions in the water break down, so do the regular surface waves associated with them. Clearly the shear flow, the unstable wave motions in the air and in the water, and the waves on the interface are all tightly coupled. There is every reason to expect this coupling to remain a characteristic of the flow after breakdown into turbulence and wind waves. Just as turbulence, wind waves are then the product and property of the air-sea shear flow.

2.1.1 Instability Theory

The origin of wind waves, according to the above story, is "explained" by instability theory. In order to accept the explanation, we must judge as realistic the idealizations the theory is built upon, know how its predictions are calculated, and be convinced of the theory's correctness by the agreement of its predictions with observation. To be useful, a theory must also be, in Einstein's words, "as simple as possible, but not simpler."

The centerpiece of instability theory, the Orr-Sommerfeld equation, derives from the equations of motion, upon a set of idealizations (see e.g., Schlichting (1960)). From one point of view, it is a generalization of the classical theory of small amplitude free surface waves on a stagnant inviscid fluid, discussed at length in Lamb's (1957) classical treatise on Hydrodynamics. As the classical theory, the derivation of the OSE postulates small-amplitude sinusoidal disturbances on a wind-blown water surface, of wavelength λ, wavenumber $k = 2\pi/\lambda$. In departures from classical theory, wave amplitude may grow or decay, and the wave motion coexists with two-dimensional parallel viscous shear flow in the (x, z) plane, both in the air and in the water, $U(z)$. "Small amplitude" means that particle velocities in the waves are much less than either the characteristic shear flow velocity or the wave propagation speed ("celerity"). The wave motions are then small perturbations on the mean flow, and linearized equations of motion and continuity describe their behavior. Instability waves necessarily outgrow the small amplitude assumption at some time: Breakdown into chaotic motions follows, but is outside the scope of the theory.

The theory portrays the instability wave motions by a streamfunction ψ, constant ψ lines paralleling the velocities, $u = \partial\psi/\partial z$, $w = -\partial\psi/\partial x$. The wave-like flow pattern propagates and intensifies or decays according to the relationship:

$$\psi = \phi(z) \exp\{ik(x - ct)\} \tag{2.1}$$

where $\phi(z)$ is the amplitude of the streamfunction, and the celerity $c = c_r + ic_i$ is complex, the real part being the propagation speed of the waves, the imaginary part portraying wave growth at the rate of kc_i. The equations of motion and continuity now require the amplitude of the instability waves to obey the following equation, primes meaning differentiation with respect to z:

$$(U - c)(\phi'' - k^2\phi) + U''\phi = -i\nu k^{-1}(\phi'''' - 2k^2\phi'' + k^4\phi) \tag{2.2}$$

known as the Orr-Sommerfeld equation. In the case of the air-water shear flow, the equation applies separately to the air and to the water, the solutions depending on the different viscosities. Boundary conditions at "infinity" (i.e., high above and deep below the interface) are that both velocity components u, w vanish. The equation is of fourth order, requiring four boundary conditions in the air, and four in the water, for a total of eight – two plus two as mentioned at plus-minus infinity, as well as four matching conditions at the interface, prescribed by the dynamics of viscous fluids.

They are: continuity of the two velocity components u, w; of the shear stress τ; and a pressure condition expressing the balance between the air-water pressure difference and the force arising from surface tension on the curved interface. Simplified forms of these conditions apply to the postulated small perturbations – they are listed in a number of publications, including Kawai (1979).

The simplified shear stress and pressure conditions at the interface are still quite complex, and so is the OSE itself for that matter, containing the velocity distribution and its second derivative. Analytical discussions of the OSE with these boundary conditions have not been very illuminating. Numerical solutions of Valenzuela (1976), Kawai (1979), and others have, however, led to important insights into mechanism of instability and revealed the properties of instability waves on the air-sea interface.

The usual numerical procedure starts with asymptotic solutions of Equation 2.2, for fixed wavenumber, at large distances above and below the interface, where U'' vanishes, and U is constant. Two such solutions exist on each side of the interface, one like the classical inviscid wave, the other rapidly varying with height or depth, known as the "viscous solution." Starting with these well above and well below the interface, with their amplitudes arbitrarily set at unity, the integration proceeds "top down" and "bottom up." Arriving at the interface, one has then two inviscid and two viscous solutions, with four calculated amplitudes, which have to satisfy four boundary conditions. The determinant of the resulting four homogeneous linear equations has to vanish, a condition yielding the complex celerity $c_r + ic_i$. The three remaining equations yield the amplitude ratios of the four solutions.

There are various practical problems in carrying out the integrations and in solving the four equations that arise from the interface boundary conditions. These are dealt with by different ingenious methods. As may be expected, the results depend on the assumed air flow and water flow velocity distributions. The solutions reveal exponentially growing waves over a limited range of wavenumbers – their growth rate peaks at some wavenumber k_m. If initial waves appearing under sudden wind have this wavenumber, then we take it for granted that they arise from shear flow instability.

The various simplification of the theory, consisting of the OSE and the boundary conditions, make it "as simple as possible," for the understanding of instability waves on the air-water shear flow. Let us now make it simpler, by neglecting viscosity, and eliminating the shear flow, $U = 0$. The remaining simple equation has the solution $\phi = const. \exp(\pm kz)$, and describes the classical inviscid wave. It neither grows nor decays, and its celerity is real, given by:

$$c^2 = k\gamma + \frac{g}{k} \tag{2.3}$$

which is the classical "dispersion" relationship (because waves of different wavelengths "disperse," i.e., travel at different speeds) valid for small amplitude waves on stagnant water. At high wavenumbers the first term on the right dominates: These are capillary waves, in contrast to gravity waves at low k. The celerity is real, waves of given wavelength travel at constant speed c and constant amplitude ϕ_w/c. The wave speed has a minimum at wavenumber $k = \sqrt{g/\gamma}$, where $c^2 = 2\sqrt{g\gamma}$.

Simplification led us to the classical theory of gravity-capillary wave propagation on the surface of a stagnant fluid, from the theory of instability waves growing on shear flow. Reversing the argument, we see that the full four boundary condition instability wave theory may be regarded as an extension of the classical theory of surface waves on water. In this light, Lock's result, that the water-side instability waves travel at much the same speed as inviscid surface waves, is not so surprising.

2.1.2 Properties of Instability Waves

A realistic specification of the mean shear flow profiles in air and water was one of the key factors in Kawai's (1979) success in reconciling theory and experiment. As mentioned before, in laboratory flumes the air flow is turbulent immediately upon starting (in the first second, at any rate, according to Larson and Wright, 1975), but the water-side shear flow remains laminar for several seconds. Kawai therefore chose an empirical turbulent air flow velocity distribution, together with a laminar shear flow profile in the water, fitted to the actual observed velocity distribution. His air flow observations did not cover the velocity distribution over that portion of the water surface where the initial waves appeared – the empirical distribution he used had two parameters, the air-side friction velocity u_a^*, and nominal wind speed, U_1. Observation yielded the friction velocity, but the nominal wind speed was uncertain with an estimated possible range of $U_1 = 5u_a^*$ to $8u_a^*$.

Numerical solution of the OSE yielded frequencies, phase speeds, and growth rates of waves of different wavelengths, at the observed friction velocities, for the two nominal air flow speeds bracketing the range of uncertainty. The phase speeds differed little from what the classical dispersion relationship predicts, indicating that they are little affected by viscosity or shear flow. Kawai (1979) showed the frequency $f = k_m c_r/2\pi$ and growth rate $\beta = 2\,k_m c_i$ of the waves with the highest calculated growth rate, together with the observed frequency and growth rate of the initial wavelets, in two dramatic illustrations (see Figures 2.1a and b). The dashed lines represent calculated quantities at the two nominal air speeds, the different circular symbols the experimental results. The agreement of theory and observation is striking, given the great difficulty of both.

Kawai's (1979) maximum growth rates are fairly high and the initial wavelets have wavelengths close to those of capillary-gravity waves of minimum celerity, in the wavenumber range of 2–5 cm^{-1}, similarly to what Larson and Wright (1975) found. This raised the intriguing question: Is there some physical mechanism holding the most unstable waves at this singular point of the gravity-capillary wave dispersion relationship? Numerical solutions of the OSE (Wheless and Csanady, 1993), in a range of wavenumbers around those of Kawai (1979) and of Larson and Wright (1975), k = 1–8 cm^{-1}, proved otherwise. The growth rate of the instability waves turned out to be sensitive to the "development time" t_d of the laminar boundary layer in

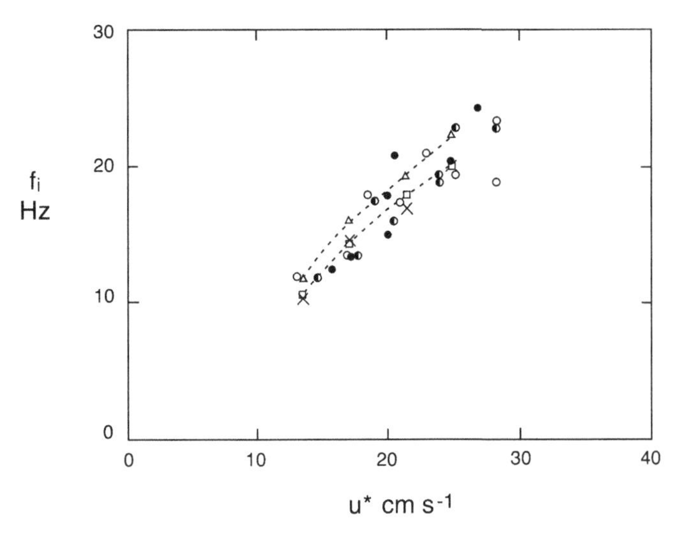

Figure 2.1a Frequency f_i of the observed initial wavelets in Kawai's experiments at fetches of 3 to 8 meters (full, half-open and open circles) versus friction velocity. Calculated growth rates (squares and triangles) are connected by dotted lines, bracketing the realistic range of U_1/u^* ratios, ranging from 5 to 8.

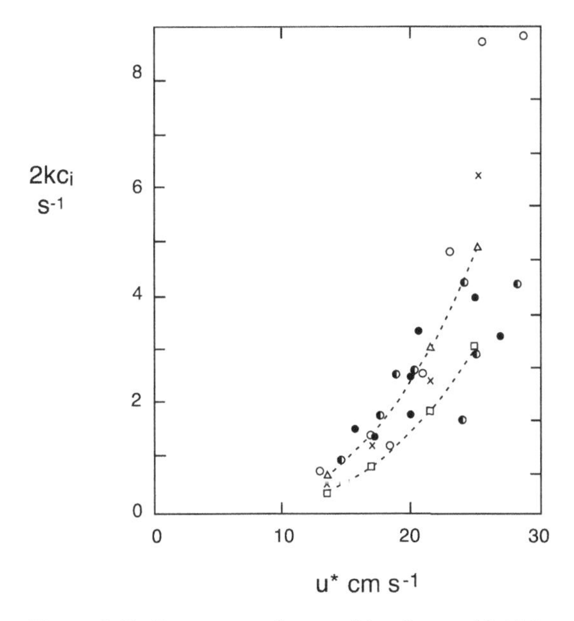

Figure 2.1b Energy growth rate of the observed initial wavelets and calculated values in Kawai's experiments. Symbols and dotted lines as in previous figure.

the water (time since starting the air flow). With increasing development time, the water-side boundary layer depth grew as $\delta = \sqrt{2\nu_w t_d}$, and the inverse wavenumber of the most unstable wave grew in proportion, approximately as $k_m^{-1} = 0.97\delta$ (Figure 2.1c). Given that laboratory observation of initial wavelets is restricted to

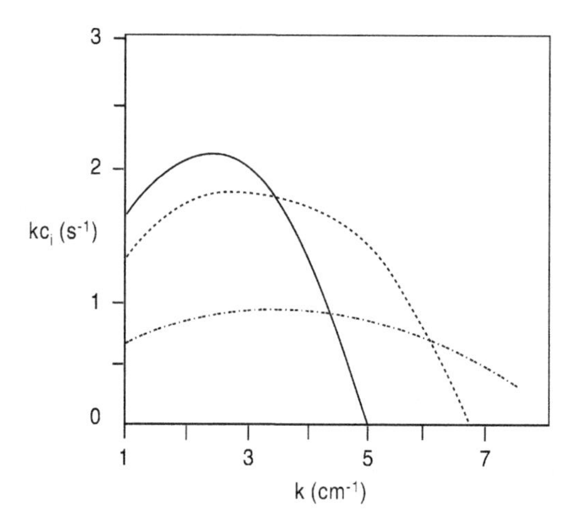

Figure 2.1c Calculated growth rate of instability waves against wavenumber, at varying thickness of the water-side viscous boundary layer. The growth rate peaks at a wavenumber k_p varying as the inverse of the boundary layer thickness δ, $k_p^{-1} = 1.35\delta$. From Wheless and Csanady (1993).

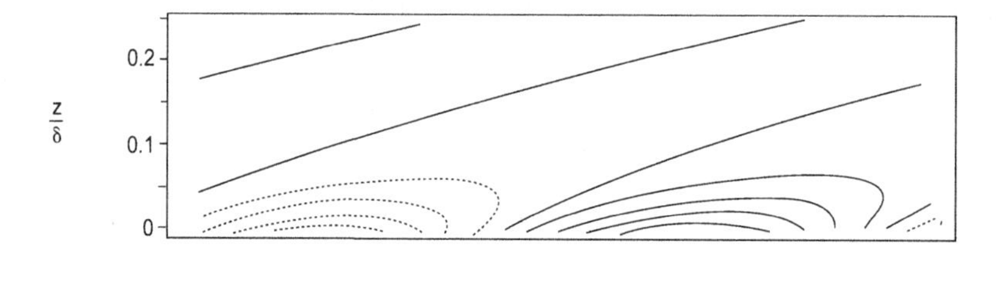

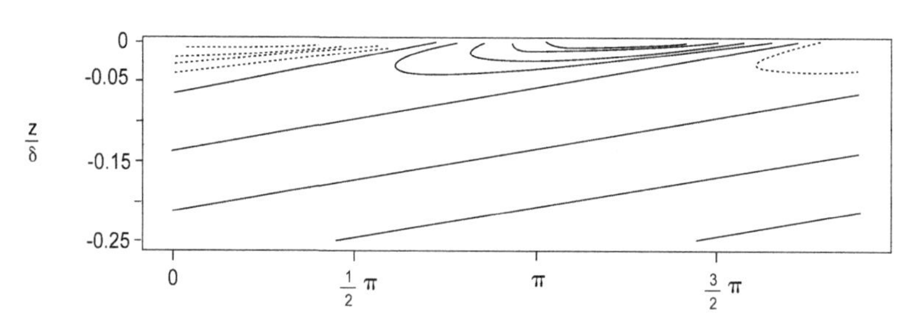

Figure 2.1d Streamlines of the instability wave of the highest growth rate on the two sides of the air-sea interface. Significant motion is confined to about $z/\delta < 0.05$, or about one millimeter from the interface. From Wheless and Csanady (1993).

development times of some 3 to 10 seconds, and that the viscosity of water is nearly constant at $\nu_w = 0.01$ cm^2 s^{-1}, the observable range of k_m has to fall between 2 and 4.2 cm^{-1}, as observations indeed showed.

Calculations also revealed the structure of the instability waves, the outstanding characteristic of which is a dense packing of streamlines on both the air and the water side, signifying high horizontal velocities (Figure 2.1d). This arises from the "viscous solution" of the OSE, the component that ensures continuity of shear stress and horizontal velocity across the interface. Lock (1953) has already drawn attention to such a viscous-wave boundary layer. Its thickness is typically a quarter of a millimeter, and its effect on the velocity distribution is to increase surface velocity in the water by about an order of magnitude, compared to the surface velocity in an inviscid free wave of the same wavelength and amplitude. This means that an instability wave outgrows the small amplitude assumption at very low amplitude, leading to early breakdown of the flow into irregular motions. The vorticity is high in the viscous-wave boundary layer: The initial wavelets thus harbor incipient eddies, which remain to contribute to the irregular motions. A final point is that the energy transfer from the mean flow to the wave motion (the product of shear stress and rate of strain) peaks in the *air-side* viscous-wave boundary layer, where the streamlines are close together and the velocity gradient is large (not in the "critical" layer).

While the laboratory experiments in which instability waves arise on a quiescent water surface under sudden wind seem remote from oceanic reality, something very similar does occur under a wind gust. The "cat's paws" of everyday experience are patches of sea surface covered by capillary-gravity waves, looking for all the world like the initial waves in Kawai's laboratory flume. It is eminently reasonable to suppose that they arise from the same causes, wind-driven surface shear and hydrodynamic instability. Within seconds they evolve into the irregular surface typical of wind waves, containing also somewhat longer waves. As the gust passes (after a typical duration of perhaps 100 seconds), the shortest waves die out until the next such episode. Each gust generates a new boundary layer at the surface, thin and strongly sheared, setting the stage for new instability waves to grow. This is not mere speculation: There can be no doubt that a wind gust greatly intensifies surface shear, nor that such a shear layer is hydrodynamically unstable. If instability waves in the gravity-capillary range occur under similar conditions in the laboratory, they can hardly fail to show up on a natural water surface. Nor can there be any doubt that turbulence generation in the water accompanies the evolution of initial wavelets into chaotic surface waves, as the highly vortical water-side instability waves break down. Thus, the natural variability of turbulent air flow over water brings about incessant generation and decay of short, steep wavelets.

2.2 The Wind Wave Phenomenon

The wind waves we encounter in nature are much higher and longer than cat's paws and their laboratory cousins arising from instability waves. What the instability theory and accompanying laboratory work has shown is that short waves arise spontaneously on the interface, and break down into irregular surface motions simultaneously with the generation of turbulent shear flow. A bold generalization of this finding is a modern view of wind waves, first articulated in recent writings of Professor Toba of Tohoku University, to the effect that the wind-blown high waves of the ocean are a complement of the air-sea turbulent shear flow from which they cannot be separated. He puts it this way (Toba, 1988):

Figure 2.2 The windsea in a condition described by the U.S. Coast Guard as "moderately high and breaking," photographed from ship. From Bigelow and Edmonson (1947).

"The wind wave is a special water wave as it is generated by the action of wind at the air-water interface. Since the wind wave is associated with air and water flows above and below the waves, its characteristics are determined by the coupling process between the boundary layers in air and water. The important elements in the wind wave are the surface wave motion, the local wind-drift and turbulence in the air and water boundary layers" (p. 263).

A precursor of the new paradigm appears in older nautical texts as a distinction between "sea" and "swell," the former meaning the irregular and rapidly changing sea surface under moderate to strong winds, the latter the smooth, regular waves propagating from some distant generation region to the observer's location. Bigelow and Edmondson (1947) describe the distinction as follows: "The characteristics of storm waves that most impress the observer are their irregularity and steepness, also their great heights in many cases, and the frequency with which their crests break." They show two Coast Guard photographs in illustration (Figure 2.2 is one of them). They then continue: "the shapes of the waves undergo wide alterations when the wind dies down, or when the waves produced by a given wind system advance to regions outside the latter, as very commonly happens. The wave train in question is then known as a 'swell'. . . " characterized by low, rounded crests and smooth surface contours. When the wind dies, the waves begin to lose energy, "the shorter ones with the least energy becoming lower and disappearing first, so that the longer ones alone are left. At the same time, the sharp peaks so frequent during a rough sea subside," the irregularities of the surface smooth out, "and the remaining crests decrease progressively in height and become more rounded. The end result is that the waves tend to approach the trochoidal profile characteristic of the so-called free wave of theory."

Toba's view quoted above ascribes the special character of wind waves to close coupling with the air-water shear flow. It conflicts with the earlier conceptual model of wind waves, that regarded them an assembly of freely propagating gravity waves, independent of and unaffected by the air-water shear flow. The new conceptual model holds the irregular wind-blown water surface to be just one manifestation of the complex coupling between air and water turbulent shear flow. A newly coined, felicitously descriptive term for the wind wave phenomenon is "windsea."

Because the wind generates both the windsea and the turbulent shear flow, the independence of the two wind products is inherently improbable. Where there is no wind, shear flow and turbulence are also absent. Any waves in such regions are smooth and more or less regular, so that they are legitimately regarded as free gravity waves, swell for short. As may be expected, in the ocean windsea and swell appear together more often than not. Properties of the windsea emerge from observations under steady winds, in the absence of swell (i.e., under rather special conditions).

Casual observation of the windsea under such conditions from ship, sailing boat, pier, or beach readily identifies "dominant" waves propagating downwind, with more or less parallel crests, more or less evenly spaced, of more or less the same height (all of these with wide error bars), together with much irregularity. The dominant waves have an identifiable mean wavelength; moreover, they propagate at a speed close to the phase speed of the classical inviscid wave of that wavelength. This is easily verified at sea, under the right conditions: one observation I can attest to, made hundreds of kilometers from shore, from a cruise ship traveling at 20 knots (10 m s^{-1}), in a tail wind blowing fortuitously at the same speed, showed dominant waves traveling together with the ship, and to have a wavelength λ (estimated using the ship as a length scale) of about 60 m. The observed wavelength approximately satisfies the gravity wave formula from equation 2.3, $k = 2\pi/\lambda = g/c^2$.

Dominant waves are not always this long, and they do not always travel at the speed of the wind. Anybody living along the shores of a large body of water has observed that dominant waves under wind blowing from the land are short, of small amplitude and of low celerity (speed of propagation). At increasing distances from shore, at longer "fetch" as this is called, the waves become both longer and higher. At some long enough distance, measured in hundreds of kilometers, the waves reach a fully developed or saturated stage. The dominant wave example of the last paragraph belongs to this stage.

Very different from the dominant waves are a great many short-scale surface disturbances of a wind-blown surface evident on casual observation. Structures of crescentic shape, in particular, of some 20 cm radius, with sharp fronts and broad backs, with a few capillary waves ahead of the front, are ubiquitous in a moderate wind, as any swimmer or sailor may easily observe. In stronger winds, bore-like fronts, of crest lengths of the order of meters, are added to the small-scale confusion, as well as many whitecaps generated by breaking long waves as seen in Figure 2.2. Looking into sunglint under a stiff breeze one sees something akin to President Bush's "myriad points of light": Like photographer's flashes, they light up for an instant and they are gone. Presumably they are reflections from sharp fronts of short waves or bores, presumably curved, because the flashes seem to originate from point-like sources. Over a larger area, the many flashes give the impression of perpetual motion, somewhat like a TV screen that lost its signal. In a "slick" (band of smooth surface caused by a surface film of organic material) the flashes are absent. Cat's paws, groups of capillary-gravity waves under a wind gust, indicate areas of locally high shear stress, and stand out in sunglint in moderate winds. In strong winds, the points of light in a sunglint apparently fuse together, and the water surface looks more like a very bright mirror.

The short surface structures are ephemeral, and their motions seem to be confined to a thin, vigorously stirred surface layer. Near Bermuda in the North Atlantic long straight rows of sargassum weed are visible from ships. Beneath the busy surface and the stirred layer, they remain apparently undisturbed as close to the surface as 0.1 m. The short and long waves clearly behave quite differently, as if they were different species; we will have to deal with their properties separately.

2.2.1 Wave Measures

First, we will discuss the long waves. Surface elevation in wind waves is an irregular function of time at a fixed location, of location at fixed time. Most of our wave observations come from instruments recording sea surface elevation above its equlibrium level, as a time series at a fixed location, $\zeta(t)$. Under steady conditions, this yields a stationary random process, analogous to the velocity record in turbulent flow that is steady in the mean. The root-mean-square elevation, $\sqrt{\overline{\zeta^2}}$, is one measure of wave height. An often used alternative measure is the average height, crest to trough, of the one-third highest waves, $H_{1/3}$. This is also known as the "significant wave height," H_s. According to Longuet-Higgins (1952), quoted by Phillips (1977), for waves in the absence of swell a good approximation is $H_s = 4.0\sqrt{\overline{\zeta^2}}$.

An important characteristic of a stationary random process is its autocorrelation function, defined in this instance by $Z(t) = \overline{\zeta(t_0)\zeta(t_0 + t)}$, measuring over what time lag t surface elevations remain related. This is similar to autocorrelation functions we have met in the previous chapter, that define the size of eddies. An autocorrelation function of surface elevation in wind waves, observed in Cheasapeake Bay, is shown in Figure 2.3, from Kinsman (1960), as quoted by Phillips (1977). Surface elevations remain related for a long time, as the slow decay of the successive peaks shows, but their sign changes rhythmically, signifying wave motion. The separation of the peaks defines the dominant period of the wave motion. An observer perceives dominant waves of much same period passing a fixed location. The envelope of the peaks decays on a time scale of some 5 periods, a measure of individual wave persistence. In casual observation, this comes across as groups of five or six extra high or extra low waves traveling together.

Another view of windsea properties emerges from another statistical measure, the frequency spectrum of surface elevation, $\phi(\omega)$, mathematically the Fourier transform

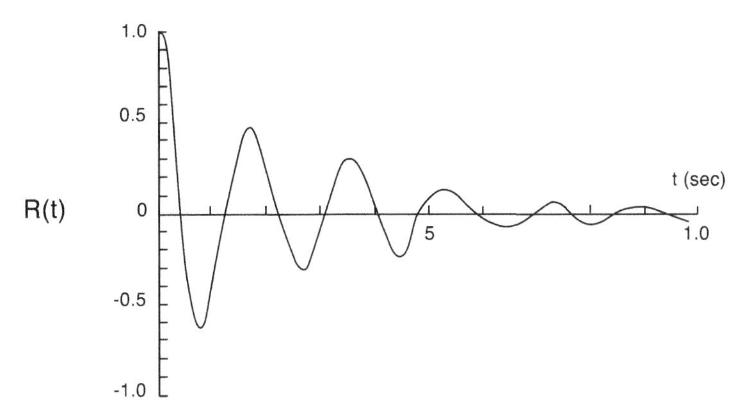

Figure 2.3 Autocorrelation function of surface elevation, observed at a location in Cheasapeake Bay by Kinsman (1960), as quoted by Phillips (1977).

of the correlation function, with ω frequency in radians. Phillips (1977) defines the connection of these two functions as follows:

$$Z(t) = \int_0^\infty \phi(\omega) \cos(\omega t)\, dt$$

$$\phi(\omega) = \frac{2}{\pi} \int_0^\infty Z(t) \cos(\omega t)\, dt \tag{2.4}$$

$$\overline{\zeta^2} = \int_0^\infty \phi(\omega)\, d\omega.$$

According to the third equation, the frequency spectrum of wind waves describes contributions to mean square surface elevation, from Fourier components of different frequencies ω. Figure 2.4, from Pierson and Moskowitz (1964), shows observed spectra, obtained in steady winds of different speeds, far from any coasts. These characterize fully developed wind waves in the absence of swell. The spectrum has a single peak, and a relatively narrow spread around the peak, indicating that most of the contributions come from the dominant waves. Equations 2.4 imply that the frequency at the peak of the spectrum is close to the reciprocal of the dominant period gleaned from the correlation function, and that the width of the spectrum is inversely related to waveheight persistence.

With increasing wind speed, the peak of the spectrum becomes higher, and moves to lower frequency, but the character of the spectrum remains the same, single peak, narrow spread. This basic similarity suggests that statistical measures of wind waves, chaotic as the waves are, nevertheless may obey simple laws.

To discover such laws, if they exist, a logical approach is to find empirically what variables affect the spectrum, apply dimensional analysis, and see if the nondimensional relationships "collapse" to simple functions. What one hopes to find are empirical laws of physics akin to the transfer laws of the air-sea interface, characterizing now the behavior of the windsea.

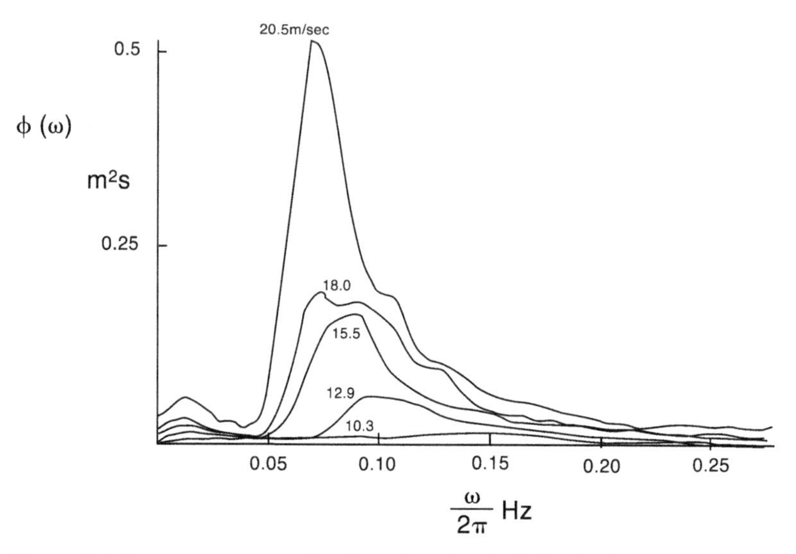

Figure 2.4 Frequency spectra of surface elevation for fully developed waves at different wind speeds, obtained by Pierson and Moskowitz (1964), as quoted by Kitagorodskii (1973).

Kitagorodskii (1973) carried out exactly this plan, relying on the Pierson-Moskowitz spectra for evidence. The observations showed that the spectral density $\phi(\omega)$ is a function of the ten-meter wind speed U, as well as of frequency ω. If, in the gravity wave range of the frequency, the only other variable of influence is the acceleration of gravity, g, the spectral density should obey a relationship of the form:

$$\phi(\omega) = func.\,(\omega, g, U) \tag{2.5}$$

or in nondimensional terms:

$$\frac{\phi(\omega)g^3}{U^5} = func.\left(\frac{\omega U}{g}\right). \tag{2.6}$$

Figure 2.5 is a replot of Figure 2.4 in these nondimensional variables, verifying the postulate of Equation 2.5. Kitagorodskii (1973) refers to this approach as similarity theory, and to the result, Equation 2.6, as a similarity law. He also notes that replacing the wind speed by the friction velocity u^* (related to $U(10)$ by Charnock's law of Chapter 1) leads to a physically more satisfactory relationship: It is the force of the wind, the wind stress, that directly drives the waves. The so-modified Equation 2.6 reveals u^* as the velocity scale of the wave motion in the windsea. That the same u^* is also the velocity scale of the shear flow and turbulence, is the strongest argument in support of the new paradigm that holds the windsea to be a byproduct of the turbulent shear flow in wind and water. We have already encountered the wind wave-turbulence connection in the transfer laws, where u^* scaled eddy velocities, while u^{*2}/g was a waveheight scale, as well as a roughness length characterizing the turbulent air flow.

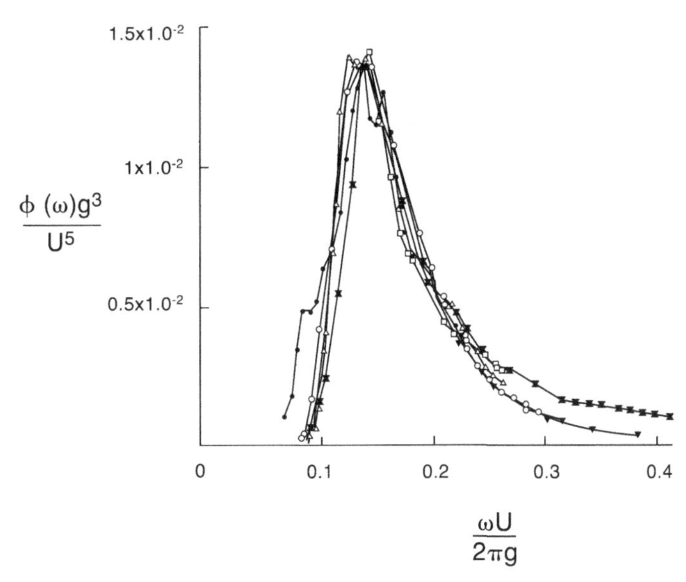

Figure 2.5 The frequency spectra of the previous figure plotted in nondimensional coordinates.

The nondimensional frequency of the spectral peak in Figure 2.5 is approximately $U\omega_p/g = 0.88$, and constitutes a key empirical characteristic of the "fully developed" windsea. Another such characteristic is the area under the spectrum, according to Equation 2.4 equal to $\overline{\zeta^2}$, the mean square surface elevation. The data yield $g\sqrt{\overline{\zeta^2}}/U^2 = 0.052$, a nondimensional measure of wave height. Taking a typical ratio $U/u^* = 27.5$, and replacing U by u^* in the formulae, Kitagorodskii's similarity laws become, for wave height and peak frequency, in a fully developed windsea:

$$\frac{g\sqrt{\overline{\zeta^2}}}{u^{*2}} = 39.4$$
$$\frac{u^*\omega_p}{g} = 0.032. \tag{2.7}$$

The relatively narrow spectrum with a single peak implies that the mean square elevation depends mainly on what we have so far called the "dominant" waves, having radian frequencies close to ω_p. These waves also propagate more or less at a phase speed given by the free gravity wave formula. The frequency at the peak of the spectrum then defines $C_p = g/\omega_p$, the celerity we may now assign to the characteristic wave, a precisely defined model of the dominant wave, of a frequency ω_p. According to the free gravity wave formula, the wavenumber is $k_p = g/C_p^2$, the wavelength $\lambda_p = 2\pi/k_p$ of the characteristic wave. Kitagorodskii's similarity laws for fully developed waves give $C_p/u^* = 31$, or $C_p/U = 1.14$, the characteristic waves barely outrunning the 10 m wind. For the significant wave height $H_{1/3} = 4\sqrt{\overline{\zeta^2}}$, we find $gH_{1/3}/u^{*2} = 160$ and $gH_{1/3}/U^2 = 0.2$. These are very similar to rules of thumb summarized earlier by Stewart (1967).

2.2.2 Wave Growth

The above laws apply to wind waves far from coasts. Near a windward (upwind) coast, waves are short in wavelength and small in amplitude, in sharp contrast to their cousins far offshore. At increasing distances from the windward coast (at longer fetch), both wavelengths and wave heights increase, and the spectrum, at short distances confined to high frequencies, spreads to lower and lower frequency. Observation revealed that these changes also obey well-defined laws under constant wind stress and in the absence of swell. Under such conditions, statistical windsea properties depend on fetch, as well as on the variables in Equation 2.5. For the spectral density, the following extension of the similarity laws applies:

$$\phi = func.\,(u^*, g, \omega, X) \tag{2.8}$$

where X denotes fetch, the perpendicular distance from the upwind shore. The nondimensional version of this relationship is:

$$\frac{\phi g^3}{u^{*5}} = func.\left(\frac{u^*\omega}{g}, \frac{gX}{u^{*2}}\right). \tag{2.9}$$

This is similar to Equation 2.6, but with the extra parameter gX/u^{*2}, identifying individual spectra at fixed fetch. Figure 2.6 shows observed spectra at varying fetch,

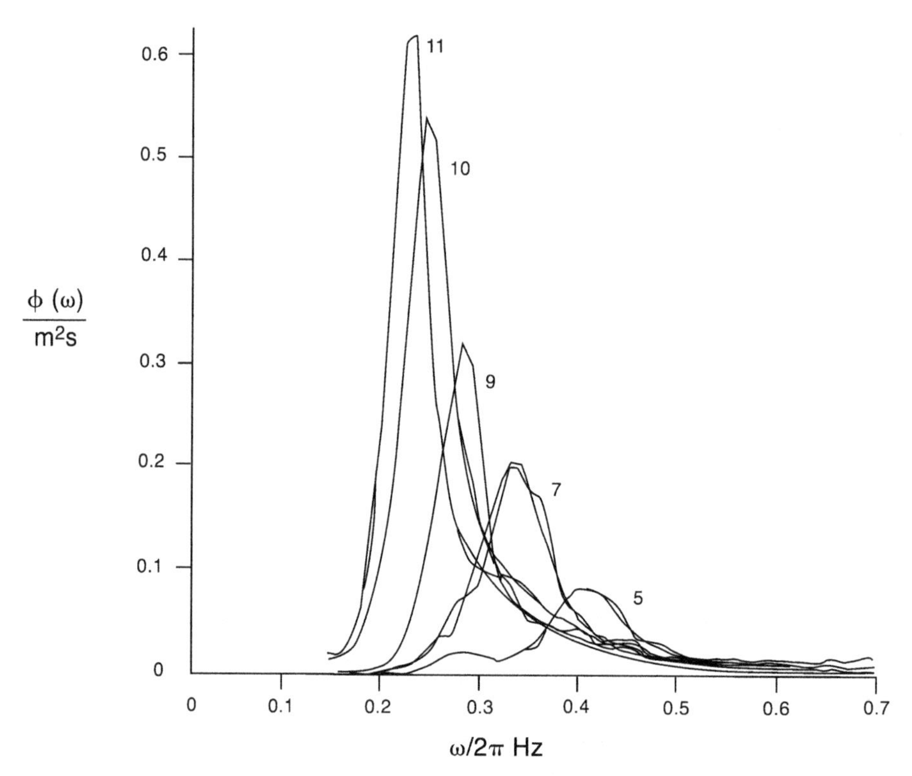

Figure 2.6 Evolution of the spectrum with increasing fetch, from 10 (label 5) to 80 (label 11) km, according to the JONSWAP observations. The labels denote successive observation platforms, the intermediate ones at 20, 37, and 52 km. From Hasselman et al. (1973).

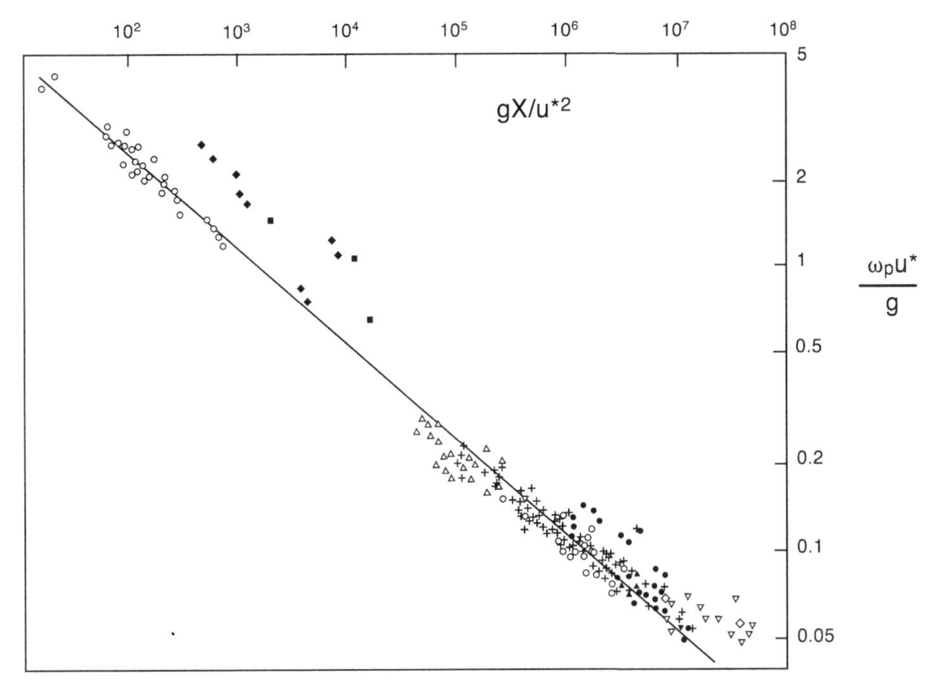

Figure 2.7 Peak frequency versus fetch in Kitagorodskii's nondimensional variables, in field and laboratory observations. From Hasselman et al. (1973).

from Hasselman et al. (1973). Although they have some common characteristics, the spectra cannot be reduced to the same shape by a rescaling of the axes. The peaks occur at lower and lower frequencies with increasing fetch, but the peaks are taller while the spreads are narrower at high than at low fetch.

The highest value of spectral density is thus a characteristic property of the wave spectrum at given fetch. At this spectral peak, the derivative of the spectral density with respect to frequency vanishes, and Equation 2.9 implies a functional relationship between nondimensional frequency at the spectral peak and nondimensional fetch:

$$\frac{u^*\omega_p}{g} = func.\left(\frac{gX}{u^{*2}}\right). \tag{2.10}$$

The particular form of this law emerged from several extensive field studies, and is, according to Hasselman et al. (1973), as shown here in Figure 2.7:

$$\frac{u^*\omega_p}{g} = 7.1\left(\frac{gX}{u^{*2}}\right)^{-1/3}. \tag{2.11}$$

Integrating the nondimensional spectra also yields a quantity depending on nondimensional fetch alone, the nondimensional mean square elevation. Hasselman et al. (1973) gave the following empirical result:

$$\frac{g^2\overline{\zeta^2}}{u^{*4}} = 1.6 \times 10^{-4}\frac{gX}{u^{*2}} \tag{2.12}$$

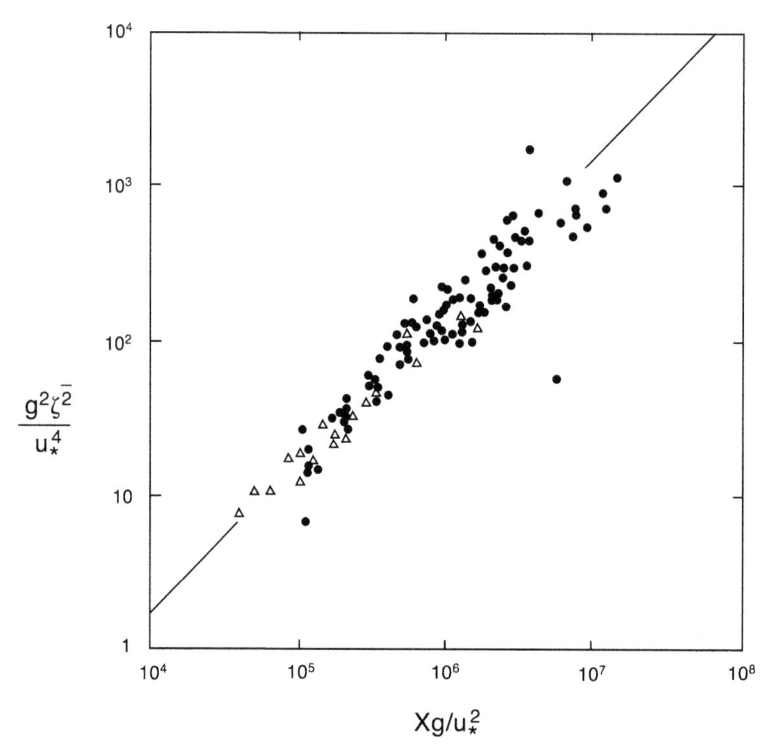

Figure 2.8 Mean square surface elevation versus fetch in nondimensional variables, from Phillips (1977).

remarkably, a linear relationship (see Figure 2.8). Eliminating the fetch from the last two equations leads to a wave-height to peak frequency relationship:

$$\frac{g\sqrt{\overline{\zeta^2}}}{u^{*2}} = 0.057\left(\frac{u^*\omega_p}{g}\right)^{-3/2}. \tag{2.13}$$

This has become known as Toba's (1972) 3/2 power law, if expressed in terms of characteristic wave period, defined as $T_p = 2\pi/\omega_p$, and significant wave height $H_s = 4\sqrt{\overline{\zeta^2}}$:

$$\frac{gH_s}{u^{*2}} = 0.061\left(\frac{gT_p}{u^*}\right)^{3/2} \tag{2.14}$$

(see Toba, 1972 and Figure 2.9).

Once the characteristic wave is long enough not to be affected by capillarity, it travels at the free gravity-wave celerity $C_p = \sqrt{g/k_p} = \omega_p/k_p$. This is true in practice on every natural surface where waves have been systematically observed. In laboratory flumes, a correction for capillarity applies. In any case, the nondimensional celerity C_p/u^* is a convenient variable to represent the stage of growth of the windsea, and has come to be known as "wave age." From the gravity wave formula, plus Equations 2.11 and 2.12, the following set of relationships emerge, expressing key wave properties in

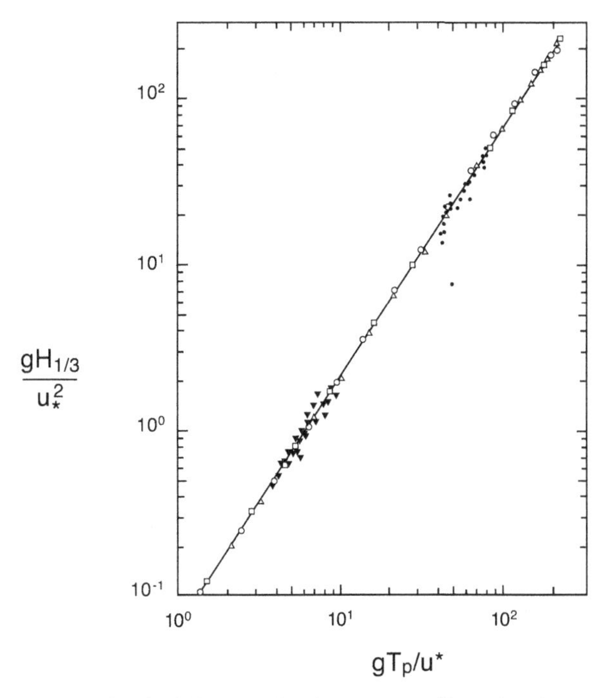

Figure 2.9 The 3/2 power law between nondimensional significant wave height and significant period. From Toba (1972).

function of wave age:

$$\frac{gX}{u^{*2}} = 358 \left(\frac{C_p}{u^*}\right)^3$$

$$\frac{gH_{1/3}}{u^{*2}} = 0.96 \left(\frac{C_p}{u^*}\right)^{3/2} \tag{2.15}$$

$$\frac{gT_p}{u^*} = 6.27 \frac{C_p}{u^*}.$$

These give the fetch, characteristic waveheight, and wave period in function of wave age. Other wave properties of interest include steepness, one measure of which is waveheight to wavelength ratio, $H_{1/3}/\lambda_p$, where $\lambda_p = 2\pi/k_p$ is the wavelength of a gravity wave of wavenumber k_p. Equation 2.15 implies:

$$\frac{H_{1/3}}{\lambda} = 0.153 \left(\frac{C_p}{u^*}\right)^{-1/2} \tag{2.16}$$

so that characteristic wave steepness decreases with increasing wave age. The steepness of a small amplitude gravity wave, $\zeta = a \sin(\omega_p t)$, is ak_p, and its amplitude a is proportional to the root mean square elevation, $a = \sqrt{2\overline{\zeta^2}}$. Using this connection to assign an amplitude to the characteristic wave, steepness as ak_p becomes:

$$ak_p = 0.34 \left(\frac{C_p}{u^*}\right)^{-1/2}. \tag{2.17}$$

For fully developed waves these formulae give 0.0275 for the waveheight to wavelength ratio, and 0.061 for ak_p.

Particles in a small amplitude gravity wave of celerity C_p execute a residual motion called "Stokes drift," because they spend slightly more time in forward than in backward motion, owing to wave propagation. The surface velocity of Stokes drift in small-amplitude, free gravity waves is $u_s = a^2k_p^2C_p$. The long waves of the windsea undoubtedly give rise to residual motion for the same physical reason as gravity waves, so that using the gravity wave formula for Stokes drift, with Equation 2.17 for steepness, should be a reasonable estimate. A substitution yields $u_s = 0.11u^* \cong 3.3u_w^*$, where u_w^* is the friction velocity on the water side, $u_w^* = \sqrt{\tau_i/\rho_w}$, smaller than the air-side u^* in the ratio of the square roots of densities. To the accuracy of this estimate, surface Stokes drift is independent of fetch or wave age, varying with wind stress alone.

Perhaps the most important property of waves is their transport of horizontal momentum. Small amplitude gravity waves transport momentum at the rate $M_t = \rho_w g a^2/4 = \rho_w g \overline{\zeta^2}/2$. The same formula presumably gives a realistic estimate of momentum transport by the windsea. A substitution of Equation 2.12 leads to:

$$M_t = 0.064 \tau_i X \tag{2.18}$$

where τ_i is the interface shear stress.

To the accuracy of the gravity wave formulae for estimating Stokes drift and momentum tranport, these results identify two invariants of wave growth under constant wind stress: the surface velocity of Stokes drift, and the divergence of momentum transport, dM_t/dX. Equation 2.18 also tells us that 6.4% of the momentum transferred from the wind to the water supports the growth of the characteristic wave, by supplying the wave momentum transport divergence. The linear growth of momentum transport with fetch comes directly from the empirical Equation 2.12, and from the gravity wave relationship of momentum transport to wave energy.

The various empirical laws governing wave growth are remarkable in their simplicity. While the functional form of the peak frequency law (Equation 2.11), and of the wave height law (Equation 2.12), follow from dimensional arguments, the power-law nature, the exponents, and the constants are all empirical and establish physical laws valid under constant wind stress and in the absence of swell, or as one might say, applying to the "perfect windsea" (à la perfect gas). Toba's law (Equation 2.14) follows from the first two, or the peak frequency law follows from Toba's law and the wave height law. The windsea properties in Equations 2.17 and 2.18 are estimates based on the hypothesis that the energy and the momentum transport of the wave field closely follow the gravity wave formulae, depending principally on the long waves.

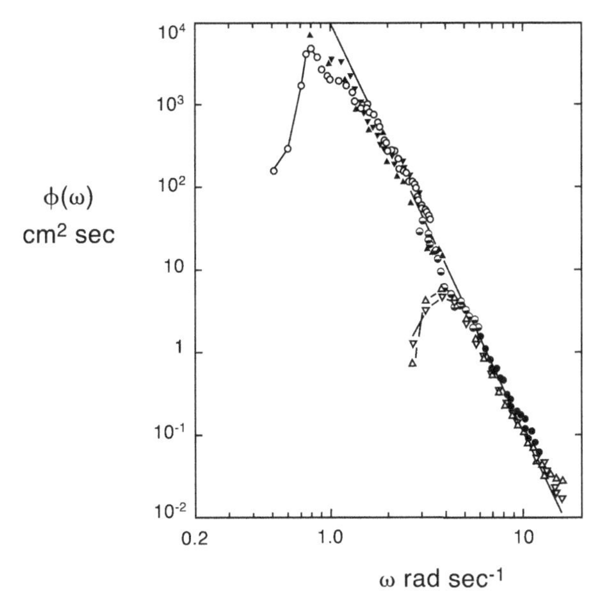

Figure 2.10 The tail of the characteristic wave at a wide range of wave ages, showing the spectral peak only in a few cases for orientation, at nondimensional fetches of $gX/u^{*2} = 5$ and 200. From Phillips (1977).

2.2.3 The Tail of the Characteristic Wave

The wind wave spectrum (Figure 2.5) shows a sharp rise ahead of ω_p, followed by a similarly sharp drop beyond, tailing off more slowly above $2\omega_p$. Remarkably, moreover, in the radian frequency range of $\omega = 1.5$–$12\ \mathrm{s}^{-1}$, spectra from short or long fetch, weaker or stronger winds, very nearly coincide at frequencies higher than their own ω_p, as Figure 2.10 shows, from Phillips (1977). In the log-log plot of this illustration, the data cluster around a straight line with a slope of -5. Phillips calls this the saturation range or equilibrium range of the wind wave spectrum, described by the formula:

$$\frac{\phi(\omega)\omega^5}{g^2} = \beta \tag{2.19}$$

where β is known as Phillips' constant, the constancy based on theoretical arguments.

Field observations showed, however, systematic variation of β with fetch, as Figure 2.11 here shows, taken also from Phillips (1977). A more appealing idea than some kind of equlibrium and $\beta = const.$ is to suppose that this "constant" is a function of the nondimensional peak frequency $u^*\omega_p/g$, as are all other characteristic wave properties. This then makes the frequency range in which Equation 2.19 applies, but now with $\beta = func.(u^*\omega_p/g)$, an appendage or "tail" of the characteristic wave. Physically, an interpretation is that many different realizations of the dominant waves, having different irregular shapes, heights and wavelengths, produce spectral densities in this range that depend on the characteristic frequency.

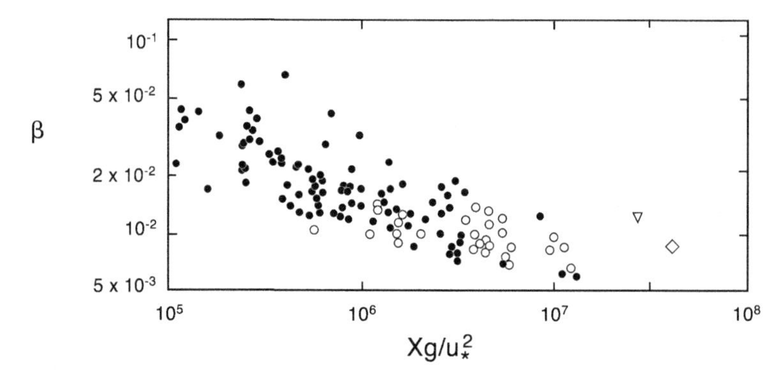

Figure 2.11 Values of the Phillips constant β from field measurements, as quoted by Phillips (1977).

If β is a function of $u^*\omega_p/g$, it is also a function of nondimensional fetch. A good match to the data in Figure 2.11 is:

$$\frac{\phi(\omega)\omega^5}{g^2} = 1.3\left(\frac{gX}{u^{*2}}\right)^{-1/3} \tag{2.20}$$

This is equivalent to $\beta = 0.183(u^*\omega_p/g)$. The so amended Equation 2.19 conforms now to the structure of the other growth laws.

Where do all these laws come from? What physical principles underlie them? Why does wave steepness decrease while waves become higher? Why are estimated momentum transport divergence and surface Stokes drift invariant? Just as thermodynamic laws arise from the chaos of molecular agitation, so we have here macroscopic laws of wave growth arising from the interface chaos of turbulence and wind waves, laws that do not follow from established laws of fluid mechanics. A "microscopic" explanation in terms of shear flow-turbulence-wave interactions may some day be found for them, but nothing like that exists today.

The theoretical literature of wind waves is replete with attempts to find such an explanation. Most postulate energy transfer between components of the wave spectrum at different frequencies or wavenumbers. The physical basis is either a tacit or explicit assumption that spectral densities arise from freely traveling waves of frequencies ω that interact dynamically, or an analogy with Kolmogoroff's (1941) ideas on the turbulent energy cascade, according to which energy flows from larger to smaller flow structures, with the dissipation concentrated at the high wavenumber end of the spectrum. Neither of these ideas is applicable to the windsea: Freely traveling waves differ in several important respects from wind waves, as discussed above (see Figure 2.2 and quotes from Bigelow and Edmondson, 1947). Energy dissipation in the windsea is dominated by wave breaking, and occurs at all frequencies, but most conspicuously near the peak of the spectrum on long waves. In a later section, we discuss the interesting physical process of wave breaking, a prominent feature of the windsea.

One well-known attempt at explaining wave growth in a windsea was Hasselman's (1962, 1963 a,b) model of the evolving wind wave spectrum through nonlinear wave-to-wave energy transfer. In this model, individual spectral components behave as inviscid gravity waves, except that their spectral densities at fixed frequency change with time or fetch, owing to energy input from wind, energy dissipation, and cross-spectral transfer. Hasselman calculated the rate of cross-spectral transfer, to or from a given frequency, from or to the rest of the spectrum, with the aid of a complex four-wave interaction theory. The model ignores the empirically well-established dependence, of just about any spectral property on the friction velocity. In a careful discussion of the Hasselman approach, Phillips (1977) shows that the calculated nonlinear transfer rates are incompatible with the empirical growth laws.

The idea that the spectral density at *any* frequency comes from freely traveling waves, does not follow from the fact that the spectrum is the Fourier transform of the observed time series of surface elevation. In Phillips's (1977) words, the spectral density so deduced from observation may come from "the presence of higher harmonics, resulting from the coherent passage of longer waves whose crests are sharper than their troughs" – or from any other distortion of the wave shape.

The dominant waves nevertheless do travel at the wave speed of free gravity waves, as we have seen. The story of wind wave origin teaches us that, even then, they may have a structure different from free waves and behave differently (e.g., grow exponentially as instability waves). The actual, observed evolution of spectral density at fixed frequency as a function of fetch also obeys similarity laws as do other wave properties. Barnett and Sutherland (1968) compared laboratory and field observations on this windsea property, and found in both a similar rapid rise first, followed by an overshoot, then a decaying oscillation and approach to a constant spectral density attained within the tail of the characteristic wave. The initial exponential growth suggests instability waves, while the overshoot and decaying oscillation are what one finds in a damped oscillator.

Spectral components of the windsea at frequencies above ω_p do not travel at their own gravity wave speed, but at the celerity of the characteristic wave, C_p, as laboratory observations of Ramamonjiarisoa (1974), quoted by Phillips (1977), have shown (Figure 2.12). Phillips remarks that the only spectral components present, at frequencies above the spectral peak, were "those associated with the deformation of the primary waves with whose speed they travel." These "do *not* [his italics] propagate as freely traveling waves .... but are bound as harmonics of the dominant waves." Phillips adds that "in the field the range of freely traveling waves is much greater," but it is not clear that any component of the windsea can be legitimately called a freely traveling wave, prior to transformation to swell. The increasing wavelength of the characteristic wave, or the spectral density overshoot of an individual spectral component are hard to reconcile with such an idea.

A further point is that the shape of the laboratory spectrum, rescaled according to Kitagorodskii's (1973) similarity laws, is much the same as spectra from field studies.

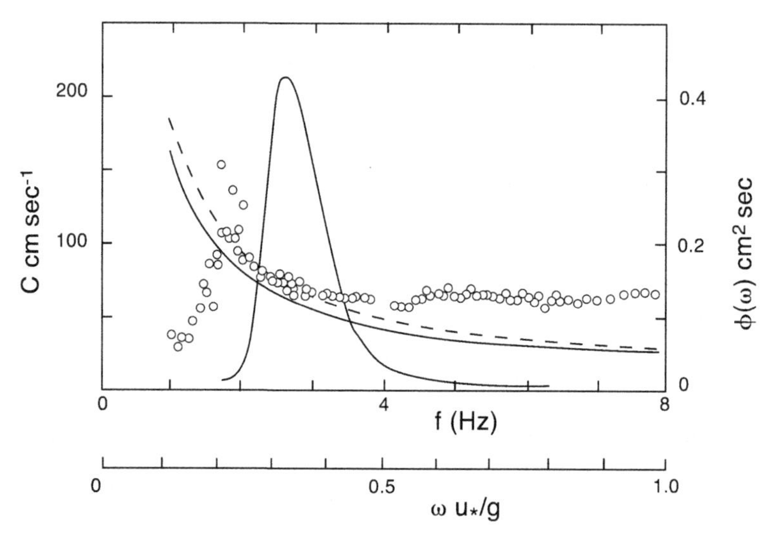

Figure 2.12 Phase speed of spectral components in a laboratory wind wave tunnel observed by Ramamonjiarisoa (1974), as quoted by Phillips (1977). Circles show observed phase speeds, the full and dotted lines close together free gravity wave celerities estimated in two ways, the other full line the frequency spectrum.

The term *similarity law* comes from the theory of hydrodynamic modeling: Model and prototype behave similarly; that is their laws are the same if stated in terms of Froude number or some other nondimensional variable, except for "scale effects" or "Reynolds number effects" attributable to viscosity. Laboratory waves model field waves in exactly the same way: the same nondimensional laws apply to both, including especially to the distribution of spectral densities over frequency, except for minor viscosity effects at high frequencies. What Figure 2.12 tells us about the phase speed of spectral components and how they vary with frequency should apply to the windsea after the appropriate similarity transformations.

Phillips (1977) finishes his discussion of wind wave spectrum development with the following:

"Many details are indistinct, and the parts played by the various dynamical processes involved are clearly not simple, changing constantly as the waves develop. Indeed, a dynamical kaleidoscope!"

2.2.4 Short Wind Waves

The tail of the characteristic wave extends to a radian frequency ω of about 10–12 s^{-1}, as shown in Figure 2.10. Already at this frequency, and more so above it, spectral density is small, and makes a negligible contribution to the mean square elevation. On the other hand, the spectral density of surface *slope*, the gradient of surface elevation, $\nabla\zeta$, has its largest values at still higher frequencies. This was revealed by pioneering laboratory observations of Cox (1958), and by field observations relying on sun glitter of

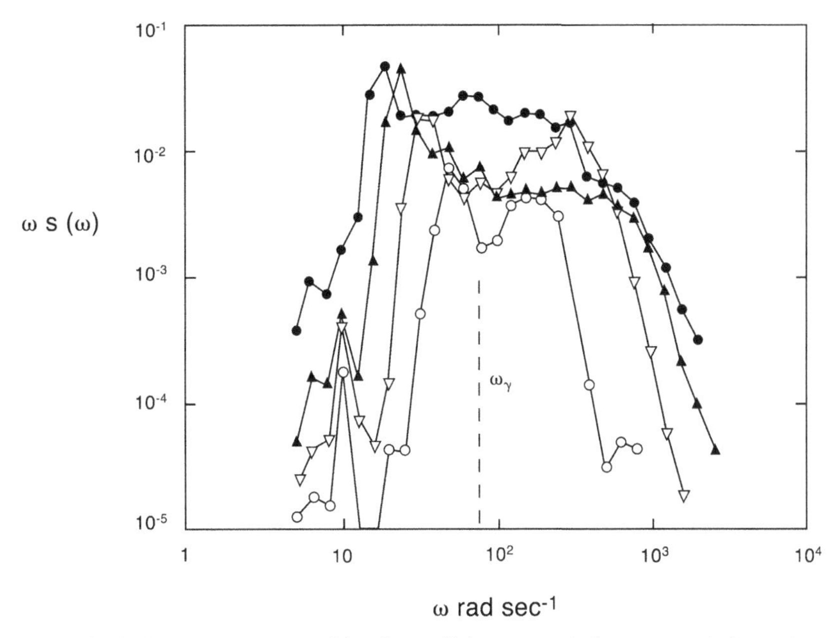

Figure 2.13 Frequency spectra of the slope of laboratory wind waves at winds varying from 3 to 12 m s^{-1}. The dotted line marks the radian frequency ω_γ that separates capillary waves from gravity waves. Data of Cox (1958), as quoted by Phillips (1977).

Cox and Munk (1954). Figure 2.13 shows Cox's laboratory data (from Phillips, 1977) on frequency-weighted slope spectral density ωS against frequency ω, with S the slope spectral density. Two striking properties of the spectra are negligible overlap with the elevation spectrum, and a nearly symmetrical distribution of frequency-weighted spectral densities around the logarithm of the frequency that separates gravity waves from capillary waves, $\omega_\gamma = (4g^3/\gamma)^{1/4}$. At this frequency, the celerity of gravity-capillary waves is a minimum. The spread around ω_γ immediately suggests that combinations of short gravity and capillary waves dominate the slope spectrum: Their presence on wind-blown natural surfaces is well-known. At any rate, the population of surface structures responsible for the slope spectrum is evidently separate from the population that produces the elevation spectrum from the characteristic wave to its tail. We will call the former population simply "short wind waves," even though they are not necessarily wave-like at all, and in nature generally three-dimensional.

Figure 2.13 also shows that both the spectral densities of surface slope, and the frequency range which they cover, increase with wind speed. So does therefore the integral of the spectrum, which is the mean square slope $\overline{(\nabla\zeta)^2}$. Results of Cox and Munk (1954) derived from sun glitter on a natural wind blown surface show the increase to be approximately linear with wind speed (see Figure 2.14 and Phillips, 1977). The field data more or less agree with the laboratory observations, except at the lowest wind speed, where the laboratory mean square slope falls below a linear trend. At these short wavelengths, corresponding to the high frequency end of the spectra in Figure 2.13, scale effects resulting from viscosity may be expected.

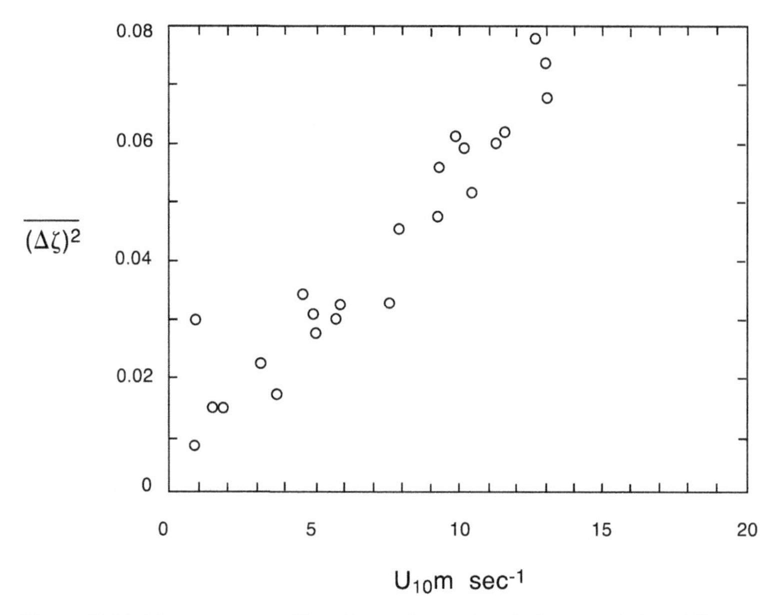

Figure 2.14 Mean square surface slope of oceanic wind waves deduced from sun glitter on a clean surface. From Cox and Munk (1954), as quoted by Phillips (1977).

Taking wind speed to be a proxy for friction velocity, the observations show that $\overline{(\nabla\zeta)^2}$ depends on u^*, as well as on the short-wave variables g, γ. A promising hypothesis is then that the mean square slope depends on those variables alone. With $\omega_\gamma = (4g^3/\gamma)^{1/4}$ a convenient proxy for surface tension, a nondimensional independent variable of familiar form is $u^*\omega_\gamma/g$. Mean square slope is already nondimensional; fitting a straight line to the data in Figure 2.14, we arrive at the relationship:

$$\overline{(\nabla\zeta)^2} = 0.017\frac{u^*\omega_\gamma}{g} \tag{2.21}$$

a similarity law for the slope spectrum. The nondimensional spectral density ωS is a function of ω/ω_γ, in addition to $u^*\omega_\gamma/g$.

2.2.5 Laboratory Studies of Short Waves

Field observation of short waves is a particularly challenging task on account of their irregularity, not made any easier by the presence of long wind waves. Not surprisingly, systematic, quantitative field observations on short waves are far less plentiful than on long waves. We would expect laboratory studies to help out. There is a rub, however: The "short wave" range starts in the windsea near a radian frequency of $\omega = 10\ \mathrm{s}^{-1}$, where the celerity of a gravity wave is near $1.0\ \mathrm{m\,s}^{-1}$, its wavelength about 60 cm. Shorter wind waves on natural wind-blown surfaces are generally three-dimensional and short-lived. This is not so with wind waves in a laboratory flume. Owing to the short fetch, laboratory wind waves of order 10 cm wavelength are more or less regular and parallel-crested, and play the role of characteristic waves. Nevertheless, laboratory observations on short wind waves have yielded a veritable treasure trove of information on how the shear flow in air and water interacts with the wave motion and the turbulence. The insights so gained helped clarify the role of short waves in air-sea transfer processes, differences between laboratory and field notwithstanding.

A whole series of pioneering observations on short wind waves have been carried out in Professor Toba's laboratory, at Tohoku University, in Sendai, Japan. That work (Toba et al., 1975; Okuda et al., 1977; Kawai, 1981, 1982; Kawamura and Toba, 1988; Yoshikawa et al., 1988; Ebuchi et al., 1993; Toba and Kawamura, 1996) provided most of the material for the following synthesis.

The short and regular initial instability waves on the laboratory air-water shear flow break down into a chaotic windsea and interior turbulence within ten seconds or so, as we discussed before. Eventually a statistically steady state of short wind waves emerges, with sharp-crested characteristic waves, similar to those visible on small lakes and ponds under moderate to strong winds. They are very "young" wind waves, of low C_p/u^*, quite steep on average; the steepness of individual dominant waves varying, however, from wave to wave.

Those variations have major effects on the air flow over the waves. In a coordinate system moving with the waves, the air flow smoothly follows the undulating surface – over waves of lesser steepness. Over some of the steeper waves, on the other hand, the air flow separates near the crest, as several careful studies have demonstrated (Banner and Peirson, 1997; Kawai, 1981, 1982; Kawamura and Toba, 1988). A "separation bubble" forms downwind of the crest, a trapped, slowly circulating air mass. A layer of strong shear separates the bubble from the free stream above, its center arching over toward the next downwind crest as Figure 2.15 illustrates (from Kawamura and Toba, 1988). The air flow reattaches again ahead of the next crest, so that the separation bubble occupies close to a full wavelength, the exact length varying from one steep wave to another. The reattaching streamlines bend upward fairly sharply where they reach the water surface, as may be seen in a photograph from Banner and Peirson (1997).

Superimposed on the above mean flow pattern is a system of large eddies attached to the waves and of wavelength dimensions, spanning the entire shear flow boundary

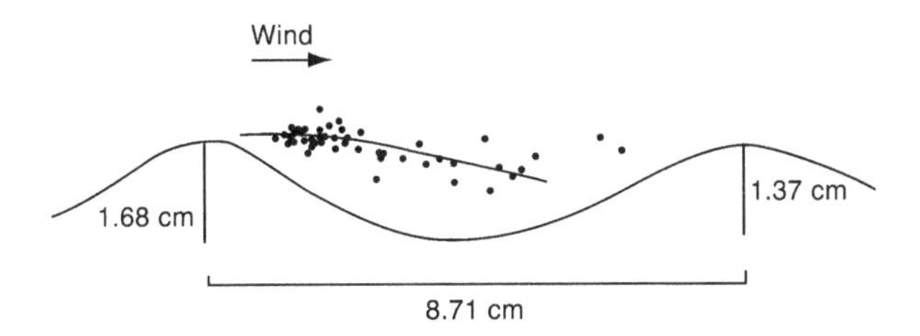

Figure 2.15 Location of high shear layer at the outer edge of the separation bubble on a steep laboratory wind wave, detected by Kawamura and Toba (1988).

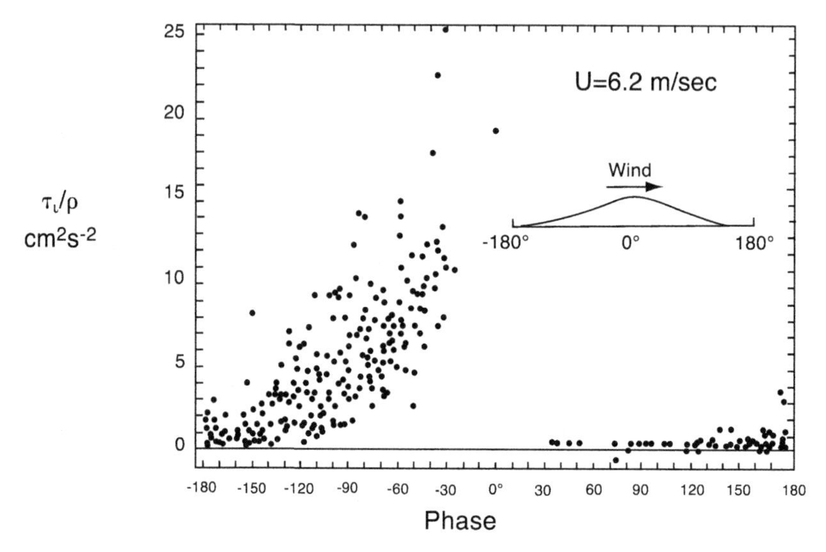

Figure 2.16 Interface shear stress distribution on laboratory wind waves, reported by Okuda et al. (1977). The peak value near the crest is some five times the area-average wind stress.

layer (Kawamura and Toba, 1988). As in wall-bound boundary layers, at a fixed point the eddies generate "bursts" of slow fluid upward, alternating with "sweeps" of fast fluid downward. Over the wavy surface, however, the bursts are statistically tied to the wave crests, the sweeps to the troughs. By way of explanation, Kawamura and Toba offer the conceptual model of a separation bubble drained by the flow, when the steep wave originating it decays, an event they call a "big burst." The slow fluid in the bubble leaves over the downwind crest, the void behind it filled by a sweep of fast fluid.

The net horizontal force exerted on the interface, driving the water or braking the air depending on your point of view, consists of viscous shear stress and pressure force acting on the inclined surfaces of the waves. The viscous shear stress is very unevenly distributed: most of it acts on the windward face of the waves, as Okuda et al. (1977) have found (Figure 2.16). In Okuda et al.'s case, the integrated shear stress force came close to accounting for the entire momentum transfer. Banner and Peirson (1997) reported a somewhat more even distribution, significant shear force also on the leeward face. Banner and Peirson (1997) furthermore determined pressures on the interface, found them high also on the windward face, and estimated the magnitude of the resultant net pressure force at 0 to 70% of the total horizontal force, the proportion increasing from low- to high-wind speeds. This leaves 30 to 100% for the shear stress force, highest in weak winds.

The flow pattern on the air side of the steeper waves explains how forces on the windward face come to be as large as they are: Where the air flow reattaches at the downwind end of a separation bubble, both shear and pressure are intense. The shear stress is high because the boundary layer is very thin near the point of reattachment, and

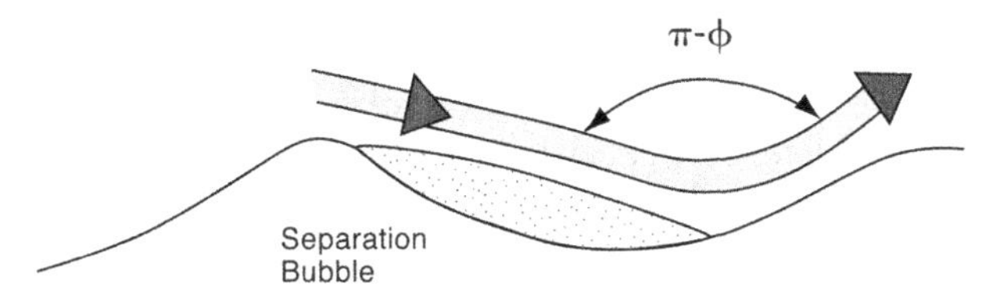

Figure 2.17 Schema of reattaching high shear layer behind a separation bubble.

the velocity outside the boundary layer is the high free stream velocity. The pressure force is high because it has to redirect the free shear layer through an angle θ.

A simple conceptual model is a two-dimensional jet (Figure 2.17), of a thickness comparable to wave amplitude a, impinging on an inclined solid surface (water being so much more massive than air, this is a good approximation). A pressure force normal to the surface deflects the jet through the angle θ, of the order of wave steepness. Acting on a wave surface inclined at a similar angle, the pressure force has a horizontal component of order $\theta^2 a U^2$ times air density, with U wind speed. The horizontal momentum transport of the air flow is diminished by this amount, in each wave with air flow separation. Divided by wave length, λ, the contribution to the resultant interface stress is of order $\theta^3 U^2$ (a/λ is of order θ). A typical value of θ for steeper laboratory waves is 0.2, so that if all waves were this steep, the drag coefficient resulting from the pressure forces alone would be $C_D = 8 \times 10^{-3}$, some four times higher than the typical observed drag.

On the water side, in the same wave-following frame of reference, the flow is upwind, counterintuitively, and also smooth under waves of low steepness. On the windward face, where the shear stress is high, a laminar boundary layer thickens and accumulates momentum deficit. The slowed down fluid shows up in pockets of vortical fluid at the crests of waves, on waves of low steepness without disorganizing the flow pattern.

On the crest of steeper waves, however, a "roller" develops from vortical fluid, a separate closed circulation, akin to, but much shorter than, the separation bubble on the air side. The presence of a roller means that the wave is "breaking," a phenomenon discussed in detail in the next section. Breaking waves transfer much of their momentum to the shear flow. The front of the roller is steep, bore-like. The air flow separates over the roller, as several observers reported. Underneath and "downstream" (i.e., upwind) of the roller a trailing turbulent boundary layer forms on the water side of the interface, containing the momentum lost from the wave to the shear flow.

Momentum transport "crosses" the air-sea interface as shear stress and pressure force. How do interface forces on laboratory waves hand over the momentum transport to fluid motions, to reappear as Reynolds stress some distance above and below the water surface? On the air side, the upward bursts of slow air over the crests, and downward sweeps of fast air over the troughs add up to the Reynolds stress. As we have seen, the slow fluid in the bursts comes mostly from separation bubbles behind steeper waves and separated shear layers slowed down by pressure forces. Thus, the interaction of waves with the air-side shear flow gives rise to ordered eddy motion, its length scale the wavelength, its reach the boundary layer depth. The ordered eddy motion carries the entire momentum transport, (i.e., windward momentum deficiency upward, and surplus downward.)

On the water side, any pressure force on the windward face of waves, exerted by impinging shear layers or otherwise, presses down on the free surface of a fluid in motion, (in a wave-following frame) and generates a trailing wave, by a mechanism to be discussed in connection with wave breaking. The momentum of the trailing wave combines with the momentum transport of the waves that follow. Those waves then steepen until they eventually break and lose their momentum to the water-side shear flow. This mode of momentum transfer on the water side is very different from what happens on the air side.

The water-side shear flow, sustained by viscous shear and wave breaking, transfers its momentum downward through large eddy motions coupled to the waves, similarly to the air side. Toba et al. (1975) reported "forced convection" from the crest of short waves, in the first of such laboratory studies at Tohoku University. They found that "the surface converges [ahead of] the crest, making a downward flow there." In the light of later studies, we may place the convergence at the front of the roller on a breaking short wave. According to Ebuchi et al. (1993): "From a point close to the crest on the leeward face of individual waves, a parcel of water with high velocity (burst) goes downward through a particular route relative to the wave phase." They showed the particular route in an illuminating illustration, (Figure 2.18): starting at the toe of the roller, the burst dives downward and backward (upwind). The compensating sweep to replace lost roller fluid is presumably less noticeable. In a further extension of the Tohoku University studies, Yoshikawa et al. (1988) and Toba and Kawamura (1996) have referred to the turbulent flow regime on the water side of the interface as the wind wave coupled "downward-bursting boundary layer." Thus, once the momentum is in shear flow momentum, the further handover process on the water side is much as on the air side, burst and sweep phases of large eddies sustain Reynolds stress, the eddies tied to rollers, therefore moving with the steeper waves. The difference

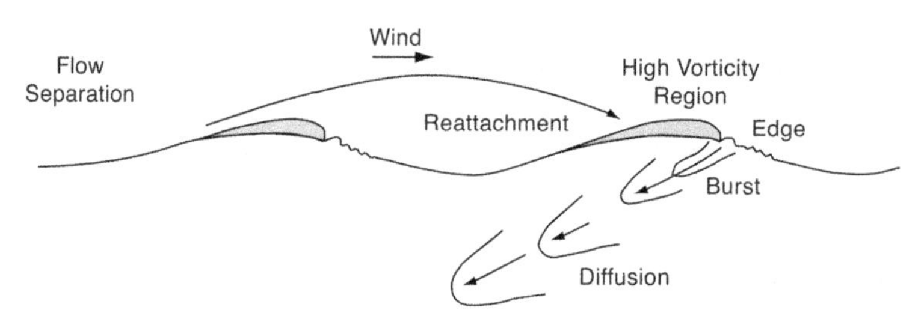

Figure 2.18 Interacting dominant flow features above and below the air-water interface illustrating: reattaching air flow, roller ("high vorticity region") on the wave crest, downward burst at roller toe, capillaries in front. From Ebuchi et al. (1993).

is at the start of this route: Pressure forces causing waves to steepen and break, and hand over momentum downward. How exactly this happens is the subject of the next section.

2.3 The Breaking of Waves

From the growth laws, we inferred that just 6% or so of the wind-imparted momentum supports the increase of wave momentum transport with fetch. A further share of the wind input must replace wave momentum continually lost and transferred to the water-side shear flow via various dissipative processes. The most important such process is wave breaking, a more or less violent overturning motion, made conspicuous by a multitude of whitecaps, particularly numerous in strong winds. Foam makes whitecaps white, air bubbles form the foam, and the overturning motion traps the bubbles. Not all waves are breakers, however, and not all breakers trap enough air bubbles to carry whitecaps. The physics of wave breaking remained obscure until quite recently, when a combination of field, laboratory, and theoretical studies elucidated its essentials.

Waves "break" when some fluid particles on their surface, usually near the wave crest, travel faster than the wave, overtaking it. From the point of view of an observer traveling with the wave, the fast particles still travel forward in the direction of wave motion, while slower particles move backward. If the motion in this frame of reference is steady, its streamlines have a stagnation point on the surface where the forward motion begins, and another such point where it ends. The streamline connecting the stagnation points forms the boundary of a separate closed circulation cell, known as a "roller," a pocket of fluid moving with the wave. It is usually located near and slightly ahead of the wave crest (Figure 2.19). The roller may be a boiling, churning mass of water, as in the "plunging" breakers often seen on beaches. At the other extreme, no more than a sharp crest marks the rollers on "microscale" breakers of short wavelength, often preceeded by a few capillary waves. At the intermediate scale,

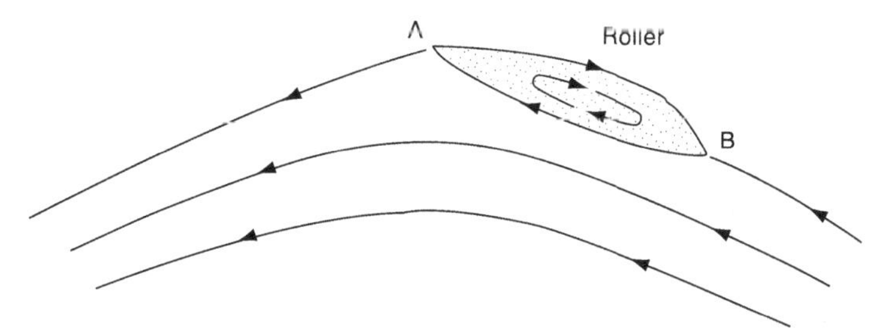

Figure 2.19 Schematic mean flow streamlines in a wave-following frame of reference near the crest of a breaking wave, showing a roller. The surface fluid travels faster than the wave (in a fixed frame of reference) between the tail A and toe B of the roller.

the "spilling" breakers of long wind waves seem to spill gently down the forward faces of the waves, from a roller ahead of the crest. Many spilling breakers are marked by whitecaps. The shape of rollers on spilling breakers resemble bores and hydraulic jumps seen on tidal rivers, in that they contain steep fronts ahead of a slightly higher surface that propagates relative to the underlying water mass. A roller similarly moves with the wave, leaving the underlying fluid behind.

The technical criterion of wave breaking is therefore $u > c$, where u is fluid particle velocity somewhere on the wave surface, and c is wave celerity. In a seminal paper, Banner and Melville (1976) pointed out that this criterion not only implies the presence of a roller, but also a separation bubble, the roller's twin on the air side of the interface. Because air and water molecules stick together at the interface, the stagnation points on the water surface are also stagnation points for the air flow, in the wave-following frame of reference.

Under what circumstances can we expect surface particle velocity in wind waves to exceed celerity? In the absence of wind, on a mechanically generated gravity wave, the maximum particle velocity is of order $u = akc$, steepness ak times celerity. At given celerity, gravity waves break at a critical steepness. Wind stress generates surface shear flow in the water in addition to waves, increasing surface velocities. Wind waves therefore break at lower steepness than gravity waves (Banner and Phillips, 1974). On the growing waves of the windsea, at given wind stress and fetch, both the surface shear flow and the steepness of the characteristic wave are fixed, and so is therefore the average propensity of long waves to breaking. Under such conditions, in a population of long wind waves, a certain fraction are steep enough to break. According to the similarity laws, that fraction should depend on wave age.

What happens to a wave when it breaks? Just as breakers on a beach, those far from shore lose some of their energy, ending up after breaking with lower waveheight and momentum transport, and a lot of turbulence. Loss of momentum transport by the wave implies momentum transfer to the shear flow, loss of wave height, and energy transfer to shear flow and turbulence. The mechanism of the momentum transfer, a puzzle until recently, has turned out to be particularly interesting.

2.3.1 Momentum Transfer in a Breaking Wave

Two classic memoirs of Duncan (1981, 1983) elucidated that mechanism, reporting on his laboratory experiments. Figure 2.20 illustrates his apparatus: laboratory flume and a towed hydrofoil, spanning the flume. A quasi-steady breaking wave appeared behind the hydrofoil "when the foil speed, angle of attack and depth of submergence were adjusted properly." The angle of attack of the hydrofoil was either 5° or 10°, in which configuration the hydrofoil exerted a strong downward force on the water. This force, moving forward with the hydrofoil at twelve different towing speeds between 0.62 and 1.03 m s^{-1}, left a wave behind, steep enough to break, and to carry a roller in a more or less steady formation. The first larger breaking wave was followed by a smaller amplitude trailing wave train. In a frame of reference moving with the hydrofoil, the

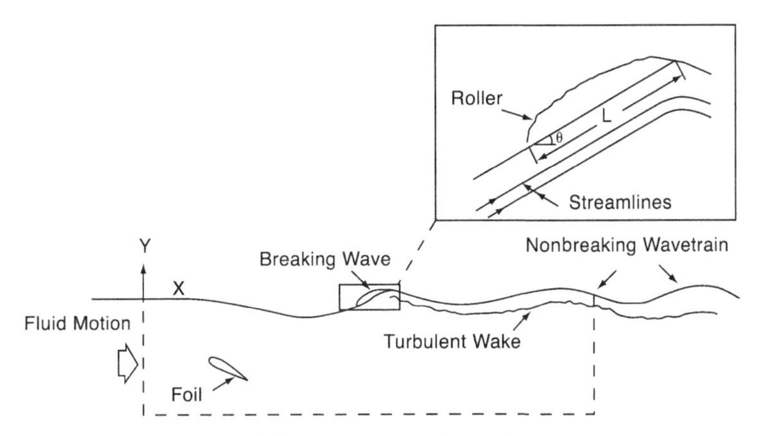

Figure 2.20 Duncan's (1981) apparatus for the study of quasi-steady breaking waves and following wave train, with details of the roller (inset) and control volume for momentum balances (broken line).

waves were stationary, while the fluid streamed by in the direction opposite to the hydrofoil's motion. The flow behind the hydrofoil thus simulated conditions along and behind a breaker on the ocean surface. Duncan's observations determined overall momentum balances, as well as the roller-wave interaction.

First, the overall momentum balances. Duncan (1981, 1983) calculated these for control volumes sketched into Figure 2.20. The undisturbed flow (in a frame of reference moving with the hydrofoil) ahead of the hydrofoil carried no waves, so that its momentum transport was simply the vertical integral of density times approach velocity, equal in magnitude to the hydrofoil velocity. Behind the hydrofoil, but ahead of the roller, the momentum transport was less by the horizontal force on the hydrofoil per unit width of the flume, the hydrofoil's so-called "wave-making resistance." In the absence of breaking, a trailing wave train moving with the hydrofoil would carry all this momentum upstream as wave momentum transport. With a breaker present, there was still a trailing wave train behind the breaker, but of smaller amplitude, as well as a turbulent "wake" (i.e., surface shear flow slowed down by the breaker). The total momentum deficit was the same before and behind the breaker, but behind the breaker it consisted of the upstream wave momentum transport of the trailing wave train, plus the momentum transport deficit in the trailing turbulent wake.

The wake started under the roller, and slowly grew in thickness with distance behind. Its growth rate was the same as of other turbulent wakes behind immersed obstacles, while its momentum transport deficit remained constant, and could be calculated from observation. That deficit had to come from a shear force that the roller exerted on the fluid stream underneath it, analogously to the resistance force of any immersed object. The turbulent wake analogy was confirmed in full detail by observations of Battjes and Sakai (1981). Furthermore, dye injection in Duncan's experiments visually demonstrated the physical presence of a turbulent wake, growing in depth "downstream" (i.e., with distance behind the hydrofoil). The important point for the physics of wave breaking is that loss of wave momentum transport in the breaker equaled wake momentum gain behind, revealing the breaker to be an instrument of "wave to wake" momentum transport conversion.

For the purpose of determining roller-wave interaction, it is necessary to regard the roller as a two-dimensional body of fluid separate from the rest of the wave motion below. This is a conceptual model of only the mean motion, superimposed on which there is vigorous turbulence, exchanging fluid between the roller and the underlying flow, and sustaining a Reynolds stress. That stress, integrated over the bottom of the roller, adds up to a shear force equaling the momentum deficit of the trailing turbulent wake. Duncan could not rigorously assign a boundary separating roller and wave motion: he approximated it by a straight sloping line, of length L, from the observed rear stagnation point (in the wave-following frame) to an apparent toe (see the insert in Figure 2.20). The length L, and the inclination angle θ against the horizontal of the supposed bottom of the roller, were the important empirical shape parameters of the roller needed in the dynamical analysis.

Pressure as well as shear forces act on roller bottom, the Reynolds stress being quite high: in a similar experiment Battjes and Sakai (1981) found it peaking at a value of $\overline{u'w'} = 0.01U^2$, where U is flow velocity relative to the hydrofoil or to the roller. The net force on the roller in its quasi-steady observed state has to vanish; the vertical and horizontal force balances are therefore:

$$\begin{aligned} W &= \tau L \sin(\theta) + pL\cos(\theta) \\ \tau L \cos(\theta) &- pL\sin(\theta) = 0 \end{aligned} \tag{2.22}$$

where W is the weight per spanwise width of the roller, τ and p are shear stress and pressure along roller bottom. The shear stress force τL generates the wake, and its horizontal component, $\tau L\cos(\theta)$, equals the observed momentum transport deficit of the wake. The same integrated shear force component, acting on the roller, balances the horizontal pressure force on the inclined bottom. For this balance to hold, the roller must sit on the forward face of the breaking wave, as it is observed to do, $\theta > 0$.

Eliminating the pressure terms from Equation 2.22, we find $W\sin(\theta) = \tau L$, tying the weight of the roller to the shear force determined from the trailing wake. This should be the roller's cross-sectional area shown in Figure 2.20, times the specific gravity of the fluid, times the aceleration of gravity. To match the shear force, the specific gravity of the fluid in the roller had to be 0.61, much less than 1.0, the specific gravity of water. Duncan attributed the difference to bubbles in the roller. This is unlikely to be the full explanation. Alternative possibilities are that the actual roller bottom was above its supposed location, so that the actual roller area (and weight) was less than estimated, or that the effective value of θ was less than estimated.

The angle θ that represents the average slope of the roller bottom was always small. If we eliminate the shear force terms from Equation 2.22, we find $W\cos(\theta) = pL$, the pressure force balancing most of the weight. This indeed is the essence of the separate roller conceptual model, the weight of fluid external to the wave motion pressing down on the fluid stream underneath the separation streamline. The inclination of that streamline against the horizontal ensures the presence of a horizontal pressure force component, that balances the shear force while diminishing wave momentum transport.

According to the momentum balances, the momentum loss of a breaking wave equals the shear force on the roller, τL. As in similar turbulent flow problems, the shear stress is likely to vary with the square of the velocity difference between roller and the fluid below, equal to wave celerity c. If we take Battjes and Sakai's (1981) laboratory value for the drag coefficient, we have the momentum loss per breaking wave of $M_d = 0.01\rho c^2 L$. Roller length is certainly less than a quarter wave, let us say $L = 0.3k_1^{-1}$. Dividing the momentum loss per breaker by wavelength we get the following rate of momentum loss per unit area:

$$\frac{M_d}{\lambda} = \frac{0.003\rho c^2}{2\pi} \cong 5.10^{-4}\rho c^2. \tag{2.23}$$

At long fetch we may put $c = U$, the wind speed. According to this result, if all waves were breakers, the total downward momentum transfer via wave breaking would be about a quarter of the wind stress, given a typical drag coefficient of $C_D = 2.10^{-3}$. Our estimates here are very "soft," however, and all we can conclude is that the breakers handle a significant fraction of the total downward momentum transfer.

The weight of the roller pressing on the surface has other effects. According to classical hydrodynamics, pressure acting on the surface of flowing water generates waves. Lamb (1957) discusses this phenomenon in detail and develops its linear theory for small disturbances. The theory shows that a concentrated force on the surface induces a gravity wave train downstream of its point of application, as well as a capillary wave train upstream. Lamb quotes the effect of a fishing line as a practical illustration, described by Russell in a delightful passage:

"When a small obstacle, such as a fishing line, is moved forward slowly through still water, or (which of course comes to the same thing) is held stationary in moving water, the surface is covered with a beautiful wave-pattern, fixed relatively to the obstacle. On the up-stream side the wave-length is short, and as Thomson [Lord Kelvin] has shown, the force governing the vibrations is principally cohesion. On the down-stream side the waves are longer and are governed principally by gravity. Both sets of waves move with the same velocity relatively to the water; namely that required in order that they may maintain a fixed position relatively to the obstacle."

For cohesion, read surface tension. The two kinds of waves having the same velocity as the moving water, $c = u$, have wavenumbers, k, satisfying the classical dispersion relationship we wrote down earlier in this chapter (Equation 2.3):

$$\gamma k^2 - u^2 k + g = 0. \tag{2.24}$$

The two roots of this equation, at $u > 0.4$ m s^{-1} are, to a good approximation $k_2 = u^2/\gamma$ and $k_1 = g/u^2$, characterizing capillary and gravity waves, respectively. The value of γ is near 8×10^{-5} m^3 s^{-2}, so that wavenumbers of capillary waves are

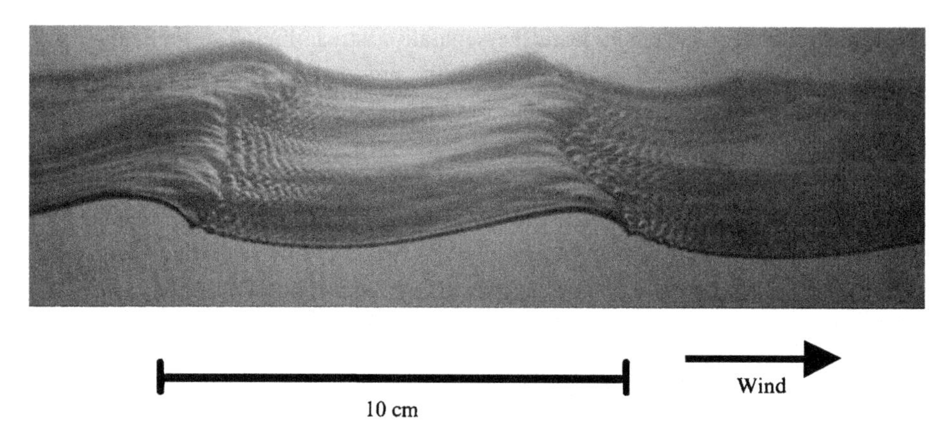

Figure 2.21 Photograph of laboratory wind wave at a fetch of 6 m, under a wind of 5.75 m s^{-1}, from Kawamura and Toba (1988). The wave shape is well-defined on the front window of the laboratory flume, the roller is marked by minor irregularities on the surface of the crest, the capillaries in front revealing the "fishing line" effect of the roller.

extremely high at celerities equal to those of long gravity waves at even moderately long fetch. Viscosity eliminates such very short capillaries, so that the upstream capillary waves of the fishing-line problem are only observable when they accompany short gravity waves, with a typical wavelength of $\lambda = 0.1$ m, or so.

Lamb (1957) gives formulae for calculating surface elevation upstream and downstream of a concentrated vertical force acting on the surface. For gravity waves with celerities of order 10 m s^{-1}, carrying a realistic size roller, the calculated downstream effect of the roller weight is negligible, compared with the effect of the horizontal force on the inclined wave surface. On short waves, however, calculations show that a typical roller generates at least upstream capillary-gravity waves of noticeable amplitude.

At the short wavelengths of laboratory wind waves, a few capillary waves should survive elimination by viscosity. They should therefore be visible at very short fetch on natural wind-blown surfaces, as well as in laboratory flumes. They are indeed commonly observed. A particularly clear illustration of a laboratory roller with capillaries ahead is a photograph by Kawamura and Toba (1988), Figure 2.21 here. The roller shape is distinct on the transparent wall of the flume, its surface ruffled by eddy motion underneath. Visible capillaries extend to some 10 wavelengths in front.

2.4 Mechanisms of Scalar Property Transfer

The ultimate objective of this chapter is to demonstrate how wind waves facilitate the transfer of scalar properties and momentum across the air-sea interface. The preceeding sections elucidated many details of the different transfer processes, and the role of wind waves in them. A synthesis is now needed, with the focus on how properties "cross the interface," rather than on waves or their interaction with the shear flow. Because the problem of scalar transfer is simpler, we start with its discussion.

Scalar properties, heat, water vapor, or carbon dioxide cross the air-sea interface via molecular conduction or diffusion. Molecular transport on its own is a slow process, however, as we have seen in the previous chapter. The resistance to transfer is proportional to rapidly growing thermal or diffusion-boundary layer thickness, and the square root of diffusivity. The lesser diffusivity, on the air or water side, controls the transfer rate. Wind waves and turbulence greatly diminish the resistance as they "feed" molecular diffusion, and hold diffusive boundary layer thicknesses small. Examination of this interplay reveals that the shortest flow structures are most effective in facilitating transfer. The interaction of the smallest eddies and the shortest waves with molecular diffusion, on the side of the lesser diffusivity, then controls the transfer rate.

An important distinction between the two sides arises from the high density ratio of water to air, owing to the boundary condition that air and water molecules stick together at the interface. The much more massive water slows down this motion, to the point where the interface appears almost like a solid surface to the air-side eddies, preventing significant eddy motion within the interface. By contrast, the air side is almost like vacuum to the water-side eddies, so that along-interface eddy motion is unhindered. This difference affects the way eddies and molecular diffusion cooperate in scalar transfer.

2.4.1 Water-side Resistance

How does the eddy-molecular diffusion interplay work? First we will discuss the water side, where the answer is fairly clear. The water side controls the transfer of gases of low diffusivity, as we have seen. The chemical engineering literature of gas transfer contains many attempts to elucidate the mechanism responsible. One intuitive idea due to Higbie (1935), and elaborated on by Danckwerts (1951), was "surface renewal." This supposes that eddies bring fluid from the interior to the surface, there expose it for a limited time to a gas constituent of the air-side, and then whisk it back into the interior again. Such a model yields the same transfer coefficient as molecular diffusion on its own, but integrated over "exposure time." The exposure time remained an arbitrary quantity: it could be deduced from observed transfer rates, but then it was simply a proxy for an empirical mass transfer velocity. Furthermore, the idea of a fluid parcel being moved from the interior to the surface conflicts with the kinematics of a fluid continuum. By definition, macroscopic parcels of fluid do not move *relative* to a free surface in the *normal* direction, only tangentially, within the interface. Spray and bubbles have their own interface that is subject to the same condition.

On account of its intuitive appeal, the surface renewal idea remains popular to this day. This in spite of Fortescue and Pearson's (1967) clear explanation many years ago of the role of eddies in scalar transfer. They noted that a "knowledge of the kinematics of any flow field should allow the mass transport to be deducible from it,

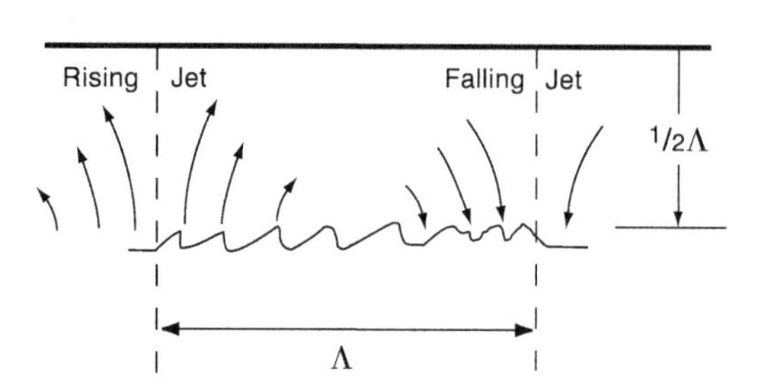

Figure 2.22 Fortescue and Pearson's (1967) schema of advection-diffusion interplay in air-sea gas transfer: Surface divergence coupled to upward moving jets keeps boundary layers thin, convergence and downward jets remove boundary layer fluid.

using standard mass conservation equations and boundary conditions." The difficulty lies in characterizing the kinematics of eddies. As a sample eddy field, Fortescue and Pearson prescribed a row of two-dimensional circulation cells at the surface, of a characteristic size Λ, velocity scale u, and numerically solved the advection-diffusion equation to find the water to air gas transfer rates. Their principal result was that fluid rising toward the interface holds the diffusion boundary layer relatively thin, the gas transfer rate relatively high. Continuity of the fluid motions dictates that the fluid rises toward the interface in those portions of the cells where the along-interface flow is divergent. Fluid returning to the interior, where flow is convergent, takes down fluid enriched in gas. Fortescue and Pearson (1967) found the gas transfer coefficient to be a constant times $\sqrt{uD/\Lambda}$, where D is the gas diffusivity in water. The fraction u/Λ is a measure of the surface flow divergence.

The important kinematic features of Fortescue and Pearson's flow field are the divergences and convergences at the surface, above rising and sinking fluid respectively (Figure 2.22). Perhaps the simplest flow pattern with surface divergence is two-dimensional stagnation point flow, described by the streamfunction (Csanady, 1990a):

$$\psi = axz. \tag{2.25}$$

Near a stagnation point on the surface $z = 0$, at $x = 0$, the velocities are $u = ax$, $w = -az$. The constant a is the divergence of the along-surface flow, du/dx. The vertical velocity at a given depth is the same at all distances from the center of the divergence. This can only apply over a limited area, the area of the rising jet in Figure 2.22, outside of which there must be a convergence somewhere. In this idealized flow field, the steady-state advection-diffusion equation for a substance present in concentration χ is:

$$u\frac{\partial \chi}{\partial x} + w\frac{\partial \chi}{\partial z} = D\nabla^2\chi \tag{2.26}$$

with D molecular diffusivity. The surface concentration of some gas in the water, χ_0, in equilibrium with the gas concentration on the air-side, set by Henry's law, (discussed in section 1.7, see Equation 1.76) drives the downward diffusive flux of the gas, while advection by the rising fluid balances it. The solution of Equation 2.26 expressing this balance is:

$$\chi = \chi_0 erfc\left(z\sqrt{\frac{a}{2D}}\right). \tag{2.27}$$

The mass transfer rate or surface flux F_0 is:

$$F_0 = -k_m \chi_0 \tag{2.28}$$

where the mass transfer velocity is $k_m = \sqrt{2aD/\pi}$, a formula of the same structure as Fortescue and Pearson's (1967) result, with $a = u/\Lambda$. (The index m is necessary to distinguish mass transfer velocity from wavenumber k.) Note that both concentration and flux are independent of x, a result only valid within some effective radius of the divergent flow region. The stagnation point flow field provides the divergence, molecular agitation the diffusivity, the two together determining the transfer rate.

A similar result is likely to apply in other flow fields, with the divergence determined by the ratio of velocity and length scales of the eddy motion near the surface. As the dependence of k_m on $\Lambda^{-1/2}$ shows, the shortest length scales are likely to dominate. Comparing the result here with the molecular diffusion example in Chapter 1, we see that the surface divergence acts as a reciprocal time scale of boundary layer growth. Or, in terms of the surface renewal model, a is reciprocal exposure time.

How large are the surface divergences on the water side, in a field of wind waves? Different divergences accompany motions of different scales, long waves, short waves, eddies tied to waves, and rollers. Gravity wave divergences are of order $a_w k^2 c = a_w k\sqrt{gk} = 0.34 g^{1/4} k^{3/4} u^{*1/2}$, with their steepness $a_w k$ taken from Equation 2.17. We have written a_ω for wave amplitude here, to distinguish it from divergence a. The shortest gravity waves have the largest divergences. At the short end of the gravity wave range their divergence is of order 1 s^{-1}.

Much larger divergences characterize capillary-gravity waves and rollers on short waves. What we might call a characteristic gravity-capillary wave, of frequency ω_γ, with all the steepness of short waves concentrated in it (a certain overestimate, $a_w k = \sqrt{2\overline{\nabla\zeta^2}}$) has a surface divergence in moderate winds of order 30 s^{-1}. Rollers on short waves originate from the rolling-up of the surface shear layer, with a vorticity of u_w^{*2}/ν_w, which is of order 100 s^{-1} in moderate winds. Surface divergence at roller fronts is of the same order.

The leading divergence producer on the water side is then the roller, with its divergence of order u_w^{*2}/ν_w. Substituting this estimate for the divergence in our formula following Equation 2.28, we arrive at:

$$k_m = const.\, u_w^* Sc^{-1/2} \tag{2.29}$$

with $Sc = \nu_w/D$, the Schmidt number. This applies to the neighborhood of a single roller generated surface divergence; if the area-density of such divergences, the fraction of the sea surface that they occupy, is φ, then φ times the above is the area-average transfer coefficient. The nondimensional version of this, $k^+ = k_m/u_w^*$, should then be proportional to the area-density of roller divergences, at a fixed value of the Schmidt number.

The gas transfer law derived from observation in the moderate wind speed range is of this form, with k^+ increasing roughly linearly with u_w^* or with mean square surface slope $\overline{\nabla\zeta^2}$, according to Jähne et al. (1987), (see the last section of Chapter 1). We may now interpret the increase as being the result of the area-density of roller divergences increasing with wind speed.

According to Equation 2.21 u_ω^* and $\overline{\nabla\zeta^2}$ are linearly related. Substituting the definition of ω_γ, their relationship is:

$$\overline{\nabla\zeta^2} = 0.017\sqrt{2}\frac{u_w^*}{(g\gamma)^{1/4}}. \tag{2.30}$$

The right-hand side contains the fourth root of the waveheight scale u_w^{*2}/g to surface tension scale γ/u_w^{*2} ratio. According to the evidence, in moderate winds, this ratio is proportional to the area-density of divergences most effective in facilitating air-sea gas transfer. A reasonable conclusion to draw is that short, sharp crested waves, also known as microscale breakers, harbor the divergences that keep the diffusion boundary layer thin on a wind-blown water surface.

The simple argument above elucidating the mechanism of gas transfer in moderate winds rests on the hypothesis that the effective divergence controlling the process is proportional to the vorticity of the surface shear flow on the water side. Accumulations of viscous boundary layer fluid in microscale breakers are the suspected divergence producers as they "roll up" into rollers. The area density of microscale breakers is then the variable controlling gas transfer as well as wave steepness.

In very light winds, surface contamination affects gas transfer, while in strong winds bubbles on breakers and spray enhance gas transfer. Current and future research efforts should elucidate the extent of the enhancement.

2.4.2 Air-side Resistance

Gas transfer is the only significant scalar transfer process controlled by the water side. The air side, with its low conductivity and vapor diffusivity, controls the important heat and vapor fluxes. A major difference compared to gas transfer is that the Prandtl number for heat transfer, and the Schmidt number for vapor transfer, are of order one: Conductivity and vapor diffusivity are of the order of kinematic viscosity. Thus, any boundary layers, for temperature, vapor concentration, or shear flow are of similar thickness. The solid surface-like behavior of the water surface, suppressing significant air-side eddy motions along the interface, interposes a viscous boundary layer between the surface and any divergent flow above, and enhances the effect of the conductive or diffusive boundary layer in impeding heat and vapor transfer. The temperature or humidity difference across the viscous cum diffusive boundary layer then constitutes a significant, but not a dominant, part of the total between the surface and a reference level in the constant flux layer. Dividing this difference by the temperature or humidity flux yields a boundary layer contribution to Resistance. Its reciprocal is a transfer velocity across the boundary layer alone. Heat transfer or evaporation therefore depends on the air-side heat conductivity or vapor diffusivity, as well as on viscosity and the waveheight length scale u_a^{*2}/g, which is also the scale of the smallest eddies.

Thicker air-side viscous cum conductive or diffusive boundary layer notwithstanding, the divergences of smallest spatial scale remain the principal conveyors of cold or dry air from above. To calculate their effects, we employ the analog of the stagnation point flow on the water side. Supposing the interface to behave as a solid surface on the air side, this is two-dimensional Hiemenz flow (Schlichting, 1960), a stagnation point flow with a viscous boundary layer. Its structure follows from boundary layer theory, and yields the mass transfer velocity across the viscous cum diffusive boundary layer (Csanady 1990a):

$$k_m = \beta\sqrt{aD_v} \tag{2.31}$$

where β is a function of Schmidt number $Sc = \nu_a/D_v$, D_v diffusivity of water vapor in air, and a is the divergence above the boundary layer. The boundary layer thickness is $\delta = \sqrt{u_a^* D_v/g}$. The heat transfer velocity is the same as k_m, with Prandtl number replacing Schmidt number, thermometric conductivity replacing diffusivity. The value of β is of order one, effectively constant.

The smallest eddies on the air side have the dimensions of short waves, with velocity scales u_a^*, length scales u_a^{*2}/g, so that the typical divergence is $a = u_a^*/(u_a^{*2}/g) = g/u_a^*$. The nondimensional transfer coefficient across the boundary layer is then:

$$k^+ = k_m/u_a^* = \beta\sqrt{\frac{gD_v}{u_a^{*3}}} \tag{2.32}$$

where u_a^{*3}/gD_v is Sc^{-1} times the air-side Keulegan number, the waveheight scale to viscous length scale ratio. This implies a rapid decrease of k^+ with increasing friction velocity. The decrease in vapor concentration across a viscous-diffusive boundary layer of thickness δ, expressed as specific humidity, is then, $\Delta q = q_s - q(\delta)$:

$$\Delta q = q^*/k^+ = \beta^{-1}q^*\sqrt{\frac{u_a^{*3}}{gD_v}} \tag{2.33}$$

which is a quantity of order $10q^*$ in moderate winds with q^* the humidity flux scale introduced in Chapter 1, $q^* = \overline{\omega' q'}/u_a$.

Again as on the water side, the result in Equation 2.32 applies to a single divergence. Multiplying by area-density ϕ we have the area-average transfer coefficient across the boundary layer. In moderate to strong winds short waves cover most of the sea surface, so that the divergences that they cause have a high area-density, ϕ lose to 1.0.

Above the boundary layer, in the constant flux layer, humidity decreases further, with the logarithm of distance above the interface, as we discussed in Chapter 1. The humidity Flux-Force relationship written as a nondimensional mass transfer velocity is $k^+ = -\overline{w'q'}/\{u_a^*[q(h) - q_s]$. Writing $q^* = -\overline{w'q'}/u_a^*$, the reciprocal of this can be split into $(k^+)^{-1} = [q(h) - q(\delta)]/q^* + [q(\delta) - q_s]/q^*$, the boundary layer Resistance plus a constant flux layer Resistance.

In Chapter 1 we found for the total Resistance to vapor transfer, (Equation 1.69):

$$\frac{1}{k^+} = C_s + \kappa^{-1} \ln\left(\frac{gh}{u_a^{*2}}\right) + \kappa^{-1} \ln\left(\frac{u_a^{*3}}{g D_v}\right)^{4/3} \tag{2.34}$$

again split into two parts: the first part depending on turbulence properties in the constant flux layer, and the second part on the same nondimensional parameter as appears in Equation 2.33, the ratio of waveheight length scale to diffusive boundary layer thickness, that governs the boundary layer Resistance. Equations 2.33 and 2.34 give two different versions of how this boundary layer/wind wave interaction affects vapor flux. Resistance to heat transfer behaves similarly. More important than the difference between them is that they both identifiy the same key nondimensional variable governing the boundary layer influence. More cannot be expected from simple models of surface divergence.

2.5 Pathways of Air-Sea Momentum Transfer

Momentum transfer differs from scalar property transfer because some of it occurs via pressure forces on the inclined surfaces of wind waves. The windsea contains two differently behaving populations of waves, one long, one short. Pressure forces generally add windward momentum to waves, long or short, increasing their momentum transport. Breaking of waves is the mechanism of transferring this momentum downward. Short waves have small amplitudes and they communicate their momentum to a shallow surface layer. Breaking long waves affect a deeper section of the oceanic mixed layer. As well as a pressure force, the airflow exerts a viscous shear force on the interface, although in a very irregular pattern, vanishing shear stress in air flow separation bubbles, high shear where shear layers reattach. We may think of these three, short waves, long waves and shear stress, as different pathways of momentum transfer "crossing" the interface, uniting again above and below into a single one of importance, the Reynolds flux of momentum (Figure 2.23). An important question on the mechanism of air-sea momentum transfer is: How wide are the individual paths?

First, momentum transfer to "long" waves. This means dominant waves of randomly varying shape, responsible for the elevation spectrum from characteristic wave to its tail. As we have seen, a constant fraction of about 6% of the wind-imparted momentum supports the increase of wave momentum transport by the windsea during wave growth from very young to mature waves. An additional fraction must balance loss of wave momentum through viscous and eddy friction plus wave breaking. Viscous friction contributes very little: The attenuation coefficient of a freely traveling classical wave is $\beta = 2\nu k^2$, yielding a half-life of the order of a year for a long wave of 60 m

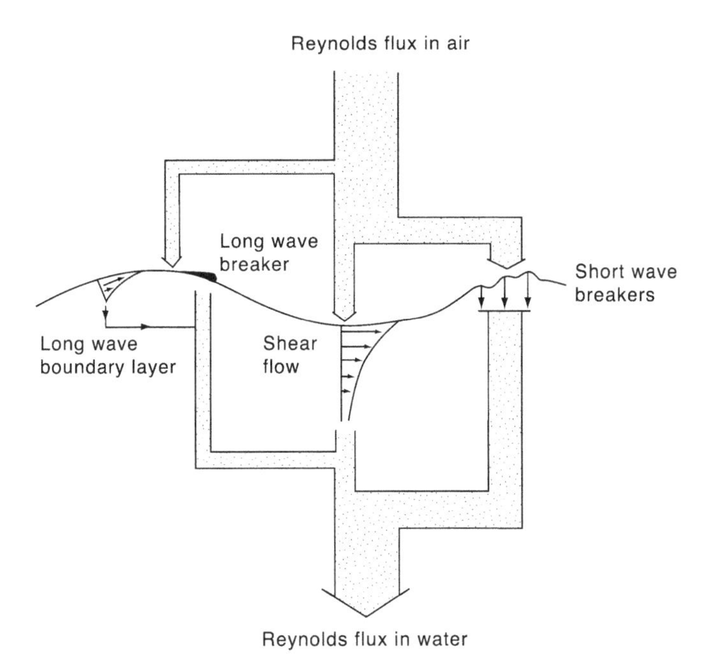

Figure 2.23 The three pathways of momentum flux crossing the air-sea interface: short waves, long waves, and viscous shear stress.

wavelength. Eddy friction on orbital motions is more effective but is still unlikely to be a significant addition to momentum transfer via wave breaking. With a typical eddy viscosity of 1 cm^2 s^{-1} in the surface layer, the half life of the same long wave is still numbered in days, or many times the wave period of 6 s. By contrast, a single breaking event transfers a large fraction of a wave's momentum downward. The dominant loss mechanism in long waves is thus wave breaking.

The details we have discussed before: the rollers on long waves and their white-caps travel with wave celerity and their turbulence transfers momentum downward via Reynolds stress. Celerity times downward momentum transfer adds up to energy transfer. Laboratory studies on short waves showed the details of downward momentum transfer: "forced convection" from roller front down, and, in a windsea with many rollers, a "downward bursting boundary layer" on the water side of the interface. As these are inertial phenomena, they are also likely to occur on long waves. If they do, on the fast moving rollers of long waves, then long wave breaking puts much Turbulent Kinetic Energy (TKE) into the water-side mixed layer. The forced convection should reach to depths of the order of waveheight.

Recent work on water-side turbulence in the windsea (Agrawal et al., 1992; Drennan et al., 1996; Terray et al., 1996) revealed that TKE dissipation associated with long wave breaking greatly exceeds TKE production and dissipation associated with the water-side shear flow. Figure 2.24 from Drennan et al. (1996) shows nondimensional energy dissipation rate $\varepsilon H_s / F$, where F is an estimate of energy input to waves, H_s the significant waveheight, versus nondimensional depth. Considering the great difficulty

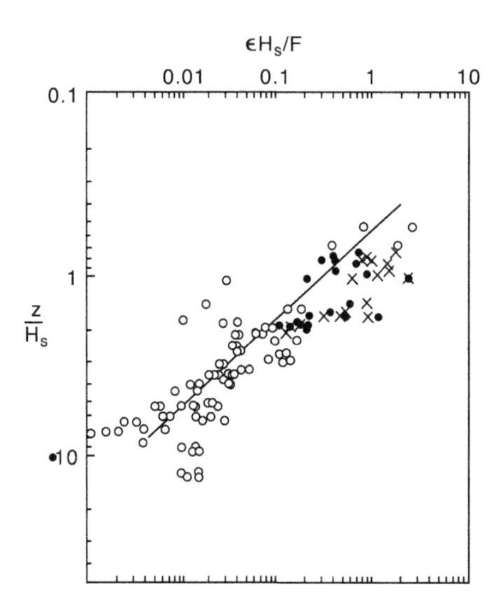

Figure 2.24 TKE dissipation rate versus depth on the water side of the interface, scaled by characteristic wave height H_s and energy input to long waves F. From Drennan et al. (1996).

of such observations in a windsea, the points line up with a clear trend. At depths comparable to wave heights, the observed dissipation rate is up to 100 times higher than what water-side shear flow generated turbulence would have produced.

These data on dissipation allow us to estimate the fraction of wind-imparted momentum that goes into long waves, traveling with a celerity close to that of the characteristic wave, C_p. Let that fraction be η, so that the momentum transfer to long waves per unit area is $\eta \rho u_w^{*2}$, the downward energy transfer C_p times that. The rest of the wind-imparted momentum goes to shear stress and short waves; these we lump together, and suppose them transmitted to the water-side shear flow, at a surface velocity of U_s. The total energy input to long waves and shear flow then equals the energy dissipation in the water-side mixed layer:

$$\eta u_w^{*2} C_p + (1-\eta) u_w^{*2} U_s = \int_{-h}^{0} \varepsilon \, dz. \tag{2.35}$$

Here $\varepsilon = u_w^{*3}/\kappa z + \varepsilon_L$, energy dissipation of the shear flow, plus the long wave-related dissipation ε_L. The latter dominates, and the shear flow portion may be neglected in a first approximation. On the left-hand side $C_p >> U_s$, so that unless η is very small, its value is nearly:

$$\eta = \frac{\int_{-h}^{0} \varepsilon_L dz}{u_w^{*2} C_p}. \tag{2.36}$$

Drennan et al. (1996) give the empirical formula for dissipation as $\varepsilon H_s/F = 0.3(z/H)^{-2}$. Integrating upward to the near-surface depth of about $0.5H_s$, to which the observations were limited, yields $0.6F$ for the dissipation rate integral. The dissipation rate above that level could hardly have been less than the maximum observed, or about $3F/H$, the integral $1.5F$, for a total of $2.1F$. The estimated energy input to the long waves was apparently low by a factor of two.

In one of the observed cases listed by Drennan et al. (1996), Run 18-09, using their estimated energy input and other listed data, the fraction of momentum going into long waves, $\eta = F/(u_w^{*2}C_p)$, was 21.3%. Multiplying by a factor of 2.1 brings this close to 50%. The ubiquitous presence of whitecaps on a windsea in moderate to strong winds certainly suggests that a substantial fraction of air-sea momentum transfer goes through the long wave pathway. High water-side TKE dissipation puts this expectation beyond a shadow of doubt.

Short waves include instability waves arising spontaneously under gusts of wind that intensify the shear flow. These very short waves (in the capillary-gravity wavelength range) grow and steepen rapidly, extracting momentum from the air-side boundary layer, until they break and transfer their momentum to the water-side shear flow. Similarly ephemeral are decimeter scale bore-like structures with steep fronts, as well as crescent shaped, sharp crested, wave-like surface disturbances of similar size; they also pop up and vanish in seconds, as casual observation shows. This very short time-scale activity stirs up a surface layer of a few cm in depth, as is easily demonstrated by dropping dye on the surface: it diffuses rapidly over a surface layer. Because of their ephemeral nature, it is stretching a point to say that such surface structures acquire wave momentum transport from the wind to promptly transfer it to shear flow. A bore-like structure, for example, may be driven by the pressure force of the wind on its rear, while the bore exerts an equal shear force on the fluid below, in the manner of a roller. What is common to these structures is that they extract momentum from the air flow in excess of what the mean shear would transfer over a smooth surface, much as roughness elements of a solid surface do. Short waves and other flow structures travel slowly, however, at speeds comparable to the surface velocity of the water-side shear flow, so that their contribution to downward TKE transfer is relatively small. This justifies lumping them together with viscous shear in Equation 2.35.

Properties of the slope spectrum suggest that momentum transfer via short waves could be a function of surface tension, in addition to u^* and g, and therefore could depend on the nondimensional variable $u^*\omega_\gamma/g$. In our survey of the momentum transfer laws, we did not find direct evidence for such dependence (the area-density of divergences controlling gas transfer depended, however on this variable). This seems to point the finger at the longer wavelengths in the short wave spectrum, perhaps the 0.1 to 0.3 m wavelength range, as those most effective in momentum transfer. The laboratory studies of short waves cover exactly this range, and they amply demonstrate the likely role of such waves in momentum transfer.

The laboratory evidence is not quantitatively transferable to the more diverse and chaotic short waves on the windsea. Therefore, we do not have a direct estimate of what fraction of momentum transfer passes through the short wave pathway in the open ocean: It must be the fraction left after accounting for long waves and mean viscous shear stress.

Viscous momentum transfer via the air-water shear flow remains an unavoidable corollary of the boundary conditions requiring continuous velocity and shear stress across the interface, even when breaking waves and associated flow separation disrupt the flow pattern. Laboratory work suggests especially intense viscous momentum transfer in locations where the air flow has a downward component, such as on the downwind face of wind waves and in regions of flow reattachment behind any separation bubbles. Whatever the details, a portion of the total momentum transfer therefore remains garden variety viscous stress.

What that portion is cannot be estimated with confidence, but it should be of the order of the shear stress on a smooth wall. In the moderate wind speed range of 7–10 m s^{-1} Charnock's law predicts a drag coefficient about twice that of a smooth wall, suggesting that close to half of the momentum transfer may proceed via shear stress. The long waves may take another half, as we have just seen, but of course these estimates are only good within a factor of two at best. That makes all three pathways of comparable importance, the exact split unknown, but known to depend on wind speed and wave age. The arbitrary split in Figure 2.23 with a preponderance of the short wave pathway applies perhaps to young waves in moderate winds.

Chapter 3

Mixed Layers in Contact

3.1 Mixed Layers, Thermoclines, and Hot Towers

The Transfer Laws of the air-sea interface link various fluxes to low-level properties of the air and to surface properties of the water. For, say, latent heat transfer to occur at all, the low level air has to be drier than the saturated air in contact with the sea surface. The resulting transfer of vapor moistens the air, however, and reduces or even eliminates the transfer. A steady supply of drier air to low levels is necessary to counteract the moistening and maintain vapor transfer.

Upper level air is much drier than the air near the interface, and if it extended downward to low levels, it would certainly maintain high vapor flux across the interface. The upper layers of the atmosphere are, however, insulated from direct contact with the interface by the Trade Inversion in the Tropics, and similar "thermoclines" poleward of the trades. Atmospheric thermoclines are layers in which the "virtual" potential temperature, θ_v, a measure of buoyancy taking into account humidity, sharply increases with height ($\theta_v = \theta + 0.61qT$, where $\theta = T + gz/c_p$ is potential temperature). Meteorologists refer to such layers as strong "inversions," because the absolute temperature T increases in them with height, instead of decreasing as it does everywhere else. The temperature gradient across an inversion would force strong downward Reynolds flux of heat, were not the associated buoyancy flux a strong sink for turbulent kinetic energy, TKE, as discussed in Chapter 1, in connection with the Transfer Laws. Thus, the sharp Trade Inversion and other such thermoclines suppress turbulence and prevent Reynolds fluxes of heat and vapor in their upper portions.

The layer between the atmospheric thermocline and the air-sea interface may with more or less justification be called a "mixed layer": it is nearly, but not quite, "mixed," in the sense that at least its potential temperature and humidity do not vary much with height. Its main characteristic distinguishing it from the quiescent upper layers of the atmosphere is vigorous turbulence, mechanical or convective, that maintains near-uniformity of properties down to low levels.

At the top of the atmospheric mixed layer, eddies intrude into the lowest layers of the thermocline, engulf thermocline air, and incorporate it into the mixed layer, a process known as "entrainment." Continuous entrainment brings down warmer and drier air, sustaining downward Reynolds flux of heat and upward flux of vapor. At some level, these fluxes have a greatest absolute value, a peak downward heat flux, a maximum of upward vapor flux. Above that level, the divergence of the vertical fluxes cools and moistens thermocline air, making it more like mixed layer air. Below that level, and over much of the mixed layer, the flux divergences heat and dry the mixed layer air, counteracting the upward flux of heat and vapor from the air-sea interface that warms and moistens the mixed layer air.

Downward Reynolds flux of heat in the lower layers of the thermocline implies downward buoyancy flux, diminished somewhat by the upward buoyancy flux resulting from vapor transfer (typically by 20% or so). Net downward buoyancy flux requires import of Turbulent Kinetic Energy (TKE), that can only come from the turbulent mixed layer. This limits the rate of entrainment and the associated Reynolds fluxes: They can only be as large as the TKE import is able to sustain. The upshot is that mixed layer turbulence controls heat and vapor fluxes across the thermocline, just as it controls fluxes across the air-sea interface.

Above the Trade Inversion and other sharp atmospheric thermoclines, the absolute temperature drops with height at about half the rate it would, were the potential temperature θ constant, i.e., at the rate of about 5×10^{-3} K m^{-1}, the potential temperature increasing at about the same rate. This is still stable enough to prevent turbulence except in strong wind-shear layers (or where radiation cooling generates convective turbulence). Reynolds fluxes above the thermocline are therefore generally small, insulating upper layers of the atmosphere from lower ones.

Mixed layer processes nevertheless manage to punch holes in atmospheric thermoclines in some locations and alter this picture. Especially near the InterTropical Convergence Zone (ITCZ, a kind of thermal equator where the equatorward trade winds of the two hemispheres collide), but also elsewhere in the Trade Wind region, sea to air transfer of sensible and latent heat produces a warm and humid air mass with high total energy content. When some local disturbance lifts such air past the level where its vapor content begins to condense (the Lifting Condensation Level, LCL), the release of latent heat makes it buoyant and promotes its further rise. In most places, this merely generates convective turbulence plus clouds, but where the energy content of the moist air is high enough, it just breaks through the inversion, and forms a "hot tower" that may reach to the tropopause and beyond, 15 km or higher. Hurricanes are extreme examples of hot towers. The horizontal dimensions of hot towers are relatively small but they transport large volumes of high energy air to the upper layers of the troposphere. In the aggregate, hot towers occupy only a small fraction of the Trade Wind region, but they are responsible for the massive total of upward transport of surface air in the tropics and subtropics.

What balances the upward transport of air in hot towers? The return flow is evenly distributed subsidence, air slowly descending through the Trade Inversion, over extensive areas of the tropics and subtropics. Heat loss through radiation to space cools the air of the troposphere between the tropopause and the Trade Inversion, and causes it to subside at a rate just high enough for the downward advection of potential temperature to balance radiation cooling. This requires a supply of air at high potential temperature near the tropopause: The hot towers deliver it.

Subsidence in the troposphere would move the Trade Inversion bodily downward, were it not balanced by continuous "erosion" (loss of lower layers) through downward entrainment of mixed layer air: subsidence and balancing entrainment maintain the Trade Inversion at a (more or less) constant height. The associated downward buoyancy flux consumes TKE. Where this energy comes from, is one of the important questions in air-sea interaction.

Similar conditions prevail on the water side of the air-sea interface. A turbulent oceanic mixed layer lies between the interface and an oceanic thermocline, underneath which lie colder stratified interior layers, with turbulence sporadic or totally absent. Mixed layer turbulence produced by shear flow or radiation cooling episodically entrains upper thermocline layers, deepening the mixed layer, unless "upwelling," slow upward displacement of the thermocline, counteracts it. Short wave solar radiation crossing the interface (the "irradiance" of the water-side mixed layer) heats surface waters, however, and the resulting stable stratification tends to suppress turbulence. Entrainment is therefore limited to periods of vanishing irradiance, mainly nights, or to stormy weather. The rate of entrainment is again governed by how large a downward buoyancy flux the mixed layer turbulence can sustain in the upper thermocline.

Hot towers are atmospheric realizations of "deep convection," air motions induced by density contrasts, extending over substantial heights. In the ocean, deep convection is due to strong surface cooling, strong enough to break down the thermocline underlying the mixed layer, and to carry surface waters in "chimneys" down to mid-depths or even right to the seafloor. Oceanographers call the process "water mass formation," because mixing in the course of descent changes the temperature-salinity characteristics of the resulting water mass. Ocean waters retain such characteristics for long periods, yielding clues to the circulation of deep layers.

A few decades ago, oceanographers thought that the mass balance of oceanic deep convection is similar to the atmosphere's: broadly distributed upwelling through the thermocline. This view now appears as too simplistic. The evidence on deep circulation shows that waters conveyed to great depths by deep convection make upward headway only in very few locations, taking a number of steps before reaching a surface mixed layer again. The physical reason is that raising heavier waters upward requires energy, in the form of TKE import, not accessible at depth. The final step in the process, entrainment of thermocline water into the oceanic mixed layer, occurs only in such singular locations as the eastern equatorial portions of the Atlantic and Pacific, or along western margins of continents where equatorward alongshore winds bring about coastal upwelling. Insulation of the deep ocean from the surface mixed layer is therefore more complete than of the upper troposphere from the air-side

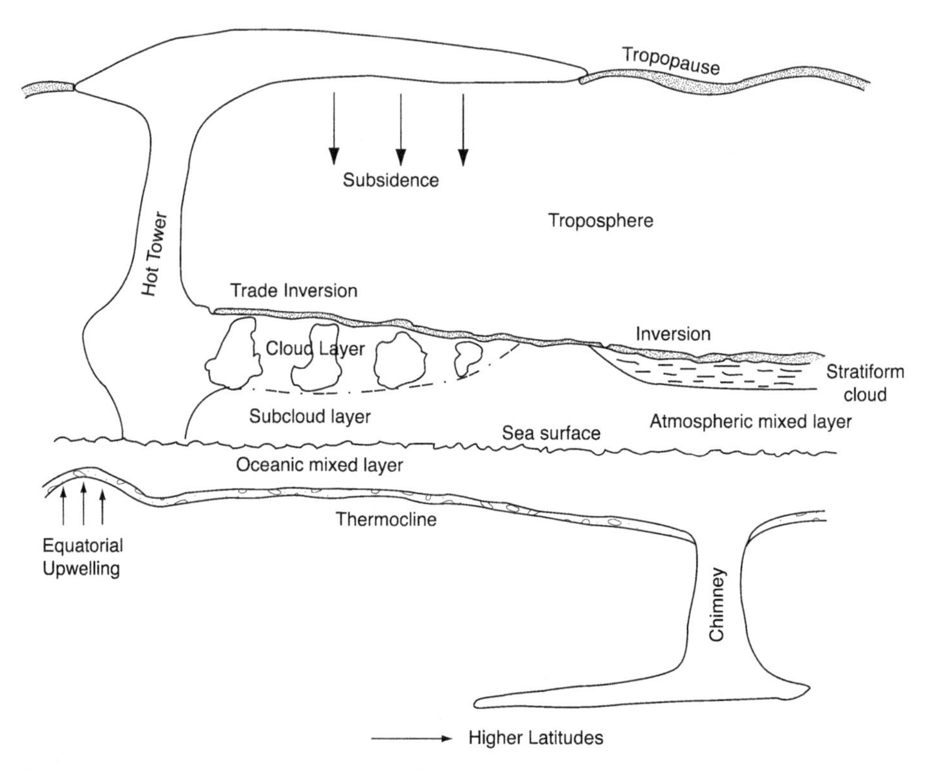

Figure 3.1 Atmospheric and oceanic mixed layers as systems separate from the high troposphere and the deep ocean, except for sites of deep convection, hot towers, and chimneys.

mixed layer. Figure 3.1 shows an artist's version of mixed layers, thermoclines, chimneys and hot towers.

Apart from sites of deep convection in either the atmosphere or in the ocean (i.e., over most of the global ocean), the two mixed layers in contact form an interacting system, tied together by interface fluxes, but insulated from upper atmosphere and lower ocean, except for limited cross-thermocline heat and mass transfer from above and below. Important components of mixed layer heat budgets are radiant energy fluxes or rather their divergences that heat or cool the air or water. Turbulence in the mixed layers controls not only the air-sea transfer of heat and vapor, but also entrainment of thermocline air or water, the heat and mass transfer across the thermoclines. Important questions relating to mixed layers are then: (1) what maintains their turbulent character, and (2) what laws govern entrainment (i.e., the mechanism of heat and mass transfer across thermoclines).

3.2 Mixed Layer Turbulence

The Turbulent Kinetic Energy (TKE) equation, familiar from the previous chapter (see Section 1.4.3) contains part of the answer to the first question. For the atmosphere, a slightly more general form of the TKE equation is, allowing for cases in which the wind direction changes with height:

$$\frac{\partial \overline{E_t}}{\partial t} = \overline{w'b'} - \left(\overline{u'w'}\frac{d\bar{u}}{dz} + \overline{v'w'}\frac{d\bar{v}}{dz} \right) - \frac{\partial(\overline{w'p'} + \overline{w'E_t'})}{\partial z} - \varepsilon \tag{3.1}$$

where E_t is TKE, overbars denote means, primes turbulent fluctuations. The terms on the right-hand side denote, in order: the Reynolds flux of buoyancy $b = g\theta_v/T = g(\theta/T + 0.61q)$, either a source or a sink term according to sign; the production of TKE by two components of shear flow, always a source term; flux-divergence of TKE because of pressure work and Reynolds flux of TKE, again either source or sink; and viscous energy dissipation, always an energy sink. This is a one-dimensional balance: Horizontal advection and Reynolds flux of TKE are generally small. Pressure is understood to be in "kinematic" units (i.e., divided by density). The vertical integral of the flux divergence term yields the difference of TKE fluxes at mixed layer top and bottom. When these boundary fluxes vanish, the flux divergence term represents internal transfers of TKE from a producing region to one where energy is dissipated and/or used up in sustaining downward buoyancy flux, negative $\overline{w'b'}$.

When all terms on the right vanish except unavoidable dissipation, turbulence quickly winds down and disappears. In laboratory flows, shear production is what usually balances dissipation. Mixed layer shear production usually peaks near the sea surface in both air and water, but is evanescent over most of the mixed layer. Thermocline shear can be strong, too, especially in the ocean, but again in relatively thin layers. Buoyant turbulence production in the atmospheric mixed layer because of sea level sensible heat flux is typically much weaker than over land. Surface cooling of the oceanic mixed layer by long wave radiation and latent heat transfer is typically much stronger than upward sensible heat flux, but the buoyancy flux on the water side remains moderate owing to the low thermal expansion coefficient of water. Of greatest global importance is probably buoyant production of turbulence in clouds.

Atmospheric mixed layers often contain clouds in their upper portion, in isolated clusters or in a full "stratiform" overcast, extending to a "capping inversion," the thermocline at mixed layer top. Liquid water in the form of a fine mist is present in clouds and makes them visible. The mist forms as water vapor condenses into small droplets in rising moist air, when air temperature drops below saturation at the Lifting Condensation Level, LCL. The droplets add up to liquid water content q_l of the cloud; this increases upward owing to further condensation under decreasing temperature.

The droplets radiate long-wave (infrared) energy to space, from a very thin sublayer at cloud top, while also absorbing short wave solar radiation over a much thicker optical depth. Entrainment of thermocline air warms and dries the top of the cloud; this causes re-evaporation of some droplets. The attendant latent heat loss further cools the air at cloud top. Figure 3.2 illustrates the various processes occurring in the cloud-topped atmospheric mixed layer.

The different heating and cooling processes in clouds, occurring at different levels, give rise to vertical Reynolds fluxes of temperature. The one-dimensional heat

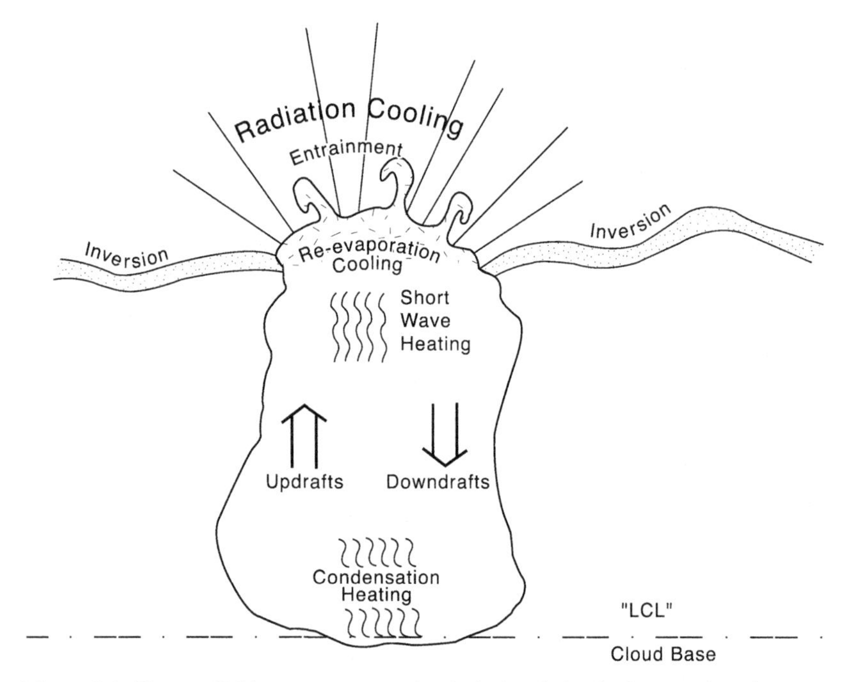

Figure 3.2 The manifold processes operating in isolated clouds. In many locations groups of such clouds fill the atmospheric mixed layer between the Lifting Condensation Level (LCL) and the Trade Inversion.

conduction equation, neglecting advection and Reynolds flux of temperature in the horizontal, but taking into account cloud processes, makes this clear:

$$\frac{\partial\overline{\theta}}{\partial t}+\overline{w}\frac{\partial\overline{\theta}}{\partial z}=-\frac{\partial(\overline{w'\theta'})}{\partial z}-\frac{1}{\rho c_p}\frac{\partial\overline{R}}{\partial z}+\frac{L}{c_p}\overline{C} \tag{3.2}$$

where θ is potential temperature, $\overline{w}$ mean vertical velocity, subsidence velocity if negative, R the vertical flux of radiant energy in W m^{-2}, L is latent heat, and C the rate of condensation (evaporation, if negative) as dq/dt. The flux divergence of radiant energy and the latent heat release rate are source terms, either increasing temperature following the fluid, or balancing the Reynolds flux divergence of potential temperature.

The results depend on the distribution of the source terms. Latent heat released by condensation heats the cloud from the bottom up, as its liquid water content rises with height, typically at a fairly steady rate. Radiation cooling of water droplets is concentrated at the top of the cloud, as just mentioned; there $d\overline{R}/dz$ is locally very high. Solar (short wave) radiation heats the cloud at the same time, but the heat gain is distributed over an optical depth comparable to cloud depth. It is therefore convenient to split the radiant heat gain into short wave and long wave components, R_{LW} and R_{SW}. In steady state, with cloud temperature unchanging, subsidence absent, integration of Equation 3.2 from the bottom of the cloud z_b to some intermediate level z in the cloud yields the flux distribution:

$$\overline{w'\theta'}(z) - \overline{w'\theta'}(z_b) = \frac{L}{c_p}\int_{z_b}^{z} \overline{C}\, dz - \frac{1}{\rho c_p}\overline{R_{SW}}(z) \tag{3.3}$$

noting that long wave cooling $\overline{R_{LW}}$ does not affect cloud interior. Over the whole cloud, if the Reynolds fluxes at cloud top z_t and cloud bottom z_b both vanish, we have the balance:

$$\overline{R_{LW}} = \rho L \int_{z_b}^{z_t} \overline{C}\, dz - \overline{R_{SW}}(z_t) \tag{3.4}$$

long wave radiation removing all the heat gained by net condensation and downward solar radiation. In the interior of the cloud, Reynolds fluxes are of the same order as cloud top radiation $\overline{R_{LW}}$: already at a short distance below the thin cloud-top sublayer, Reynolds flux must replace all of the long wave radiation $\overline{R_{LW}}$ in Equation 3.4. Similarly, re-evaporation at cloud top requires Reynolds flux of heat from below, even if depth-integrated condensation and re-evaporation nearly balance, as is usually the case in a mixed layer capped by an inversion.

In the absence of Reynolds fluxes of temperature, Equation 3.2 shows that radiant and latent heat sources raise or lower the mean temperature, or balance temperature advection through subsidence. Above the atmospheric thermocline, subsidence-heating and radiant heat loss balance, as we have mentioned. Where this is not the case, suppose that a local heat source produces a bell-shaped distribution of excess temperature over height. Above the height of the maximum heating, the temperature distribution becomes unstable: more intensely heated and therefore lighter air comes to underlie heavier air. Underneath the maximum the opposite is true, the arrangement is stable. Convective turbulence arises above the maximum, while the layer below becomes a TKE sink. A similar distribution of cooling generates convective turbulence underneath the level of maximum cooling, TKE sink above. When a heat sink above accompanies a heat source below, the layer in between becomes turbulent on two counts, heating from below and cooling from above. In steady state, Reynolds flux of temperature then connects the source and the sink, the associated buoyancy flux sustaining turbulence and usually exporting TKE.

Condensation affects the humidity balance, just as cooling affects temperature:

$$\frac{\partial \overline{q}}{\partial t} + \overline{w}\frac{\partial \overline{q}}{\partial z} = -\frac{\partial(\overline{w'q'})}{\partial z} - \overline{C}. \tag{3.5}$$

In steady state, with mean humidity unchanging, and subsidence absent, condensation over most of a cloud requires upward Reynolds flux of humidity from the subcloud layer to near cloud top. Re-evaporation, on the other hand, generates downward flux of humidity at cloud top.

As we pointed out in Chapter 1 already, indispensable for significant Reynolds flux of temperature or humidity is a significant mean-square value of temperature or humidity fluctuation: one cannot have finite $\overline{w'\theta'}$ or $\overline{w'q'}$ unless $\overline{\theta'^2}$ or $\overline{q'^2}$ is nonzero. We refer to these important mean squares as Turbulent Temperature Variance, TTV, and Turbulent Humidity Variance, THV. A brief look at the balance equations for these variances in clouds is instructive.

The TTV equation is, as in Businger (1982), but with radiant and latent heat release terms added:

$$\frac{d(\overline{\theta'^2}/2)}{dt} = -\overline{w'\theta'}\frac{d\overline{\theta}}{dz} - \frac{d(\overline{w'\theta'^2}/2)}{dz} - \frac{1}{\rho c_p}\overline{\theta'\frac{dR'}{dz}} + \frac{L}{c_p}\overline{C'\theta'} - \varepsilon_t. \tag{3.6}$$

This has a structure similar to the TKE equation, the flux-gradient product being a positive definite source term, the second term on the right the flux divergence of TTV, the last term ε_t is dissipation of temperature variance, always a sink for TTV. The radiant and latent heat terms are sources or sinks. If, for example, $dR'/dz > 0$ mostly occurs together with $\theta' > 0$ (owing to radiation cooling of warmer air parcels, for example), then the radiant term is a sink, damping TTV changes. On the other hand, if the warmer parcels come from the thermocline, they are also drier, without liquid water content, so that they radiate less than the local average, $dR'/dz < 0$, and they act as TTV sources. The latent heat term gets negative contributions from rising parcels condensing their vapor, positive contributions from drier and warmer thermocline air causing re-evaporation. If the source terms win, they may not only maintain TTV locally, but also give rise to export of TTV.

We have met with the flux-gradient source term of Equation 3.6 in Section 1.6.7, and identified it there as entropy transfer to turbulence. The flux-gradient terms in the TKE Equation 3.1 are also handover to turbulence terms. The actual entropy source terms in W kg^{-1} K^{-1} are ε/T, associated with TKE dissipation, and $c_p\varepsilon_t/T^2$, with heat flux irreversibility.

Finally, the THV equation is as follows, with cloud processes taken into account:

$$\frac{d(\overline{q'^2}/2)}{dt} = -\overline{w'q'}\frac{d\overline{q}}{dz} - \frac{d(\overline{w'q'^2}/2)}{dz} - \overline{C'q'} - \varepsilon_q. \tag{3.7}$$

Flux-gradient THV production (alias entropy handover to turbulence) and THV flux divergence are followed here by a condensation-humidity correlation. Because excess humidity at constant temperature causes condensation, the correlation may be positive, yielding a damping term, a sink for THV. The actual entropy source is $c_p\varepsilon_q$.

Entropy sources associated with turbulence imply that export of TKE from a production region is limited to a fraction of production. This also limits downward buoyancy fluxes that turbulence can sustain via entrainment, and calls for an investigation of the entrainment process and of the laws governing entrainment.

3.3 Laws of Entrainment

This sounds like Terms of Endearment, but it does not convey satisfaction with the current state of knowledge about the laws that govern the rate of entrainment. They are well-established only in some simple cases, and then only with considerable error bars. One difficulty is that the entrainment laws of the two kinds of turbulence – mechanical and convective – differ, and there is not much evidence on how their effects combine when both are present.

The concept of entrainment originated in laboratory studies of air and water jets and plumes. A classical paper by Morton et al. (1956) elucidates the underlying physics and discusses some applications. Turner's (1973) great monograph develops the subject in detail and gives futher insight. Briefly, in the absence of vertical density gradients, a jet increases its mass transport as its turbulence spreads to neighboring fluid, dragging the latter along and incorporating it in the jet mass. The increase rate of the mass transport is the entrainment rate, represented usually by an "entrainment velocity" w_e such that the volume transport increase is w_e times jet perimeter, the latter defined somewhat arbitrarily. The magnitude of the entrainment velocity is of the order of eddy-velocities in the jet.

Entrainment into an atmospheric or oceanic mixed layer is a similar phenomenon, but it has to overcome stratification and sustain downward buoyancy flux, negative $\overline{w'b'}$, in the lower portion of an atmospheric thermocline, upper part of an oceanic one. In these locations, upward acceleration because of buoyancy acts predominantly on fluid moving downward, so that the fluid is doing work against gravity, some of its kinetic energy changing into potential energy. A local source, or import from an external source, must then provide TKE for conversion into potential energy. Furthermore, reconversion of chaotic TKE into organized potential energy turns out to be possible only partially, in analogy with Carnot's principle: Viscosity acting on small-scale shear converts to heat much of the TKE produced by shear or convection.

In atmospheric or oceanic thermoclines, the buoyancy flux is negative only in a relatively shallow layer, and it peaks at some level, where its magnitude is $B_d = (\overline{w'b'})_{\text{min}}$. In the oceanic thermocline, buoyancy depends on temperature alone, in the atmosphere also on humidity, but in either medium B_d translates into cross-thermocline Reynolds flux of temperature or of temperature and humidity. Above the level of B_d, the buoyancy flux is divergent, it takes heat and vapor from the thermocline and distributes it in the mixed layer where it is convergent, modifying mixed layer temperature. Therefore, a relationship between B_d and external variables is in effect a cross thermocline transfer law for heat and vapor, a Law of Entrainment. Because buoyancy flux is an entry in the TKE balance, the TKE equation serves as the basis for Laws of Entrainment for convective as well as shear flow turbulence, under different boundary conditions.

3.3.1 Entrainment in a Mixed Layer Heated from Below

There are two sources of TKE: shear flow and buoyancy. If only one of these is present, and if boundary conditions are uncomplicated, the problem of how much negative buoyancy flux the turbulence can sustain is relatively simple: the sink to source ratio can only be a function of few external variables. Dimensional argument plus well conceived experiments might then yield a useful law of entrainment. Convection in the atmospheric mixed layer heated from below, with shear production insignificant, is such a case.

Buoyancy flux at the surface, B_0, associated with surface fluxes of vapor and sensible heat, drives convection in that case. "Thermals," the elements of convective turbulence, extend throughout the mixed layer height and reach into the lower thermocline. The buoyancy flux is positive (upward) over most of the mixed layer, but changes sign a short distance below the top. In a thin top portion, the buoyancy flux is negative and increases in absolute value to a peak, before declining again rapidly to insignificant amplitude. The level of the peak downward buoyancy flux serves as a convenient choice for mixed layer top, at a height h. With no other external variable affecting the TKE balance, the peak downward buoyancy flux, B_d, should depend only on the two variables B_0 and h.

In a seminal paper entitled "Control of inversion height by surface heating," Ball (1960) advanced the hypothesis that, in the simple case of surface heating-driven convective turbulence, the entrainment rate depends only on the upward buoyancy flux integrated from the surface to the level of vanishing buoyancy flux. This is the total source strength in the TKE equation, and the hypothesis is eminently reasonable. Ball (1960) also surmised incorrectly, however, that the total source strength equals the total sink strength, the depth integrated negative buoyancy flux. This means vanishing TKE dissipation, an unrealistic assumption, as Lilly (1968) has noted.

Later work, notably laboratory and numerical studies of Deardorff (1970) and Willis and Deardorff (1974), and parallel theoretical contributions of Tennekes (1970) and others, developed a similarity theory of convective turbulence and of the fluxes it sustains. At any level z, the theory postulates the buoyancy flux or other local variables such as TKE intensity or dissipation to be functions only of the surface buoyancy flux B_0 and the height h of the mixed layer. Dimensional analysis then yields the similarity law for the distribution of the buoyancy flux:

$$\overline{w'b'} = B_0 \, func\,(z/h). \tag{3.8}$$

Strictly speaking, the zone of downward buoyancy flux in the lower thermocline lies outside the range of the similarity theory, because the turbulence there may be affected by such thermocline properties as the potential temperature gradient. Equation 3.8 nevertheless applies to a good approximation up to the level of the peak downward buoyancy flux B_d, at $z/h = 1$ by the above chosen definition of mixed layer height. The ratio B_d/B_0 is then a constant, and so is, to a fairly good approximation, the ratio of the depth-integrated positive and negative buoyancy fluxes. More detailed dimensional argument therefore supports Ball's main hypothesis.

Other useful results of the similarity theory are that the scale velocity of convective turbulence is $w^* = (B_0 h)^{1/3}$ and that buoyancy fluctuations b' scale with $b^* = B_0^{2/3} h^{-1/3}$. When buoyancy is due to heating alone, $b' = g\theta'/T$, and $B_0 = (g/T)\overline{w'\theta'}(0)$. Nondimensional TKE dissipation, $\varepsilon h/w^{*3}$, is also a function of z/h.

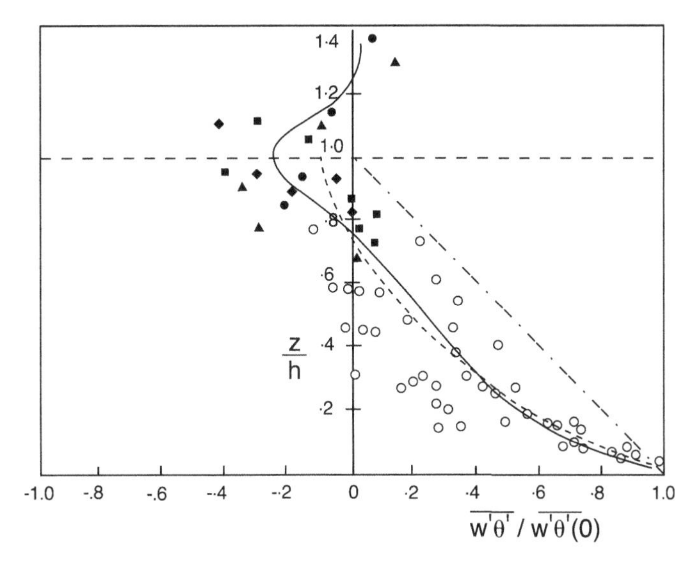

Figure 3.3 Reynolds flux of temperature versus height in an atmospheric mixed layer heated from below, normalized with ground level flux and the height of the peak downward flux. The circles and other dots are values observed in two field experiments, the full line a mean distribution drawn through the data. From Caughey and Palmer (1979).

Several major cooperative experiments established the validity of the similarity laws above a thin surface layer where shear production is also important. They have confirmed that the various properties of convective turbulence depend on the two variables B_0 and h only (see the detailed account of Caughey, 1982). Observations of the atmospheric boundary layer over land, under clear skies, where high surface buoyancy fluxes coupled with vanishing shear production are no rarity, also yielded fairly conclusive data on the various empirical constants and distributions of the similarity theory. Figure 3.3 from Caughey and Palmer (1979) shows the distribution of the Reynolds flux of temperature in the convective mixed layer heated from below, in different locations, at different values of the ground level heat flux and of mixed layer height. Under the conditions of these observations, the Reynolds flux of heat alone is responsible for buoyancy flux.

The observations determine the peak downward buoyancy flux, albeit only with rather large scatter, and place its magnitude between 20 to 25% of B_0. Carson (1973) analyzed observations of the heat flux distribution in convective mixed layers and recognized the linear relationship between B_0 and B_d as an empirical law of physics:

$$-B_d = \alpha_c B_0 \tag{3.9}$$

although he gave a somewhat broader range for α_c, up to 0.4. The larger database of later observations show $\alpha_c = 0.2$–0.25. For later convenience, we will refer to this result as "Carson's law," the law of entrainment for the convective mixed layer heated from below. In terms of the energy budget, Carson's law stipulates constant efficiency

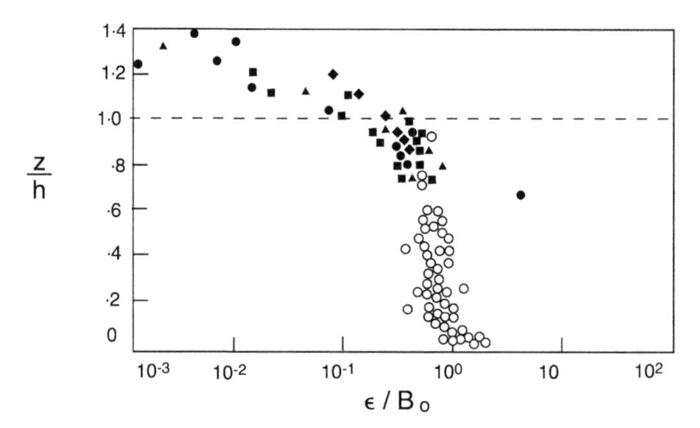

Figure 3.4 Energy dissipation rate normalized with ground level buoyancy flux, in the same experiments as the previous figure. From Caughey and Palmer (1979).

for TKE conversion into potential energy, for the particular case of heating from below.

Carson's law implies that 75 to 80% of the energy input to convective turbulence is dissipated. Figure 3.4, also from Caughey and Palmer (1979) verifies this: the rate of TKE dissipation, determined from the same field observations as the buoyancy flux distributions in the previous figure, is comparable to the rate of buoyant production throughout the mixed layer.

3.3.2 Mixed Layer Cooled from Above

The mirror image of the mixed layer heated from below is the mixed layer cooled from above. In the atmosphere, full or partial cloud cover under an inversion radiates heat to space, while the entrainment of dry upper air induces re-evaporation (Figure 3.2). Both phenomena cool the top of the mixed layer. Moreover, these cloud-top cooling mechanisms occur in a layer so thin that observations only show a large jump in net long wave radiation and liquid water content. Data of Duynkerke et al. (1995) from a recent major cooperative experiment ASTEX (Atlantic Stratocumulus Transition Experiment) in Figure 3.5 show this very clearly. Cooling at cloud top forces upward Reynolds flux of heat and buoyancy in the body of the mixed layer, as we discussed in connection with Equation 3.3, much as heating from below does. In contrast to the case of surface heating, however, entrainment of warmer air from above, and the associated downward buoyancy flux, occurs only a short distance above the cloud top cooling. Cloud top cooling is also often a stronger TKE source than sea level buoyancy flux.

Although the source-sink distribution of TKE in cloud top cooling differs from that in sea level heating, a similarity law is again likely to govern the distribution of buoyancy flux and TKE dissipation. If this also applies to the downward buoyancy flux region, then Carson's law, Equation 3.9, should again quantify the entrainment rate, albeit possibly with a different constant α_c.

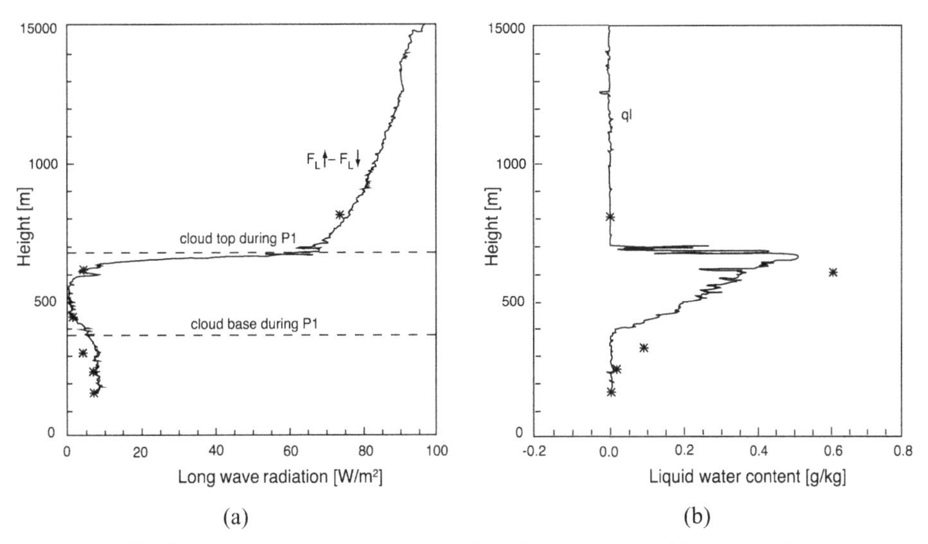

Figure 3.5 Net long-wave radiation flux (a) and liquid water content (b) in a cloud layer over the ocean observed in ASTEX. Points are results from level flights, lines profiles taken in ascent or descent. From Duynkerke et al. (1995).

Experimental evidence in support of this conjecture is of recent vintage and not as robust as in the case of surface heating, nor is the empirical value of α_c well established. Nichols and Leighton (1986) have observed and analyzed the structure and properties of mixed layers under stratiform cloud decks, in several locations over the seas around the British Isles. Their observations extended to radiation fluxes and liquid water content at cloud top, and they were also able to estimate an entrainment rate w_e from the heat balance of the cloud. Their data determine the total upward Reynolds flux of heat required to support the radiant plus re-evaporative heat loss. It equals the net upward long wave radiation flux R_{LW} (directly determined) plus latent heat L times the rate of liquid water re-evaporation $w_e q_\ell \rho$, with w_e the entrainment rate, q_ℓ liquid water content. The total upward buoyancy flux just below cloud top, B_0, is g/T times the Reynolds flux of temperature, $\overline{w'\theta'}$, to cloud top. The peak downward buoyancy flux B_d is g/T times $w_e \Delta\theta_v$, with θ_v the virtual temperature jump across the inversion, also given by Nichols and Leighton (1986). Table 3.1 lists the upward and downward

Table 3.1. *Cloud-top buoyancy fluxes, as equivalent heat flux in W m^{-2}*

Flight	$\rho_a c_{pa} \overline{w'\theta'_v}$	$\rho_a c_{pa} w_e \Delta\theta_v$	α_c
511	86	41.4	0.48
526	95	45.1	0.48
528	55	21.4	0.39
620	62	39.1	0.63
624	82	49.7	0.60
DUY	82	36.0	0.44

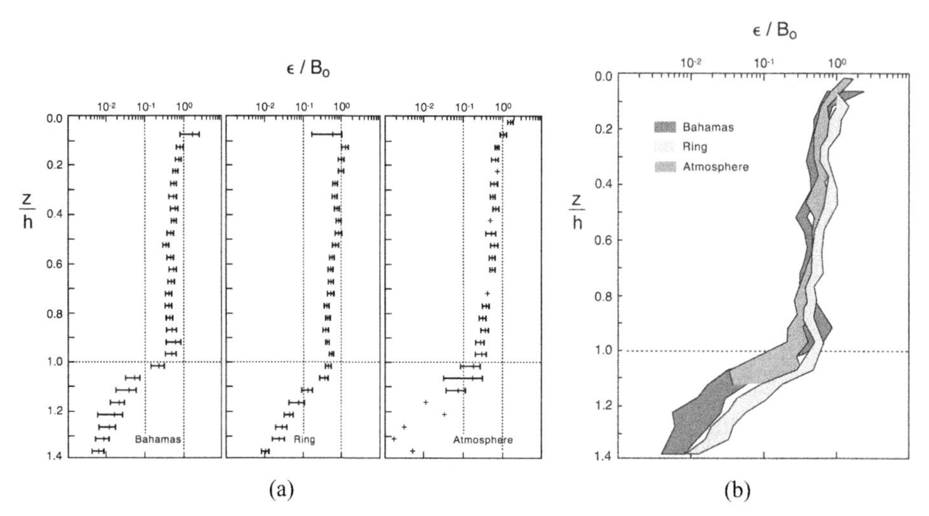

Figure 3.6 Energy dissipation in two oceanic convective mixed layers (near the Bahamas and in a Gulf Stream ring) under night-time cooling from above, compared with energy dissipation in the convective atmospheric mixed layer shown in Figure 3.4. Dissipation is scaled by surface buoyancy flux, height by mixed layer depth in either medium. From Shay and Gregg (1986).

virtual temperature fluxes for the five flights of Nichols and Leighton (1986), expressed as heat flux in W m^{-2}. The empirical values of the constant $\alpha_c = -B_d/B_0$ scatter from 0.4 to 0.6. Similar data of Duynkerke et al. (1995), with the entrainment rate taken from Bretherton et al. (1995), yield the sixth row in the table. According to these data $\alpha_c = 0.5$ is a reasonable empirical value for the present, twice the value applying to the case of heating at the bottom of a mixed layer.

In the ocean, long wave radiation and latent heat transfer cool a very thin layer at the surface, creating an unstable density distribution there, as well as convection to at least some shallow depth. Convective turbulence then entrains cooler water from below. When convection extends over mixed layer depth, the situation is analogous to the atmospheric mixed layer heated from below, entraining warmer air from above. In a study of such cases, Shay and Gregg (1986) compared TKE dissipation rates in two oceanic mixed layers under night-time cooling with dissipation in the convective atmospheric boundary layer, both scaled by surface buoyancy flux ε/B_0. Figure 3.6 shows the comparison with the data of Caughey and Palmer in Figure 3.4. The agreement could hardly be better. Although no direct evidence on entrainment is available in these cases, it is safe to suppose that Carson's law applies to them.

Oceanic entrainment may be the result not only of shear production of TKE, which affects the atmospheric mixed layer no less, but also of TKE input from breaking long waves, a phenomenon discussed in Chapter 2.

3.3.3 Shear and Breaker Induced Entrainment

In the classical laboratory studies of entrainment described in detail by Turner (1973), shear production was the source of the turbulence that engulfed nonturbulent fluid.

A relatively simple case of purely shear-induced entrainment is when the peak shear production of TKE takes place in an inversion or a thermocline, very close to where downward buoyancy flux also peaks. In analogy with Carson's law, peak downward buoyancy flux should then be proportional to peak shear production:

$$B_d \equiv -(\overline{w'b'})_{\min} = \alpha_s \left(\overline{u'w'} \frac{d\overline{u}}{dz} + \overline{v'w'} \frac{d\overline{v}}{dz} \right)_{\max} \tag{3.10}$$

with α_s an empirical constant, presumably different from α_c

What does observation say? Lofquist (1960) carried out a particularly careful and extensive series of laboratory experiments on turbulent flow along a sharp and stable density interface between fresh and saline water in a flume, with the saltwater in forced motion underneath the interface. Each of the two layers was deep enough to insulate the flow along the interface both from the free surface and from the bottom. The turbulent flow near the interface had therefore velocity and length scales depending on local variables, the friction velocity u^* derived from the peak interface stress, the length scale $\ell = u^*/(d\overline{u}/dz)$ from the peak mean velocity gradient at the interface. Lofquist conducted forty-six experiments with different flow velocities and density contrasts, and determined the downward density flux, expressing it as the product of an entrainment velocity w_e and density contrast $\Delta\rho$. The buoyancy contrast $\Delta b = g\Delta\rho/\rho$ times w_e was then the peak downward buoyancy flux $-B_d$.

An analysis of Lofquist's results (Csanady, 1978) revealed that the nondimensional entrainment velocity w_e/u^* varied in direct proportion with the nondimensional parameter $P = u^{*2}/\Delta b\ell$ (see Figure 3.7). Substituting the definition of ℓ, the parameter P turns out to be proportional to $u^{*2}(d\overline{u}/dz)$, the peak TKE production rate. Rounding off the constant in the linear relationship between entrainment velocity and the parameter P, the results imply the following entrainment law:

$$-B_d = 0.03 \frac{u^{*3}}{\ell} = 0.03 u^{*2} \frac{d\overline{u}}{dz} \tag{3.11}$$

which is of the form of Equation 3.10. Thus, approximately 3% of peak TKE production is recovered as peak downward buoyancy flux alias potential energy gain, in this case when TKE shear-production and downward buoyancy flux overlap.

Turner (1973) and Thompson and Turner (1975) reported on experiments in which a mechanical stirrer produces the turbulence near a density interface, bringing about entrainment and potential energy gain. TKE production in this case equals the energy input to the stirrer. Remarkably, the results again showed the same 3% TKE recovery rate, appearing as potential energy gain. The mode of TKE production apparently does not matter, as long as it is mechanical, not convective. Thompson and Turner (1975) made the further important observation that with the stirrer further away from the interface, entrainment decreased. We conclude that Equation 3.11, to be called here "Turner-Lofquist law," applies to shear-induced or other mechanical turbulence-induced entrainment at sufficiently small distances between TKE source and sink.

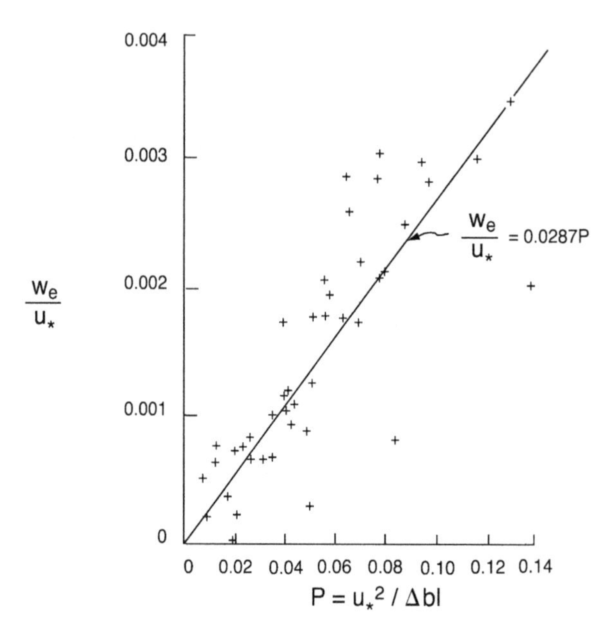

Figure 3.7 Nondimensional entrainment velocity across a density interface versus a stability parameter P proportional to the peak TKE shear-production rate, in laboratory observations of Lofquist (1960). The linear relationship implies proportionality of shear production and entrainment. From Csanady (1978).

More difficult to interpret are laboratory experiments in which shear production peaks at the surface, while entrainment takes place at the bottom of a mixed layer. This realistically simulates an oceanic mixed layer under wind stress, but is a more complex situation than the Lofquist experiment, because shear production of TKE may sport two peaks, one at the surface another at or near the density interface, where a second high-shear zone develops as the surface stress accelerates the mixed layer. The depth of the mixed layer is then another variable influencing entrainment. Phillips (1977) discusses such observations and points out the difficulties of applying the results to an oceanic mixed layer.

When TKE production at the bottom of an oceanic mixed layer is negligible, so that production peaks only at the surface, the Turner-Lofquist law should govern the entrainment rate, at any rate while mixed layer depth remains moderate, the TKE source-sink separation less than some limit. The peak shear production rate at the surface is $u_w^{*2}(d\bar{u}/dz)_0$. On the windsea, Charnock's law, plus much evidence on wind waves, suggests that surface shear should be proportional to u_w^* divided by the wavelength scale u_w^{*2}/g, that is to g/u_w^*. The peak TKE production rate resulting from the surface shear flow is then:

$$u_w^{*2}(d\bar{u}/dz)_0 = \lambda u_w^* g \tag{3.12}$$

with λ an empirical constant.

Observations on the peak surface velocity gradient are almost nonexistent. One exception is the study of Churchill and Csanady (1983), consisting of just two series of drogue and drifter determinations of the near-surface velocity distribution, with details on the centimeter scale close to the surface. One series was carried out in Lake Huron, the other in Cape Cod Bay, in weak to moderate winds, u_w^* ranging from 0.005 to 0.011 m s^{-1}. The data were listed as water-side friction velocity and eddy viscosity for 19 cases in Lake Huron and 5 in Cape Cod Bay. These data readily yield the parameter λ, scattering within a factor of two around $\lambda = 0.67 \times 10^{-3}$. The corresponding peak TKE production rate just below the surface, calculated from shear stress and velocity gradient, was of the order of 5×10^{-4} m^2 s^{-3}.

Applying now the Turner-Lofquist 3% law to the peak TKE production according to Equation 3.12, with λ as just estimated, we have the following entrainment law expressed as peak negative buoyancy flux:

$$-B_d = 2 \times 10^{-5} u_w^* g \tag{3.13}$$

a simple enough formula, predicting typically negative buoyancy fluxes of order 10^{-6} W kg^{-1}.

As we have seen in the last chapter, wave breaking dominates observed TKE dissipation rates in oceanic mixed layers under moderate to strong winds, not shear flow related dissipation. The study of Drennan et al. (1996), code named SWADE (Surface Wave Dynamics Experiment), reported observations of TKE dissipation close to the sea surface in the presence of open ocean wind waves at wind speeds near 10 m s^{-1}, at relatively short fetch. In an illuminating illustration (Figure 3.8), they showed that the water-side TKE dissipation rate exceeded shear flow dissipation (calculated from the logarithmic layer formula, $\varepsilon = u_w^{*3}/\kappa z$) by a large factor, typically 30, at depths

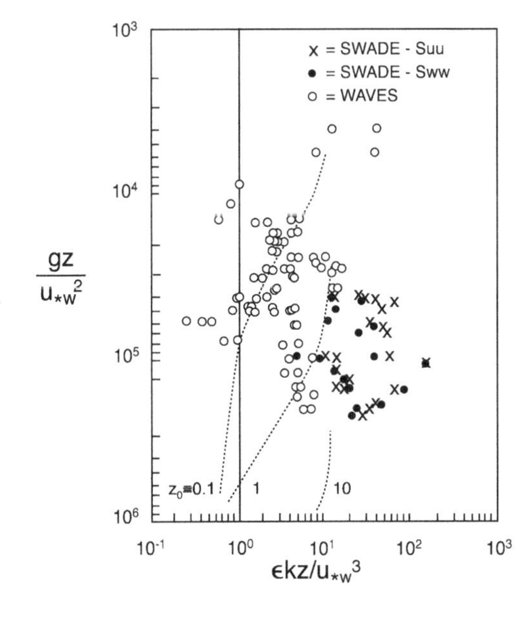

Figure 3.8 Energy dissipation in the wave zone of an oceanic mixed layer, in "wall layer" coordinates. Shear flow dissipation would group around the vertical line at $\varepsilon\kappa z/u_w^{*3} = 1$. The dots and crosses are from the SWADE experiment, the circles from a previous experiment. The SWADE data scatter around a vertical line at $\varepsilon\kappa z/u_w^{*3} = 30$. From Drennan et al. (1996).

of the order of one waveheight. The dissipation rate decreased with the square of waveheight-scaled depth z/H_s, with H_s the characteristic wave height.

As we have discussed in the last chapter in connection with the observations of Drennan et al. (1996), the downward energy input per unit mass via breaking waves was $\eta u_w^{*2} C_p$, with ηu_w^{*2} the fraction of the wind stress that goes into wave momentum transport, C_p the celerity of the characteristic wave. Typical values of η are 0.3 to 0.5 or so. The peak dissipation rate should then be proportional to the energy input divided by the wave height, $\lambda_b \eta u_w^{*2} C_p / H_s$, with λ_b an empirical constant. Relating wave height to wind stress and wave age with the aid of equations 2.19, we have the estimate:

$$\varepsilon_{\max} = \lambda_b \eta g C_p \left(\frac{u_w^*}{u_a^*} \right)^2 \left(\frac{C_p}{u_a^*} \right)^{-3/2} = \lambda_b \eta g u_w^* \sqrt{\frac{\rho_a u_a^*}{\rho_w C_p}}. \tag{3.14}$$

For a fully developed windsea the square root has a value of about 6×10^{-3}, so that with $\eta = 0.3$, the peak dissipation rate estimate is $\varepsilon = 2 \times 10^{-3} \lambda_b g u_w^*$.

The SWADE data of Drennan et al. (1996) do not reach closer to the surface than about $0.7 H_s$, where they are decreasing with the square of the depth. At that level (at the smallest depth of the dots and crosses in Figure 3.8) $gz/u^{*2} = 4 \times 10^4$ while $\varepsilon \kappa z / u_w^{*3}$ scatters around 30, implying a dissipation rate of $\varepsilon = 2 \times 10^{-3} u_w^* g$. The peak breaker-related dissipation rate should be higher than this, somewhere closer to the surface, but we don't know how much higher. If $\lambda_b = 1.0$ in Equation 3.14, $\varepsilon_{\max} = 2 \times 10^{-3} g u_w^*$, or typically 2×10^{-4} m^2 s^{-3}, equal to the observed maximum dissipation. This is four times our estimate of the peak TKE production rate by surface shear, and is very likely an underestimate. If entrainment in an oceanic mixed layer is proportional to the peak TKE production or dissipation rate, whether because of shear or breakers, then breaker related entrainment clearly dominates. This is indeed what one would expect simply from the preponderance of TKE dissipation via breakers over shear flow by a factor of about 30.

To portray the effect of entrainment, it is usual to convert peak negative buoyancy flux into an entrainment velocity $w_e = -B_d / \Delta b$, with Δb the buoyancy jump across the thermocline. One must be careful with this substitution. In the Lofquist experiment, the directly observed quantity was the actual density flux (i.e., effectively $B_d = -w_e \Delta b$), so that the derived 3% rule is solid. In other cases, specifying B_d as entrainment velocity times the total buoyancy change across the thermocline may be misleading, because the relevant buoyancy jump is not necessarily the total across the thermocline: Turbulence may very well only be able to penetrate some of the thermocline layers. With this caveat, we nevertheless estimate typical values of w_e from our results on shear and breaker related entrainment, valid for turbulence engulfing the entire thermocline. A typical oceanic thermocline buoyancy jump is (in the shallow seasonal thermocline) $\Delta b = 10^{-2}$ m s^{-2}. The corresponding shear-related entrainment velocity is $w_e = 2 \times 10^{-4}$ m s^{-1}, implying moderately rapid mixed layer deepening (0.72 m/hr). Typical breaker related entrainment velocity may, on the other hand, be higher than 6×10^{-3} m s^{-1}, or 20 m/hr, almost instant adjustment of thermocline depth to a level outside the range of breaker-induced turbulence. The reader should not forget the tentative nature of all these estimates, as they are based on recent and scant evidence. The very idea of breaker related entrainment is new and based on analogy, not direct observation.

This completes our discussion of mixed layer turbulence and the laws of entrainment. In the next section, we take a look at how these laws work out in the real world of observed mixed layers.

3.4 A Tour of Mixed Layers

As may be expected, the structure of mixed layers varies greatly with latitude and season both in the ocean and in the atmosphere. Their influence on the circulation of either medium is greater in some locations than in others. Key locations are the tropical and subtropical oceans, where the interplay of the two mixed layers decisively influences the global climate through latent heat fluxes. Midlatitude oceans are of special interest to weather forecasters in proximity to population centers, as off western Europe, Japan, or the California coast. Among the oceanic mixed layers, those at the equator and at high latitudes play an important role in the global heat balance. Not surprisingly, these are the sites of major observational studies on mixed layers carried out in the past three decades or so. They provide the empirical basis for the laws entrainment, as we have already seen. To put those laws in perspective, we take a detailed look in this section at a few of the better explored mixed layers, their thermodynamic structure and its relationship to turbulence and entrainment.

The distribution of absolute temperature T and humidity q over height z defines the thermodynamic structure of atmospheric mixed layers. Useful derived quantities are:

$$\theta = T + \frac{gz}{c_p}$$

$$\theta_v = \theta + 0.61Tq$$

$$\theta_e = \theta + \frac{Lq}{c_p}$$

with θ potential temperature, θ_v virtual potential temperature, a measure of buoyancy, and θ_e "equivalent potential temperature," a measure of total energy including internal and gravitational energy and latent heat.

In the oceanic mixed layer, comparable properties are temperature and salinity (salt concentration), the latter of order 35 parts per thousand. The density of seawater ρ_w is a function of these two properties (pressure not significantly affecting density in the mixed layer), and is usually expressed as the excess of density over freshwater, known as "sigma-tee," $\sigma_t = \rho_w - \rho_f$, with $\rho_f = 1000$ kg m^{-3}.

3.4.1 The Atmospheric Mixed Layer Under the Trade Inversion

The Trade Inversion, the atmospheric thermocline over the Trade Wind regions of the world ocean, covers a substantial portion of the globe, from the equator to about latitudes 30°N and S, excepting only the small area fraction under hot towers, located mainly in the ITCZ (InterTropical Convergence Zone). This very large area is the main source of atmospheric water vapor, the "fuel" that drives atmospheric circulation and sustains the hydrologic cycle.

The typical height of the mixed layer under the Trade Inversion is 2 km near the equator, dropping to 1–1.5 km at 30° latitude. The first hard data on the horizontal distribution of inversion height over the Atlantic came from the Meteor expedition of 1925–27. Von Ficker (1936) prepared an area map of it (Figure 3.9) from Schubert et al. (1995).

Clouds under the Trade Inversion are isolated "trade cumuli," similar to fair weather clouds over land, of horizontal dimensions of order 5 km, covering typically 30% of the area. They follow a cycle of development, rise and collapse; their liquid water content rises with height to a maximum of order 1 g/kg at cloud top. Our information on their structure and behavior comes from classical studies of Malkus (1954) and others in the 1950s, as well as later work.

The mean large-scale atmospheric circulation in the Trade Wind region consists of weak subsidence across the thermocline over most of the region, divergent equatorward flow in the mixed layer toward the hot towers of the ITCZ, the rise of air there to great heights, and the return flow poleward to feed the subsidence. The schema in Figure 3.10 (from Schubert et al., 1995) illustrates this. As Figure 3.9 has shown, the actual distribution of mixed layer height is not as simple, but conforms in principle to the schema.

Major cooperative field experiments beginning in the 1960s documented the internal structure of the Trade Wind atmospheric mixed layer. An early study dating from 1969 was the Atlantic Trade wind EXperiment (ATEX). Augstein et al. (1974) described

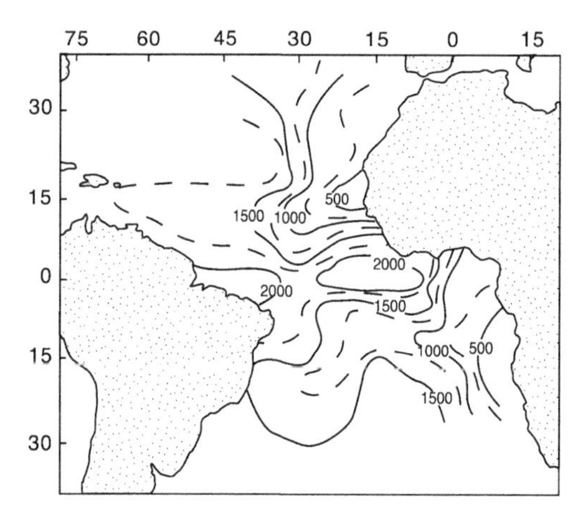

Figure 3.9 Trade inversion height in meters over the Atlantic, data of the Meteor I expedition. From von Ficker (1936).

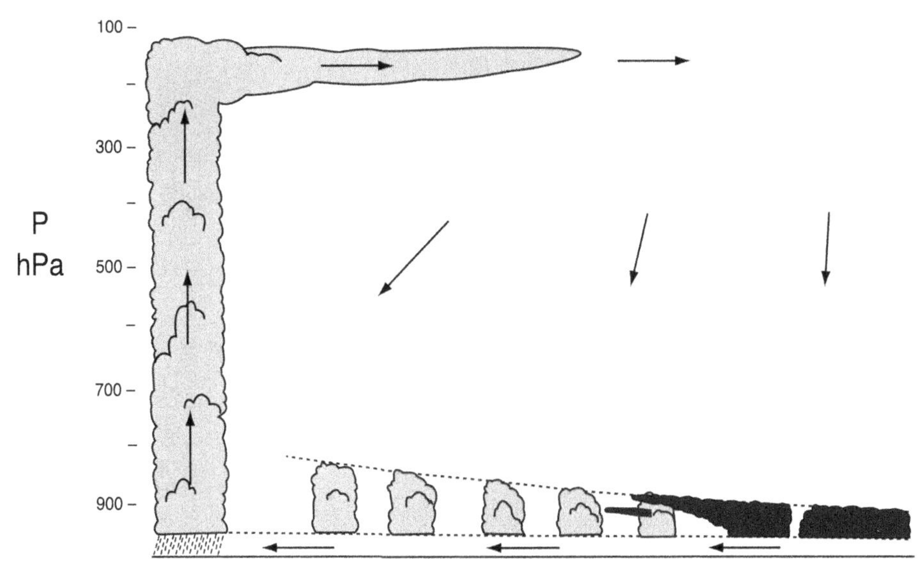

Figure 3.10 Schema of the cloud structure in the tropics and subtropics, and of the large-scale meridional circulation, showing the subcloud layer, the Trade inversion, the hot towers of the ITCZ, and the subtropical subsidence. From Schubert et al. (1995).

this experiment and analyzed its results. The data included a section of the lowest 4 km of the atmosphere taken as the research ship "Meteor" steamed along 30°W longitude from 32°N to 6°S, as well as soundings from three research ships – the Meteor, Planet and Discoverer – when they formed a drifting equilateral triangle near 30°W, 10°N. The drifting observations went on for 15 days, divided into two periods of roughly equal length. Winds in the first period remained near 9 m s^{-1}, but dropped to near 6 m s^{-1} in the second.

Augstein et al. (1974) showed the mean thermodynamic properties of the mixed layer for the two periods at the three ships in two illustrations: Figures 3.11 and 3.12. They noted that minor changes in the thermodynamic properties divide the mixed layer into a well defined structure of several sublayers:

1. A surface sublayer of some 50–100 m height, with a barely noticeable decrease of humidity with height, implying weak gravitational instability.

2. A truly mixed layer or subcloud layer of constant potential temperature and humidity, hence constant θ_v and θ_e, of about 600 m thickness.

3. A 100 m thick "transition layer" at the Lifting Condensation Level, marked mainly by a drop in humidity, but with a small rise in potential temperature, resulting in slight stability.

4. The cloud layer extending upward to 1300–2000 m height, with weakly increasing potential temperature and decreasing humidity, but nearly constant θ_e as some of the vapor condenses but remains in the cloud as fine mist.

5. The trade inversion, where the humidity drops in the case shown by $\Delta q = 7.10^{-3}$ and the temperature T rises by 5 K.

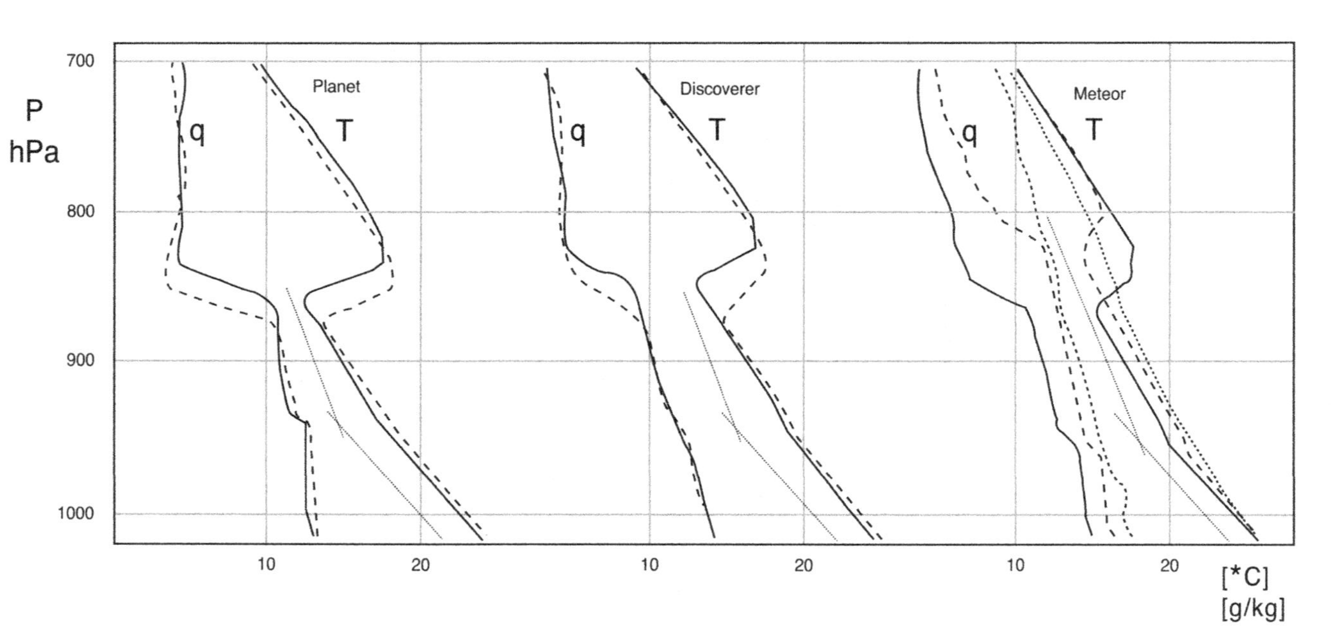

Figure 3.11 Temperature and humidity distribution over height at three ships in the tropical Atlantic. From Augstein et al. (1974). Typical heights corresponding to pressures are 100 m at 1000 hPa, 1000 m at 900 hPa, and 2000 m at 800 hPa.

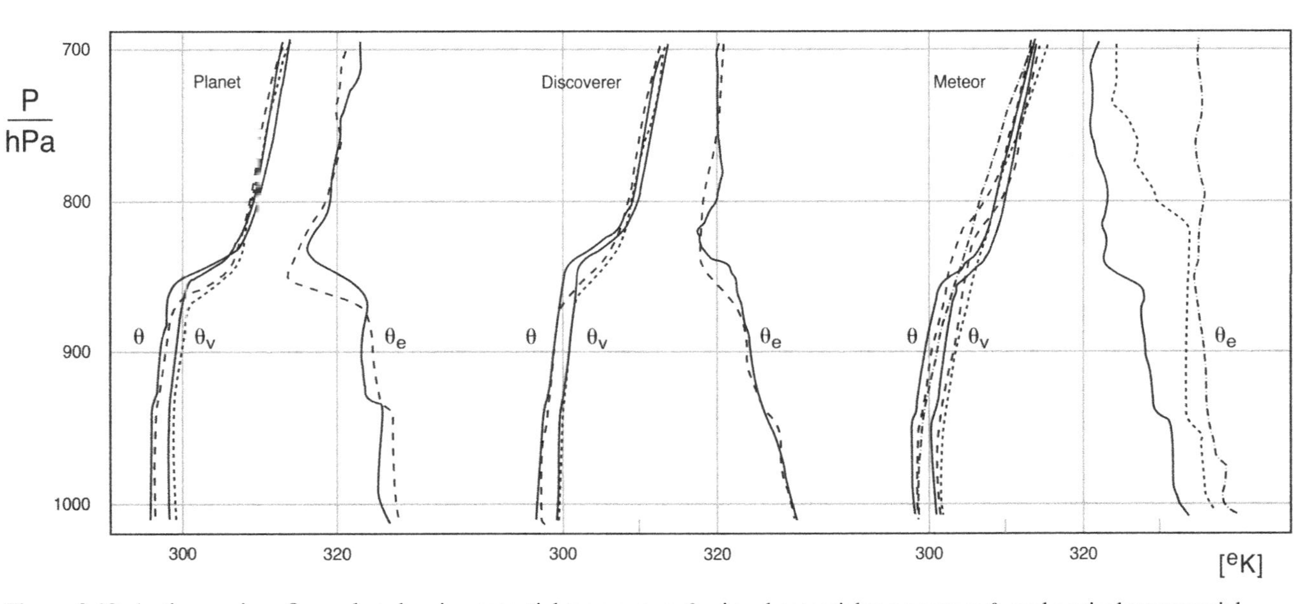

Figure 3.12 As the previous figure, but showing potential temperature θ, virtual potential temperature θ_v and equivalent potential temperature θ_e, in two observation periods, full and dashed lines. At the Meteor, the dash-dotted lines show data from a third, rainy, period. From Augstein et al. (1974).

As the illustrations show, in spite of the divisions, the differences in potential temperature and humidity within the mixed layer remain small compared to the differences across the trade inversion. The slight stability of the transition layer may legitimately be attributed to condensation and short wave heating at the bottom of the overlying cloud layer. Cooling of cloud tops presumably sustains convective turbulence in the cloud layer in spite of a weakly stable gradient of θ_v. The subcloud layer under the transition layer is no doubt well-mixed owing to convective turbulence, in this layer sustained by surface buoyancy flux. The two well-mixed layers are therefore independent, in the sense of having their own separate sources of TKE.

At the Meteor, stationed near the equator, no inversion was present in about 40% of the soundings. The dash-dotted lines in Figures 3.11 and 3.12 show the structure of the lower atmosphere under one of those conditions. The revealing distribution is of θ_e, constant to great heights above a thin surface mixed layer. As we will see later, this is the hallmark of hot towers, moist surface air of high energy content reaching great heights.

North-south cross sections of the lower atmosphere taken by the Meteor while steaming south along 30°W (Figure 3.13) show sloping inversion heights interrupted by two "trough lines," alias hot towers, contours of high θ_e reaching to the tops of the soundings, one at the ITCZ at 10°N, the other at 23°N, associated with a "decaying depression," as Augstein et al. (1974) remark. Away from the hot towers, the Trade Inversion covers the mixed layer at varying heights, presumably determined by the balance of subsidence advecting warm and dry air downward, and entrainment opposing it. As we have seen, radiation cooling at the tops of cumulus clouds forces convection, which is then responsible for cloud top entrainment.

3.4.2 Stratocumulus-topped Mixed Layers

Another large class of atmospheric mixed layers has a solid deck of stratiform cloud for a top, typically only 200 m or so thick, butting against a strong inversion. They occur over subtropical and midlatitude oceans, off the Pacific coast of California and Mexico, over the North Sea and other seas off western Europe, as well as over warm ocean currents such as the Gulf Stream and the Kuroshio.

In the course of their five flights in British coastal waters, that we mentioned in connection with cloud-top cooling induced entrainment, Nichols and Leighton (1986) obtained detailed data on the thermodynamic structure of this type of mixed layer. Figure 3.14 shows the structure found on one of those flights. Nichols and Leighton (1986) chose to call the lowest, evidently turbulent sublayer (by the evidence of the fluctuating humidity trace) "surface or Ekman layer." The substantial humidity gradient, coupled to strong Reynolds flux of humidity, shows this layer to be a source of humidity variance, THV, from where convective turbulence is exported upward into what is clearly a well mixed layer between 250 and 500 m height. On top of that layer, the humidity gradient is again well marked, while the temperature traces shows a slight rise of potential temperature, and weak stability. This layer is only 100 m or so thick,

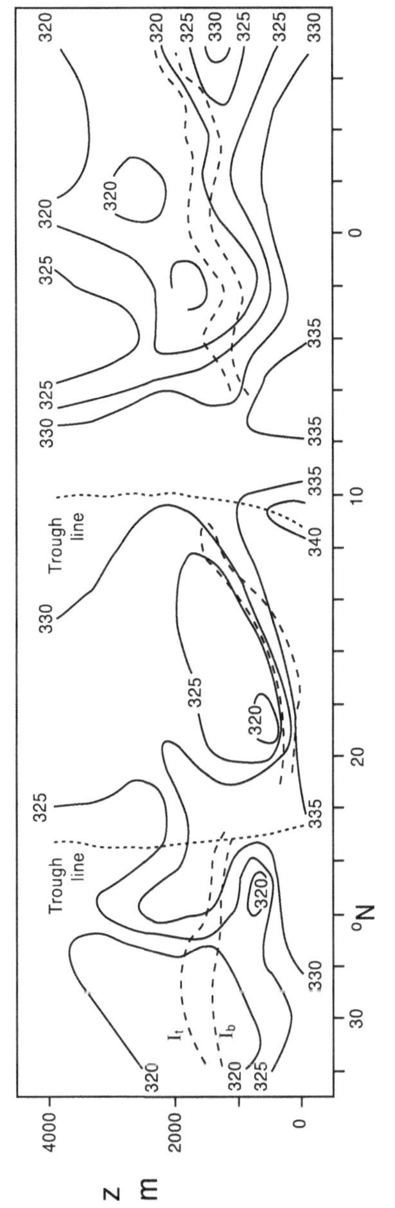

Figure 3.13 North-south section of equivalent potential temperature, along 30° longitude over the Atlantic. From Augstein et al. (1974).

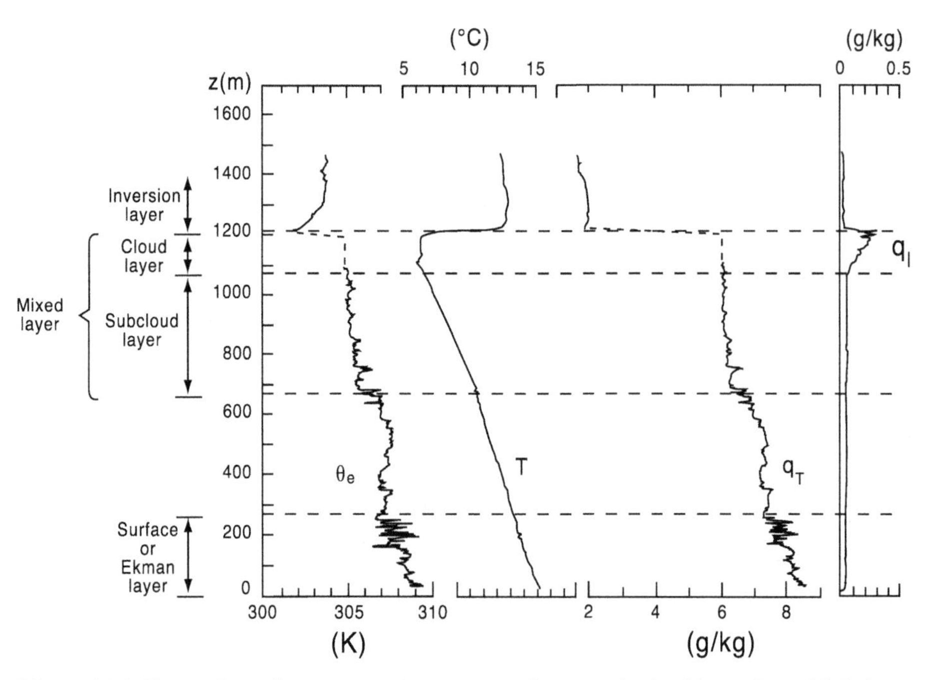

Figure 3.14 Thermodynamic structure of a stratocumulus-topped mixed layer. From Nichols and Leighton (1986).

and strongly resembles Augstein et al.'s "transition layer." The same terminology should surely be retained here.

Next upward lies what Nichols and Leighton (1986) designate the subcloud layer, with convective turbulence originating from cloud top cooling, and spanning the stratiform cloud deck as well as the subcloud layer. The two layers together thus substitute for the isolated-cloud layer of the Trade Inversion. A new piece of information is the liquid water content: its typical linear increase with height identifies the cloud deck.

Nichols and Leighton (1986) do not differentiate between the well-mixed layer between 250 and 500 m and the transition layer, referring to the whole of it as a stably stratified layer that "decouples" the surface or Ekman layer from the "mixed" (cloud plus subcloud) layer. The decoupling is, however, better attributed to the thin transition layer that forms a floor for the upper convectively mixed layer, as well as a ceiling for the lower one. From another point of view, decoupling is a consequence of having two separate sources of TKE: one at the top of the cloud deck, another just above the sea surface.

Further light on decoupling comes from the Atlantic Stratocumulus Transition EXperiment, ASTEX. As a contribution to that experiment, Miller and Albrecht (1995) recorded and analyzed the structure of the atmospheric mixed layer at the island of Santa Maria in the Atlantic, about 38°N, 25°W. Figure 3.15 shows one of their daytime soundings of humidity, and temperatures θ, θ_v, and θ_e (as well as θ_{es}, saturation equivalent temperature, a measure of total energy at a hypothetical 100% relative humidity). The stratus deck was here 160 m deep, the subcloud mixed layer extended

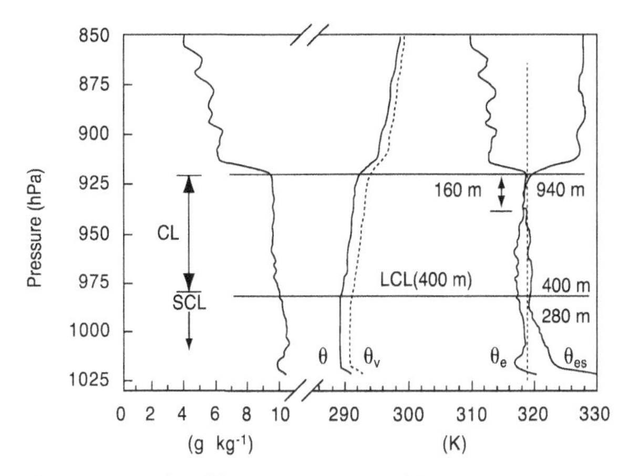

Figure 3.15 The different sublayers of the stratocumulus-topped mixed layer over the Atlantic, as shown by the humidity and temperature distributions. From Miller and Albrecht (1995).

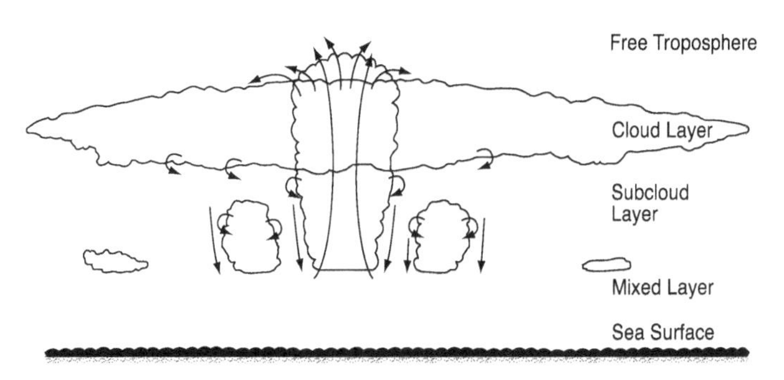

Figure 3.16 Cumulus clouds interacting with an overlying stratiform cloud. From Martin et al. (1995).

downward to the transition layer near 400 m height, which is also the Lifting Condensation Level. Cumulus clouds rise from this level as part of the convective turbulence, much as they do under the Trade Inversion, but here they penetrate, and combine with, the stratus deck. Underneath the LCL lies what these authors designate the "surface moist layer," a layer with convective turbulence sustained by surface fluxes of heat and humidity (a shallow surface sublayer with strong gradients of temperature and humidity is barely visible in the figure). This corresponds to the truly well-mixed layers just above surface sublayers in the previously discussed atmospheric mixed layers, whether under isolated or stratiform clouds.

The interesting new feature here is the presence of cumulus clouds in addition to the stratiform deck, and the formers' role as tracers of convective motions in the subcloud layer. Figure 3.16, due to Martin et al. (1995), shows a schema of cumulus clouds breaking through, and interacting with, the overlying stratiform deck. The convective turbulence in the cloud and subcloud layers between LCL and stratus top owes it

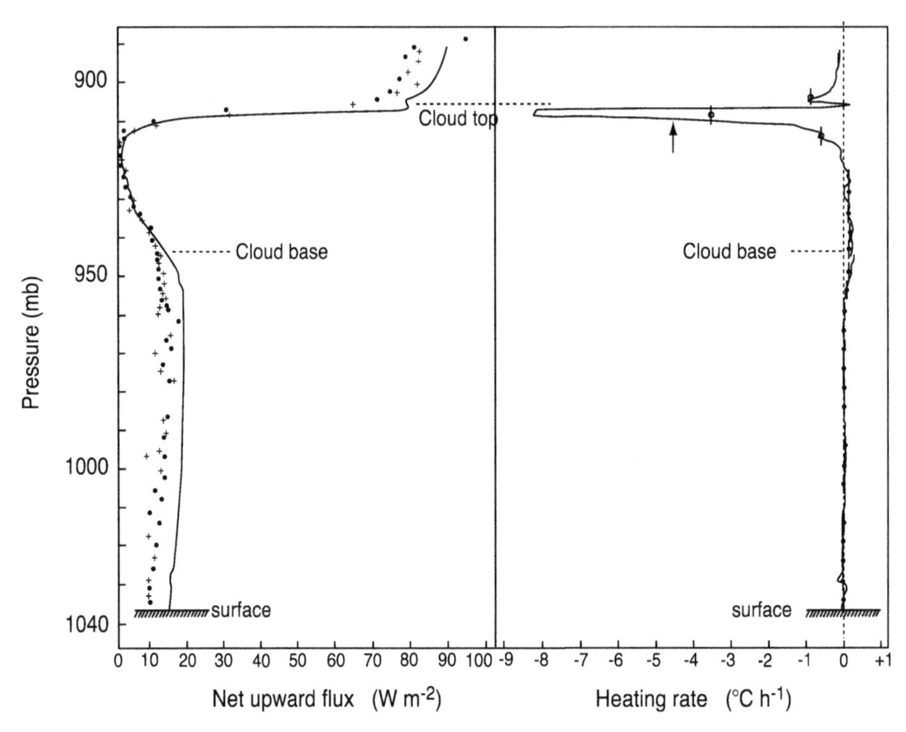

Figure 3.17 Stratiform cloud blocking radiation from below, while massively radiating upward at its top, from a thin layer cooled at a very high rate. From Slingo et al. (1982).

existence to cloud top cooling, while the surface mixed layer draws its convective TKE from surface buoyancy flux. This structure is seen now as the common feature of cloud-topped mixed layers, whether the clouds are trade cumuli or stratiform deck. "Decoupling," in the sense that turbulence and entrainment above the transition layer has nothing to do with sea level fluxes, characterizes all of the surveyed cloud-topped mixed layers, and presumably all such layers with significant area-coverage by cloud. The physical reason is simply that long wave radiation from clouds, or rather the upward buoyancy flux it induces, is a potent source of TKE, that easily equals or exceeds the typical sea level buoyancy flux. Figure 3.17 from Slingo et al. (1982) drives the point home: 80 W m^{-2} upward long wave radiant flux above the cloud, zero immediately below cloud top. This constitutes a large fraction of the typical radiant heat loss of the entire atmosphere, which is about 180 W m^{-2}. The typical sea level buoyancy flux corresponds to a heat flux of only about 50 W m^{-2}, although that is mostly due to vapor flux.

3.4.3 Oceanic Mixed Layers

Short wave solar radiation penetrates the ocean surface to depths measured in tens of meters, heating the oceanic mixed layer at mid-day under clear skies at rates that dwarf other components of the heat budget. The short wave heat flux $-R_{SW}$, or "downward irradiance," crossing the sea surface at mid-day under clear skies can reach 1000 W m^{-2} or more. Ohlmann et al. (1998) report peak two-hour average fluxes of 950 m^{-2} in the Western Pacific "warm pool," daily averages ranging from 47 to 270 W m^{-2}, in a location where the climatic average is 220 W m^{-2}. This large heat gain stands in dramatic contrast to the atmospheric mixed layer's radiation balance, which involves much smaller if any gain, and losses exceeding gains.

The distribution of day-time heating in the water column, i.e., the divergence of the radiant heat flux, consists of rapid heat absorption in the top ten meters, and exponential decay of flux with depth below, where only the blue-green light is left (Simpson and Dickey, 1981). The rates of decay depend on the turbidity of the water. "Type I" or relatively clear water absorbs radiation according to the formula of Paulson and Simpson (1977):

$$R_{SW} = -I_0 \left(0.58 \exp\left[-\frac{z}{\zeta_1}\right] + 0.42 \exp\left[-\frac{z}{\zeta_2}\right] \right) \tag{3.15}$$

where I_0 is the surface irradiance, $\zeta_1 = 0.35$ m and $\zeta_2 = 23$ m are extinction coefficients for red and blue light. Ohlmann et al. (1998) find that a similar formula with $\zeta_1 = 0$ works well below 10 m depth, and point out that some heat escapes a shallow mixed layer downward. At the average mixed layer depth of 30 m in their experiment, the downward radiation flux was 9.2% of surface irradiance.

Solar radiation thus heats the ocean from the top, and tends to generate a stable density distribution in tropical and mid-latitude seas, that insulates the surface from deeper layers. At the sea surface and immediately underneath, however, long-wave radiation and evaporation change this picture. Here, conditions resemble the top of a cloud-topped atmospheric mixed layer: The sum of $R_{LW} + LE$ is high and positive right at the interface, but vanishes at a very small depth. Continuity of heat flux requires high upward Reynolds flux to supply the surface heat loss, even under maximum day-time surface heating. This then also implies strong upward flux of buoyancy, hence convective TKE production and turbulent mixing in a near-surface zone. The Reynolds flux of heat remains positive to a "compensation depth" such that the radiant heat gain above that depth equals the surface heat loss. Convective turbulence may be expected to extend further downward, but not very much further, because it has to contend with the stabilizing influence of radiant heating below the compensation depth.

All this applies in daytime. Nocturnal conditions are a high rate of cooling from the surface, convection, entrainment of cooler water from below, and rapid expansion of the convectively mixed layer downward. A diurnal cycle of daytime stratification and night-time convectively mixed layer deepening thus comes to characterize oceanic mixed layer behavior. Brainerd and Gregg (1993) illustrate typical conditions by the schema of Figure 3.18. When convection reaches a strong seasonal or permanent thermocline, the negative buoyancy flux it can support is only able to slowly erode thermocline top, until the next sunrise generates a new "diurnal" thermocline at the bottom of a shallow convectively mixed layer. In another illuminating illustration (Figure 3.19), Lombardo and Gregg (1989) show surface heat and buoyancy fluxes and mixed layer

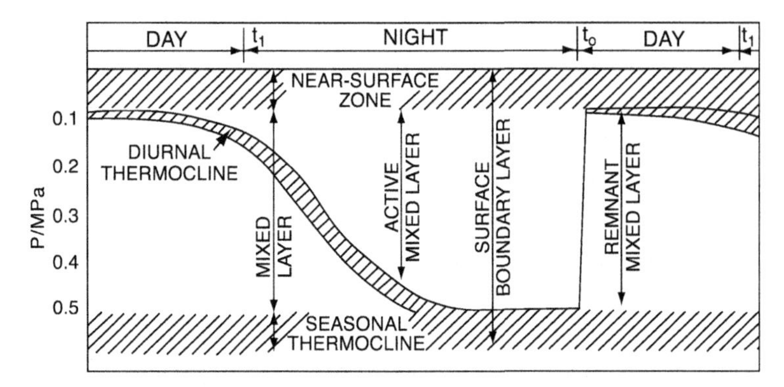

Figure 3.18 Diurnal cycle of changes in the oceanic mixed layer. From Brainerd and Gregg (1993).

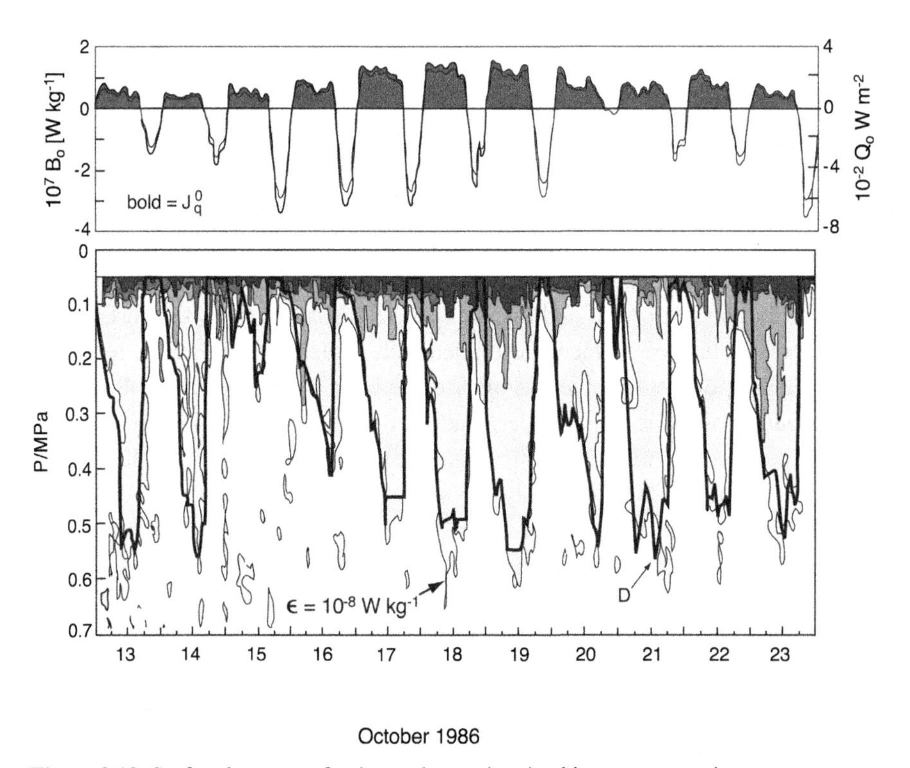

Figure 3.19 Surface buoyancy forcing and oceanic mixed layer response in energy dissipation rate. From Lombardo and Gregg (1989).

response, the latter as TKE dissipation rate contours, and mixed layer depth defined as "the middle of the entrainment zone." The data come from the PATCHEX experiment in the outer reaches of the California Current, 35°N, 127°W, 500 km west of Point Conception. Daytime heating rates were up to 600 W m^{-2}, night-time cooling rates up to 250 W m^{-2}. The mixed layer depth roughly coincides with the $\varepsilon = 10^{-8}$ W kg^{-1} contour. Figure 3.20 shows a blow-up of a single diurnal cycle, entrainment deepening

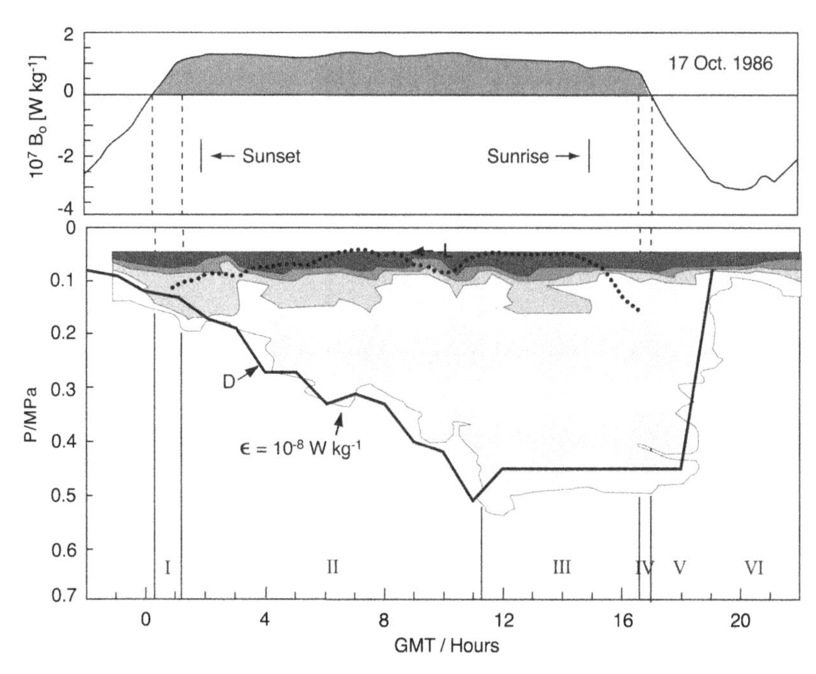

Figure 3.20 Details of the diurnal cycle of mixed layer energy dissipation. From Lombardo and Gregg (1989).

of the mixed layer down to about 50 m (MPa = 100 m in water), where it "encountered a large increase in stratification," according to Lombardo and Gregg (1989). At that level it remained until three hours after sunrise, when the turbulence died in an hour, and the diurnal thermocline reappeared at a shallow depth.

In the PATCHEX case the density jump across the diurnal thermocline was quite small. Brainerd and Gregg (1993) show temperature differences across it of the order of 0.01 K, or a buoyancy jump of $\Delta b = 2 \times 10^{-5}$ m s^{-2}. The mixed layer entrained water from below and descended at a rate of about $w_e = 10^{-3}$ m s^{-1}, driven by a surface buoyancy flux of $B_0 = 10^{-7}$ W kg^{-1} $\equiv 10^{-7}$ m^2 s^{-3}. Putting $B_d = w_e \Delta b = 2 \times 10^{-8}$ m^2 s^{-3} we find Carson's law satisfied to the accuracy of these estimates. The buoyancy jump across the diurnal thermocline was very small, yet it clearly separated a zone of active turbulence above from a zone of decaying turbulence below, the remnant mixed layer.

The same minuscule density jumps characterized the thermocline at the bottom of the convective mixed layer in Shay and Gregg's (1986) observations near the Bahamas and in a Gulf Stream ring, which we quoted earlier to demonstrate the equivalence of night-time convection in the ocean to daytime convection in the atmosphere: TKE dissipation in the oceanic mixed layer mirrored that in the convective atmospheric boundary layer. Lombardo and Gregg (1989) also found essentially the same ε distribution in PATCHEX, in cases when shear production of TKE was small. They further found a scaled depth-average mixed layer dissipation of $\varepsilon / B_0 = 0.44$ while the mixed layer was deepening, versus 0.65 when the mixed layer reached the stable thermocline

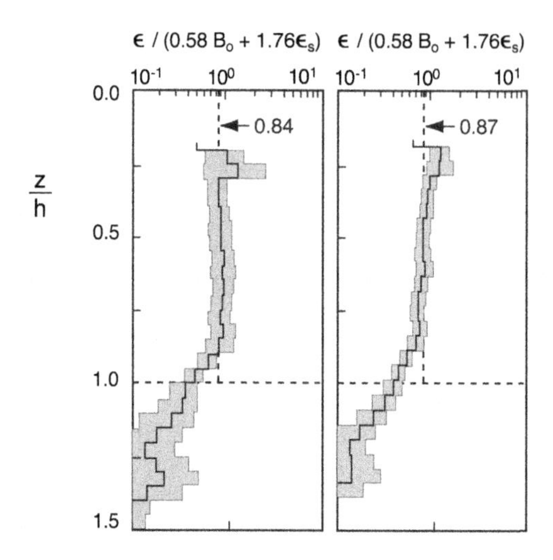

Figure 3.21 Energy dissipation when both surface cooling and shear are present. From Lombardo and Gregg (1989).

and its depth remained constant. The difference, $0.21B_0$, is what one might surmise went into entraining thermocline fluid according to Carson's law.

Yet another interesting finding of Lombardo and Gregg (1989) was that when significant shear production of TKE was present, along with surface cooling, the mixed layer dissipation rate was simply the sum of the rates resulting from cooling and shear flow (Figure 3.21). The part resulting from cooling was again about $0.6B_0$, while the shear flow related dissipation was $\varepsilon = 1.76(u^{*3}/\kappa z)$, the same as they found in cases when shear production was the only source of TKE. That simple superposition works on the dissipation rate suggests that entrainment rates should be additive, too.

The minuscule density jumps in the diurnal oceanic thermocline resemble the similar weakly stable "transition layer" in the atmospheric mixed layer, in that they also mark the boundary of turbulence sustained by surface fluxes. Underneath the small density jumps is a weakly stratified layer extending downward to a much stronger, longer-lived seasonal or permanent thermocline. The density jumps of the more permanent oceanic thermoclines are typically of order $\Delta\rho/\rho = 10^{-3}$, the buoyancy jumps $\Delta b = 10^{-2}$ m s^{-2}. Surface cooling driven entrainment with the typical buoyancy flux of $B_0 = 10^{-7}$ is only able to entrain fluid of comparable density excess at the rate of order $w_e = 2 \times 10^{-6}$ m s^{-1}, 23 cm/day, which would be imperceptible in the diurnal cycle. Entrainment by such a weak TKE source can only eat slowly into the seasonal thermocline.

In some locations, dynamically imposed surface divergence dictates the mass rate of upwelling into the oceanic mixed layer. If a strong thermocline underlies such a mixed layer, the laws of entrainment only allow the incorporation of upper thermocline fluid into the mixed layer, at an average buoyancy such that the peak buoyancy flux is what the TKE supply can sustain. In this case, the mixed layer turbulence "peels off" a thin slice from the top of the thermocline, incorporating it in the mixed layer. This cools the mixed layer, while the thermocline moves upward at a rate dictated by the surface divergence, less the slices peeled off. A steady state sets in only when surface heating balances the entrainment flux of heat, and when a cool enough mixed layer temperature has reduced that flux to a sustainable level. Equatorial upwelling is a situation of this kind, in which large volumes of upper thermocline water enter the surface mixed layer, to be exposed to solar heating.

3.4.4 Equatorial Upwelling

Easterly winds prevail over the tropical oceans and drive their surface waters westward. Earth rotation deflects that flow into "Ekman transport," poleward in both hemispheres, northward north, southward south of the equator (see e.g., Gill, 1982). The upwelling velocity that feeds this divergence is proportional to the wind stress, but has a typical magnitude of order 10^{-5} m s^{-1}. Where the mixed layer is deep, convergence at lower levels balances surface divergence and the upwelling circulation all takes place in the mixed layer without affecting the thermocline. Where the surface mixed layer is shallow, however, upwelling has to come from the thermocline and entrainment has to balance upward advection to maintain steady state. This is the case in the eastern portions of both the Pacific and the Atlantic Oceans: the thermocline in both oceans tilts down westward, so that a surface pressure gradient balances the westward wind stress. This keeps the mixed layer deep in the west, shallow in the east. Surface divergence in the eastern Pacific and Atlantic thus generates upwelling from the thermocline over a band several hundred kilometers wide North to South, several thousand kilometers long West to East. In the eastern Pacific, according to mass balance estimates of Wyrtki (1981), the average upwelling rate, over the north-south extent of 400 km of the upwelling zone, is a little over 10^{-5} m s^{-1}, the maximum on the equator about three times this high.

Upwelling depresses sea surface temperature, as we just pointed out, and with it, the air-sea transfer of sensible and latent heat. Occurring as it does over a large expanse of the ocean, equatorial upwelling and its wind-induced variations significantly influence large-scale atmospheric circulation. The global importance of this became clear following a seminal paper of Bjerknes (1966), in which he documented weather changes in North America correlated with sea surface temperature changes in the eastern equatorial Pacific Ocean, a phenomenon now known as El Niño. Serious oceanographic investigations of equatorial upwelling and its consequences followed some two decades later, notably in the course of the major cooperative experiment "Tropic Heat" of 1984–1987.

As part of this experiment, Peters et al. (1988) collected and analyzed data on equatorial turbulence, beginning in late 1984, and presented a particularly full catalogue of the different variables playing a role in the equatorial mixed layer balances of heat and momentum. Figure 3.22 shows mean (4.5 day average) temperature, salinity, and potential density distributions. The illustration shows density as σ_θ instead of σ_t, potential density anomaly over fresh water, with the in situ density reduced to surface pressure, supposing isentropic expansion, in the standard presentation of oceanographers. In the surface mixed layer of the ocean, the difference between in situ and potential densities is insignificant. Averaging over several days smoothes out the

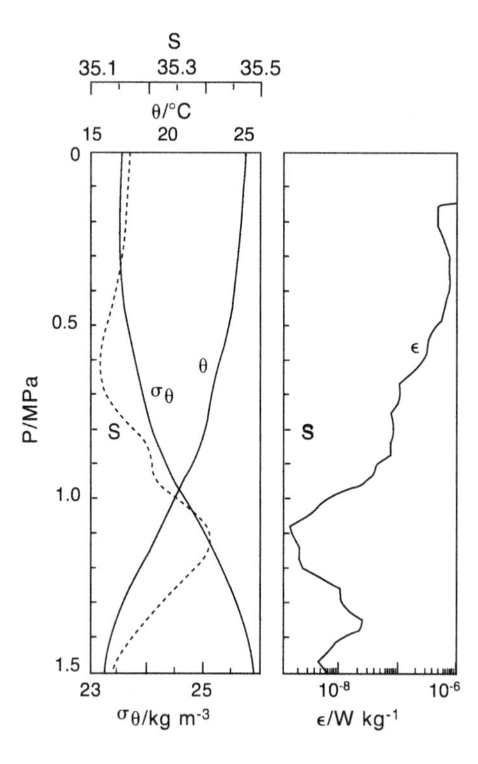

Figure 3.22 Temperature, salinity, and density (σ_θ), and energy dissipation, in the oceanic mixed layer of the eastern equatorial Pacific. From Peters et al. (1988).

diurnal cycle. Figure 3.22 also contains the dissipation rate of TKE, ε, while Figure 3.23 shows component momentum fluxes τ_x, τ_y and heat fluxes Q below a well-mixed surface layer.

What one calls the mixed layer in this situation is a matter of definition. In our atmospheric examples, we had structured mixed layers with moderate density changes up to a sharp inversion. What was present throughout, and universally characterized the structured mixed layer, was turbulence. An analogous choice for mixed layer bottom here would be the level of the sharpest density gradient at about 100 m depth (1.0 M Pa) coincident with a sharp drop of energy dissipation, signifying the limit of active turbulence.

The momentum and heat fluxes at the bottom of the mixed layer so defined vanish. There is a truly mixed layer in the top 20–30 m. Solar radiation dominates the daily average heat flux at the surface, so that it is downward, $R(0) = -115$ W m^{-2} (the night-time buoyancy flux, however, is upward at 1.5×10^{-7} W kg^{-1}, corresponding to a heat flux of $R(0) = 200$ W m^{-2}). At the bottom of the truly mixed layer, at about 20 m depth, the average heat flux is still downward, $R(20) = -81$ W m^{-2}. There is furthermore high downward momentum flux and high energy dissipation below the truly mixed layer. The high energy dissipation at levels below 50 m is due to the Equatorial UnderCurrent, EUC, flowing eastward under the westward drift of the surface. The EUC speed peaks at 110 m s^{-1}, the shear (velocity gradient) in it is high where the momentum flux is high, and so is therefore TKE shear-production.

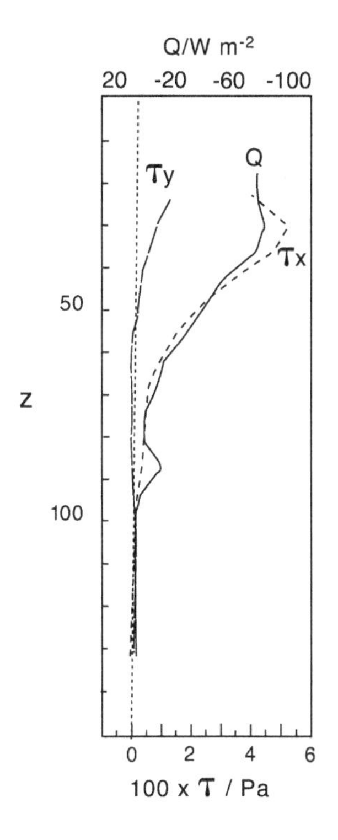

Figure 3.23 Surface wind stress components, τ_x eastward, τ_y northward, and heat flux in the oceanic mixed layer of the eastern equatorial Pacific. From Peters et al. (1988).

The downward buoyancy flux at the bottom of the truly mixed layer is the kinematic heat flux $R(20)/\rho c_p$, multiplied by αg, with α the thermal expansion coefficient of seawater, which at 25°C is about 3×10^{-4} K^{-1}, yielding $B_d = 6 \times 10^{-8}$ m^2 s^{-3}. The TKE shear production term has a maximum of about 1.2×10^{-6} m^2 s^{-3}, close to 50 m depth. Shear production alone, by the 3% rule, accounts for about 60% of the downward buoyancy flux. The nighttime upward buoyancy flux at the surface, 1.5×10^{-7} m^2 s^{-3}, with Carson's law and $\alpha_c = 0.2$, is sufficient to support the other half of the downward buoyancy flux. Within the limited accuracy of these estimates, the two TKE sources seem simply to add up, as in the observations of Lombardo and Gregg.

A new insight gained from Figure 3.23 is that the divergence of the downward heat flux is significant well below the bottom of the truly mixed layer, from 40 m down to about 70 m depth. Upwelling thus comes from a slice of the upper thermocline between 40 and 70 m depth, where the average temperature is only about 1 K lower than in the truly mixed layer. Putting $w_e \Delta\theta$ for the kinematic version of the downward heat flux of about 80 W m^{-2}, with $\Delta\theta = 1$ K, we calculate an entrainment (upwelling) velocity of $w_e = 2 \times 10^{-5}$ m s^{-1}. This is the same value as found by Gouriou and Reverdin (1992) for equatorial upwelling in the Atlantic, and similar to what Wyrtki (1981) inferred from surface divergence in the Pacific.

All of the entrainment according to these results takes place within what by one definition would be the equatorial mixed layer, extending downward to the limit of vigorous turbulence, as far as dissipation remains high. Mixed layer bottom by this definition is at 100 m. An alternative choice of mixed layer bottom, also consistent with our treatment of the atmospheric mixed layer, is the level of peak downward buoyancy flux, found at about 40 m depth. What is different in the equatorial oceanic mixed layer is that these two levels are far apart, owing to shear-production of turbulence in the EUC. Because the peak downward buoyancy flux quantifies entrainment from the thermocline, the level where it occurs is still the most sensible choice of mixed layer bottom. We will return to this question in the next section.

Other oceanic examples of dynamically forced upward entrainment into the mixed layer are coastal upwelling, and upwelling in regions of the ocean with "cyclonic" wind stress curl (tending to produce anticlockwise circulation in the northern, clockwise circulation in the southern hemisphere). Where Ekman transport conveys mixed layer fluid offshore from a coast, a band of divergent surface flow must exist near the coast, and upwelling must maintain mass balance. The upwelling has to be sustained by shear flow turbulence, if steady state is to be maintained. A typical Ekman transport magnitude at midlatitude is $u_w^{*2}/f = 1 \text{ m}^2 \text{ s}^{-1}$ (f is the Coriolis parameter), a typical width of the divergent band 10 km, so that entrainment velocity is typically $w_e = 10^{-4} \text{ m s}^{-1}$, rather higher than typical of equatorial upwelling, but confined to a narrow band. Coastal upwelling depresses sea surface temperature just as equatorial upwelling does, and it also brings to the surface waters from the upper thermocline. These waters tend to be rich in nutrients, nitrates, and phosphates, and they fuel high biological productivity.

In a region of cyclonic wind stress curl, Ekman transport is also divergent, so that the mixed layer becomes shallow and cool in steady state. In the North Atlantic, for example, north of the peak westerly wind stress, the cyclonic curl is of order $d(u^{*2})/dy = 10^{-10} \text{ m s}^{-2}$, so that the divergence of the Ekman transport is some 10^{-6} m s^{-1}. This is also the magnitude of the upward entrainment velocity needed to maintain a mixed layer in steady state, an order of magnitude less than in equatorial upwelling. Nevertheless, entrainment at this rate still keeps mixed layers much shallower and sea surface temperatures lower than where Ekman transport is convergent, as it is over large areas of the subtropical ocean.

3.5 Mixed Layer Interplay

The oceanic mixed layer absorbs the large short-wave downward irradiance mostly in the top ten meters. The two mixed layers in contact handle this heat gain, retain some of it in the ocean, transfer much of the rest to the atmosphere, and lose the remainder through long-wave radiation from the sea surface to space. The exact proportions of the split are the outcome of an interplay between the mixed layers, subject to the heat and vapor budgets of the atmospheric mixed layer, and the heat budget of the oceanic one. Lumping the two mixed layers into a single system and analyzing its interaction with the rest of the ocean and the atmosphere yields further insight.

In the mixed layer budgets, exchanges across the thermoclines play an important role. Thermoclines are, however, somewhat arbitrary boundaries of the system we call a mixed layer, as the case of the equatorial ocean's mixed layer demonstrated. What we choose to call the top of the atmospheric mixed layer or the bottom of the oceanic one, requires some thought.

3.5.1 Mixed Layer Budgets

In our survey of mixed layers above, we have defined the upper boundary of the atmospheric mixed layer, or the lower boundary of the oceanic one, as a surface in the air or water fixed by some rule such as a surface of specified constant temperature or constant TKE dissipation rate, or as a surface where the downward buoyancy flux peaks. This kind of boundary is "open," in the sense that the fluid may freely flow through it, or to put it another way, the boundary may move relative to the fluid. Thus, the inversion that marks the upper boundary of a midlatitude atmospheric mixed layer over land moves upward during the day, when surface heating generated convection entrains air from above. The Trade Inversion does not move, not at least on a long-term average, but air subsides through it continuously. The heat or vapor budget of such a mixed layer has to take into account any movement of the boundary, along with the fluxes across the moving boundary, both advection and Reynolds fluxes.

Let χ be an arbitrary scalar property, temperature or water vapor concentration, subject to the conservation law:

$$\frac{\partial \chi}{\partial t} + \frac{\partial(u\chi)}{\partial x} + \frac{\partial(v\chi)}{\partial y} + \frac{\partial(w\chi)}{\partial z} = -\frac{\partial(\overline{w'\chi'})}{\partial z} + S \tag{3.16}$$

where S is any source term; u, v, w are mean fluid velocities; and the primed quantities are fluctuations, their mean product the vertical Reynolds flux of the property. Horizontal Reynolds fluxes are negligible, and neither the horizontal velocities, nor the property χ depart more than a small amount from their average value in a mixed layer of height h. The fluid velocities satisfy the continuity equation:

$$\frac{\partial u}{\partial x} + \frac{\partial v}{\partial y} + \frac{\partial w}{\partial z} = 0. \tag{3.17}$$

The depth-average value of the property χ is of interest, how it varies with boundary movement and Reynolds fluxes from above and below.

We first take the atmospheric mixed layer, which has a fixed lower and a moving upper boundary. In integrating these equations over variable height h, Leibniz's rule has to be observed, e.g.:

$$\int_0^h \frac{\partial(u\chi)}{\partial x}\, dz = \frac{\partial}{\partial x}\int_0^h u\chi\, dz - u(h)\chi(h)\frac{\partial h}{\partial x} \tag{3.18}$$

where $u(h)$, $\chi(h)$ designate values at mixed layer top. The depth-integrated horizontal velocities and scalar property χ are:

$$\begin{aligned} \int_0^h u\,dz &= u_m h = U \\ \int_0^h v\,dz &= v_m h = V \\ \int_0^h \chi\,dz &= \chi_m h \end{aligned} \tag{3.19}$$

the index m designating depth-average quantities. Depth-integration of the continuity equation yields:

$$\frac{\partial U}{\partial x} + \frac{\partial V}{\partial y} = w_e - \frac{\partial h}{\partial t} \tag{3.20}$$

with $w_e = dh/dt - w(h)$, the rate of advance of the mixed layer top relative to the air, or entrainment velocity, $w(h)$ being mean vertical air velocity at mixed layer top, and the total derivative is $d/dt = \partial/\partial t + u(h)\partial/\partial x + v(h)\partial/\partial y$.

Depth-averaged values of the products $u\chi$, $v\chi$ are $U\chi_m$, $V\chi_m$; the source term average is S_m. The depth-integrated conservation law is then:

$$\frac{\partial(\chi_m h)}{\partial t} + \frac{\partial(U\chi_m)}{\partial x} + \frac{\partial(V\chi_m)}{\partial y} = w_e\chi(h) - \overline{w'\chi'}(h) + \overline{w'\chi'}(0) + S_m h \tag{3.21}$$

neglecting products of departures from the depth-average values of velocities and the property χ. Making use of the continuity equation to eliminate the transports U, V, we arrive now at the budget equation for the atmospheric mixed layer:

$$h\frac{d\chi_m}{dt} = w_e\,[\chi(h) - \chi_m] + \overline{w'\chi'}(0) - \overline{w'\chi'}(h) + S_m h \tag{3.22}$$

where $\overline{w'\chi'}(0)$ is Reynolds flux at the surface, $\overline{w'\chi'}(h)$ Reynolds flux at mixed layer top $z = h$, and the total derivative is defined as $d/dt = \partial/\partial t + u_m\partial/\partial x + v_m\partial/\partial y$.

The square-bracketed term on the right of Equation 3.22 is the property difference between mixed layer top and the depth-average. Its value depends on the exact choice of mixed layer top. Possibilities are illustrated in Figure 3.24, a schema of the two mixed layers in contact (Deardorff, 1981). If h is to be fixed at the level of the peak downward buoyancy flux, then it is where turbulence is active, near $h = h_1$ in the figure, and the difference in χ between that level and the mixed layer mean χ_m is small. The downward Reynolds flux, $-\overline{w'\chi'}(h)$, is then responsible for practically all of the downward transfer of the property χ, because the "entrainment flux" containing the square-bracketed term vanishes. Alternatively, if the mixed layer is supposed to extend further upward into the thermocline, to where the downward Reynolds flux just becomes insignificant, somewhere between h_1 and h_2 in the figure, $\chi(h)$ differs from χ_m, and the entrainment flux is the total flux. If the two levels are only a small distance apart, and if there are no concentrated sources of the property in between, other terms in the equation do not change significantly, and the sum of the two fluxes must remain (nearly) the same. The

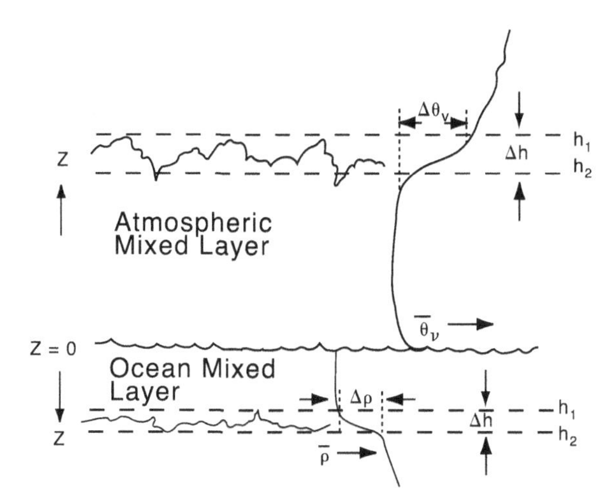

Figure 3.24 The two mixed layers in contact and the thermoclines separating them from the rest of the atmosphere and ocean. From Deardorff (1981).

entrainment flux with the second choice of mixed layer top then (nearly) equals the Reynolds flux with the first choice. Either formulation describes the total downward transfer of the property χ. The mixed layer mean of the property changes in response to that flux, plus the upward Reynolds flux at sea level and any source terms, most importantly radiation flux divergence in the case of temperature. This balance holds whether the mixed layer boundary moves or is stationary, and whether there is subsidence or not.

A particularly simple version of the above balance, with the entrainment flux the total flux, applies when the mixed layer height and the mixed layer average value of the scalar property χ_m do not change in time. The entrainment velocity then equals the divergence of horizontal transport, and the mixed layer budget may be written as:

$$\left(\frac{\partial U}{\partial x} + \frac{\partial V}{\partial y}\right)[\chi(h) - \chi_m] + \overline{w'\chi'}(0) + S_m h = 0. \tag{3.23}$$

Expanding the square bracket, we may interpret the first term as the influx of the property from above, $w_e \chi(h)$, the second term as outward transport of mixed layer property χ_m via the divergence of horizontal mass transport. The difference between influx and outward transport is the interface Reynolds flux of the property and any interior source input. In other words, the circulation consisting of subsidence and divergent horizontal transport takes away the sea level flux, plus any interior source input.

Much the same results hold on the water side, *mutatis mutandis*. There, mixed layer bottom is at variable depth, the top fixed. In place of Equation 3.22, we have:

$$h\frac{d\chi_m}{dt} = w_e[\chi(h) - \chi_m] - \overline{w'\chi'}(0) + \overline{w'\chi'}(h) + S_m h \tag{3.24}$$

with $w_e = dh/dt + w(h)$. Equations 3.22 and 3.24 embody mixed layer budgets of any scalar property χ. On the water side, only the temperature is of interest in mixed layer interplay.

The depth of the mixed layer is not always large compared to thermocline depth, as we have seen in the equatorial case. The Reynolds flux and the advective flux then do not simply exchange roles, and both have to be included in the budgets.

3.5.2 Atmospheric Temperature and Humidity Budgets

Choosing for χ in Equation 3.22, the potential temperature of the atmospheric mixed layer, let its depth average be θ_a, and let the upper boundary of the layer be the locus of the peak downward buoyancy flux, at a level $z = Z$. The Reynolds flux of temperature at this level is $\overline{w'\theta'}(Z)$ in m s^{-1}, a negative quantity, and the downward heat flux is $-\rho_a c_{pa}\overline{w'\theta'}(Z)$ in W m^{-2}. The turbulence accomplishing the transfer originates either from shear flow or from upward buoyancy flux, the latter owing to heating from below or cooling from above, and the peak downward buoyancy flux is subject to the Laws of Entrainment. The heat flux and vapor flux implied by a given peak downward buoyancy flux B_d depends on the ratio of differences in humidity and temperature between entrained air and mixed layer air, $\Delta q/\Delta\theta$. These differences are not necessarily the total "jumps" of humidity and temperature across the thermocline; the humidity jump is negative, however. Expressed in terms of the differences, the peak downward buoyancy flux is $B_d = w_e\Delta b = w_e(g\Delta\theta/T + 0.61g\Delta q)$, from which the individual fluxes follow with given $\Delta q/\Delta\theta$.

The surface Reynolds flux of temperature $\overline{w'\theta'}(0)$ is subject to the Transfer Laws of the air-sea interface, varying in a complex way with wind speed and sea level buoyancy flux.

Outside hot towers condensation and evaporation more or less cancel, and radiant fluxes of heat are the only source terms affecting the atmospheric mixed layer heat budget. They enter the integral balances of Equation 3.22 as boundary fluxes, in the case of the atmospheric heat balance as $S_m h = -[R(Z) - R(0)]/\rho_a c_{pa}$, say, with $R(Z)$ total short wave plus long wave upward radiant flux in W m^{-2}, $S_m h$ as K m s^{-1}.

The temperature budget of the atmospheric mixed layer is then:

$$Z\frac{d\theta_a}{dt} = \overline{w'\theta'}(0) - \overline{w'\theta'}(Z) - \frac{R(Z) - R(0)}{\rho_a c_{pa}}. \tag{3.25}$$

With no net evaporation or condensation, there are no sources of humidity within the mixed layer. Writing for the depth-average specific humidity in Equation 3.22 simply q, the humidity budget is then:

$$Z\frac{dq}{dt} = \overline{w'q'}(0) - \overline{w'q'}(Z). \tag{3.26}$$

3.5.3 Oceanic Temperature Budget

The temperature budget of the oceanic mixed layer, Equation 3.24 with depth-average water temperature θ_w replacing χ_m, is a little longer. The dominant term in it is the short-wave downward irradiance, $I_0 = -R_{SW}(0)$ in W m^{-2}, diminished by the long-wave radiant heat loss, $R_{LW}(0)$, for a net downward surface heat flux of of $-R(0) = I_0 - R_{LW}(0)$. The mixed layer does not absorb all of the irradiance, a fraction I_h penetrating layers below the depth h, where the total radiant flux is $-R(h) = I_h$. Other debit entries are the sensible and the latent heat loss. The former is the surface flux of temperature in the atmospheric budget, multiplied by the heat capacity of air, $SH = \rho_a c_{pa} \overline{w'\theta'}(0)$ (the heat flux in W m^{-2}, not the temperature flux, is continuous across the interface). The latent heat flux is latent heat times rate of evaporation, *LE*. The rate of evaporation is the surface vapor flux in Equation 3.26, times air density, thus $LE = \rho_a L\overline{w'q'}(0)$. The final entry is heat loss downward across the oceanic thermocline, at the depth of the peak downward buoyancy flux, $z = -h$, where the heat flux is $\overline{w'\theta'}(h)$, always negative, toward colder water. The temperature budget of the oceanic mixed layer is then:

$$h\frac{d\theta_w}{dt} = \frac{-R(0) + R(h) - SH - LE}{\rho_w c_{pw}} + \overline{w'\theta'}(h) \tag{3.27}$$

with $\rho_w c_{pw}$ the heat capacity of water. Another way of writing this equation is instructive: Let the terms signifying heat retained by the ocean be collected into $A_w = \rho_w c_{pw} \left(h d\theta_w/dt - \overline{w'\theta'}(h)\right) - R(h)$. Then Equation 3.27 takes on the form:

$$-R(0) = A_w + SH + LE \tag{3.28}$$

with the left-hand side the downward surface radiation, the right-hand side heat retained by the ocean, plus the heat handed over to the atmosphere as sensible and latent heat flux.

3.5.4 Combined Budgets

Now let us add to this last balance equation the atmospheric mixed layer's heat balance, 3.25 multiplied by $\rho_a c_{pa}$, plus Equation 3.26, multiplied by $\rho_a L$, turning the latter thus into a latent heat balance. The terms $SH + LE + R(0)$ cancel, leaving:

$$A_a + A_w = -R(Z) - \rho_a c_{pa}\overline{w'\theta'}(Z) - \rho_a L\overline{w'q'}(Z) \tag{3.29}$$

where $A_a = \rho_a L Z(dq/dt) + \rho_a c_{pa} Z(d\theta_a/dt)$ is latent and sensible heat storage and advection in the atmospheric mixed layer. The right-hand side contains fluxes at the top of the atmospheric mixed layer, downward radiant flux and Reynolds heat flux, and a heat loss term, $\rho_a L\overline{w'q'}(Z)$, the heat used up in moistening the dry air entrained from above. The Reynolds heat flux is usually downward, primarily entrainment flux, also known as "subsidence heating."

This is the heat balance of the system encompassing both mixed layers in contact: The net downward radiation at the top of the atmospheric mixed layer plus subsidence heating is the total income, some of which moistens the dry entrained air, the rest is stored or advected in the two mixed layers, with a small fraction heating the waters below the oceanic mixed layer.

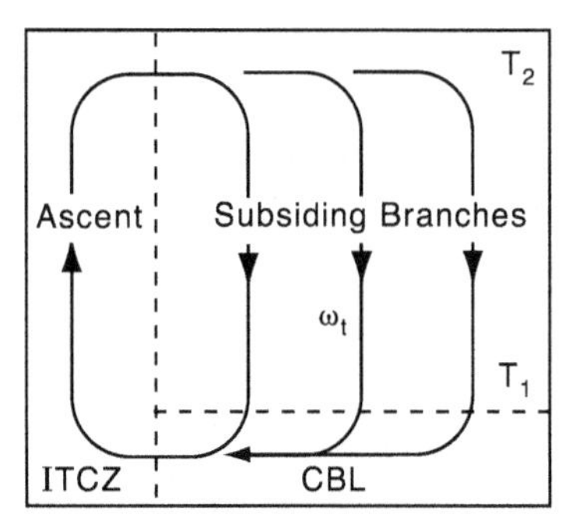

Figure 3.25 Betts and Ridgway's (1988) schema of the tropical and subtropical oceans' overturning circulation.

In the equilibrium case, with storage and advection vanishing in both media, the balance of radiation $-R(Z)+R(h)$ plus subsidence heating $-\rho_a c_{pa}\overline{w'\theta'}(Z)$ is all taken up by the latent heat required to moisten the dry subsiding air. The internal transfers of radiant, sensible, and latent heat, have all dropped out. The latent heat absorbed in moistening the subsiding air is exported to the hot towers, which wring out the moisture and deliver the latent heat to the upper layers of the troposphere.

Our results for the equilibrium case of the two-mixed layer heat balance, and the conclusion that the latent heat takeup absorbs the entire radiant heat gain, parallels the arguments and conclusions of Betts and Ridgway (1988) on the large-scale circulation of the tropical atmosphere. These authors took the Trade Wind region subsidence to be the descending branch of the large-scale, time and area-average, tropical circulation (Figure 3.25). They argued that the ascending branch is the aggregate of hot towers that transport all of the warm and moist air from the mixed layer to the tropopause. This branch covers only a small fraction of the tropical ocean, so that the area of the descending branch nearly equals the total. The descent of the air would lower the trade inversion, were it not for entrainment from below. Because the average height of the trade inversion does not change, the average entrainment rate must equal the average velocity of descent. The steady-state, area-average vapor balance of the entire tropical atmospheric mixed layer is then, by an application of Equation 3.22 to humidity, $\chi = q$:

$$w_e \Delta q + (\overline{w'q'})_0 = 0 \tag{3.30}$$

which is the equilibrium case of our Equation 3.26.

Betts and Ridgeway (1988) take the heat balance of the entire atmosphere over the tropics to be, area-average surface heating balances radiation cooling. Equation 3.22 invoked again, now applied to temperature:

$$\rho c_p \overline{w'\theta_e'} - R_{LW}(\infty) = 0 \tag{3.31}$$

where θ_e is equivalent potential temperature, the low level Reynolds flux of which includes sensible and latent heat gain at sea level, and $R_{LW}(\infty)$ is radiant heat loss at the top of the atmosphere. The latent heat gain is much the larger component of the surface heating, so that to a good approximation the last equation is:

$$L\rho_a \overline{w'q'}(0) = R_{LW}(\infty). \tag{3.32}$$

The entrainment velocity can now be calculated from the last equation and Equation 3.30, given the radiant heat loss and the humidity jump across the trade inversion. Typical quantities according to Betts and Ridgeway (1988) are 180 W m^{-2} radiant heat loss, Δq of order 0.015, requiring an average subsidence velocity of 4×10^{-3} m s^{-1}. This is of the order of entrainment velocities calculated from typical cloud top radiation flux and Carson's law, with $\alpha_c = 0.5$.

Equation 3.32, meant to be a reasonable approximation not an exact result, is another version of our principal finding for the equilibrium case, that the latent heat of moistening the dry air subsidence absorbs most of the radiation gain.

The elimination of internal transfers in the equilibrium heat balance, with storage and advection set to zero, is an obvious point physically, but has an interesting mathematical interpretation. Equations 3.25, 3.26, and 3.27, with the storage and advection terms deleted, describe the equilibrium properties of the two-mixed layer system. We express sensible and latent heat transfer through the Transfer Laws:

$$\begin{aligned} SH &= \rho_a c_{pa} c_T u_a^*(\theta_w - \theta_a) \\ LE &= \rho_a L c_q u_a^* (q_s(\theta_a) - q) \end{aligned} \tag{3.33}$$

where c_T and c_q are modified transfer coefficients applying to mixed-layer average temperatures and humidities in place of 10 m level ones. Substituting these into the equilibrium budgets of the two mixed layers in contact, three linear equations for the three equilibrium properties θ_a, θ_w, and q result. Collecting the homogenous terms on the left, we find that their determinant vanishes, indicating that only two of the equilibrium equations are independent. Those may be taken to be the equations for the temperature and humidity differences $\theta_w - \theta_a$ and $q_s - q$, expressing them in terms of the fluxes imposed. An equilibrium sea surface temperature cannot be determined from these equations, connecting as they do only the externally imposed fluxes.

In place of the humidity difference in equations such as the last one above, the dew point depression often proves convenient. If λ is the rate of change of saturation humidity with temperature, then the dew point temperature in air of temperature θ, humidity q, is:

$$\theta_d = \theta - \frac{q_s(\theta) - q}{\lambda}. \tag{3.34}$$

As Equation 3.33 shows, the latent heat flux is proportional to the dewpoint depression, $\theta - \theta_d$, or dry-bulb wet-bulb temperature difference, a directly observed quantity. At a reference temperature of 25°C, the saturation humidity is 0.02 and $\lambda = 1.20 \times 10^{-3}\ K^{-1}$.

Having established the relationships between the different fluxes to and from the two mixed layers, we would like to know now which dominate and which are negligible, under the different conditions prevailing in various parts of the world ocean, in the different seasons. For answers we have to turn to observation.

3.5.5 Bunker's Air-Sea Interaction Cycles

In a remarkable contribution, Bunker (1976) described the seasonal progression of the four terms in the heat budget of the oceanic mixed layer, Equation 3.28, at a number of locations in the Atlantic Ocean, based on his analysis of 8 million ship observations. There is no better way to portray the behavior of the two mixed layers in contact than to take Bunker's locations one by one and point out the sometimes surprising features of their heat budgets.

1. ITCZ region of the Trades, 9°N, 45°W. Figure 3.26 shows the yearly variation of the net monthly average radiant heat gain of the ocean, $-R(0)$, that Bunker designated R, the latent heat flux LE, the nearly vanishing sensible heat flux SH, designated S, and the heat retained by the ocean, A_w, called just A, plus a few observed quantities that underlie the calculation of the fluxes, notably the temperature differences, including the dewpoint depression.

   At this location, the principal balance is between radiation gain and latent heat transfer, the equilibrium case discussed above, with a small residue retained by the ocean, arising from a small gain in late northern summer, loss in winter. Bunker (1976) points out that the easterlies are strong here in winter when the ITCZ is far to the south, causing high latent heat flux, to the point that

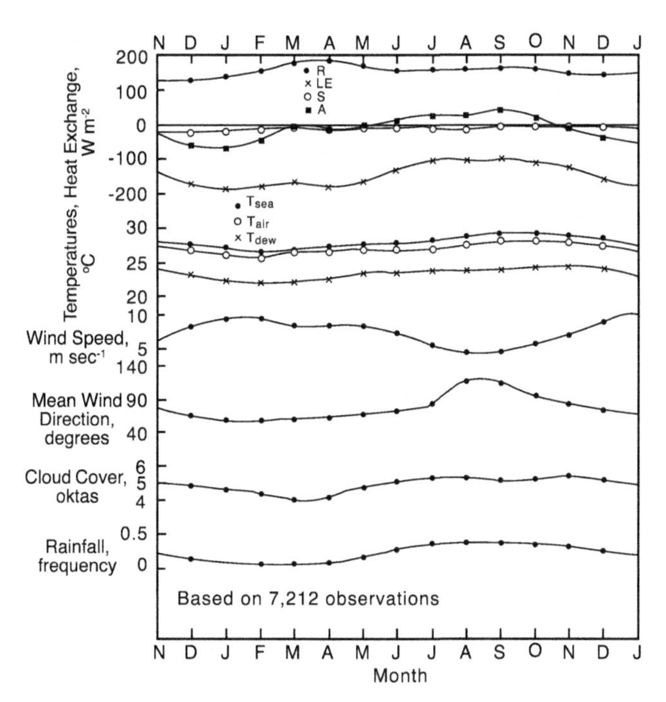

Figure 3.26 Annual march of the components of the surface heat balance in the ITCZ region of the Trades: R is net radiant heat gain of the ocean, LE and S latent and sensible heat transfer to the atmosphere, A net oceanic heat gain. From Bunker (1976).

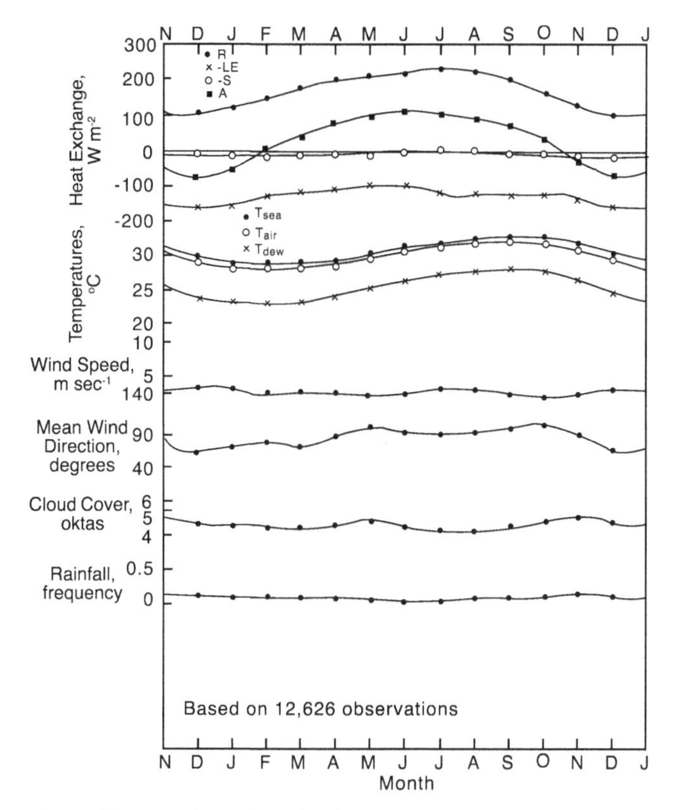

Figure 3.27 As Figure 3.26, but in the trade wind region away from the ITCZ. From Bunker (1976).

the ocean loses heat at up to a monthly average rate of 50 W m^{-2}. In summer, the ITCZ arrives, cloudiness increases, reducing heat gain by radiation somewhat, but the winds weaken, and latent heat loss drops more than the increase in radiation, so that the ocean loses little heat on the yearly average. The constancy of the dew point depression shows that the seasonal change of latent heat transfer is entirely due to the wind speed.

2. Trade winds further north, 23°N, 52°W (Figure 3.27) Bunker (1976) remarks that this location is typical of the "fresh and steady trade winds which cover a huge expanse of the North Atlantic from Spain to the Caribbean Sea." The sensible heat flux is again small, the latent heat flux fairly steady throughout the year, and comparable to the previous location's. The major difference is the seasonal variation of the radiant heat gain, double in summer of the winter gain. The corresponding changes in the heat retained by the ocean are a quite large gain in summer, with a long heating season, adding up to a significant yearly gain. In the yearly average, however, the principal balance is still between radiation and latent heat flux, the latter implying large vapor flux to the atmosphere.

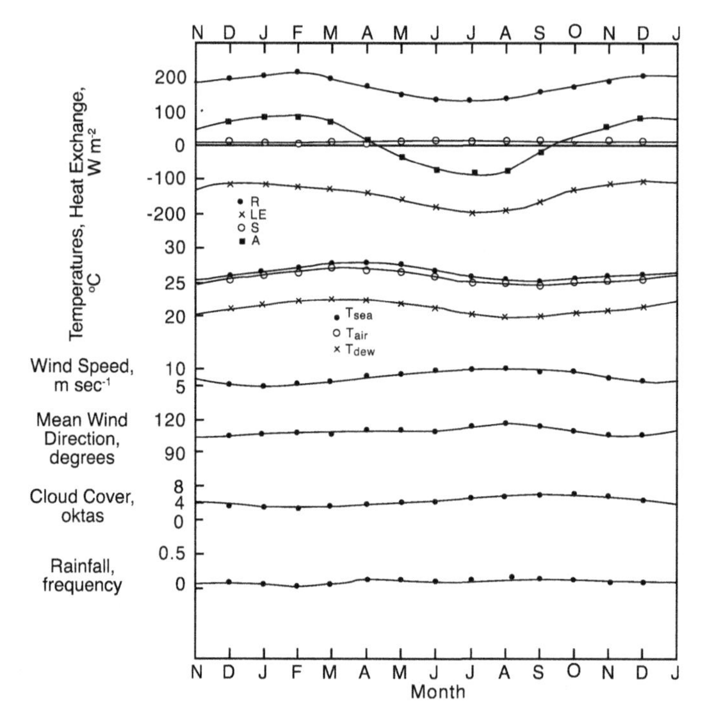

Figure 3.28 As Figure 3.26, but in the South Equatorial Current. From Bunker (1988).

3. South Equatorial Current, 11°S, 25°W (Figure 3.28). This comes from Bunker's (1988) posthumously published work on the South Atlantic. Again large latent heat transfer throughout the year, especially in southern winter, owing to strong of winds. The radiation gain varies less with the seasons than in the previous figure. The seasonally varying oceanic heat gain and loss therefore more or less balance, the large latent heat transfer responsible for winter loss of oceanic heat.
4. Spanish Sahara upwelling region 23°N, 17°W (Figure 3.29). In this region, steady NNE winds of 5–8 m s^{-1} parallel to the African coast cause upwelling of cold water. Subsidence of dry air through the Trade Inversion keeps the cloud cover low, rainfall rare. Solar radiation reaching the sea surface is thus high, varying regularly with the seasons. A major difference compared to other Trade Wind regions is that the dewpoint depression $\theta_w - \theta_d$ is low, and with it evaporation and latent heat transfer. Bunker blames this on the cold water temperatures, but that is not the whole story. Our analysis above reveals that the proximate cause of low dewpoint depression is low entrainment flux of dry air (Equation 3.26). Because cloud-top radiation sustains entrainment into the atmospheric boundary layer of the Trades, the root cause is presumably the low cloud cover. In any case, in this location the ocean retains most of the radiant heat gain, in midsummer at rates exceeding 200 W m^{-2}. The heat gain serves

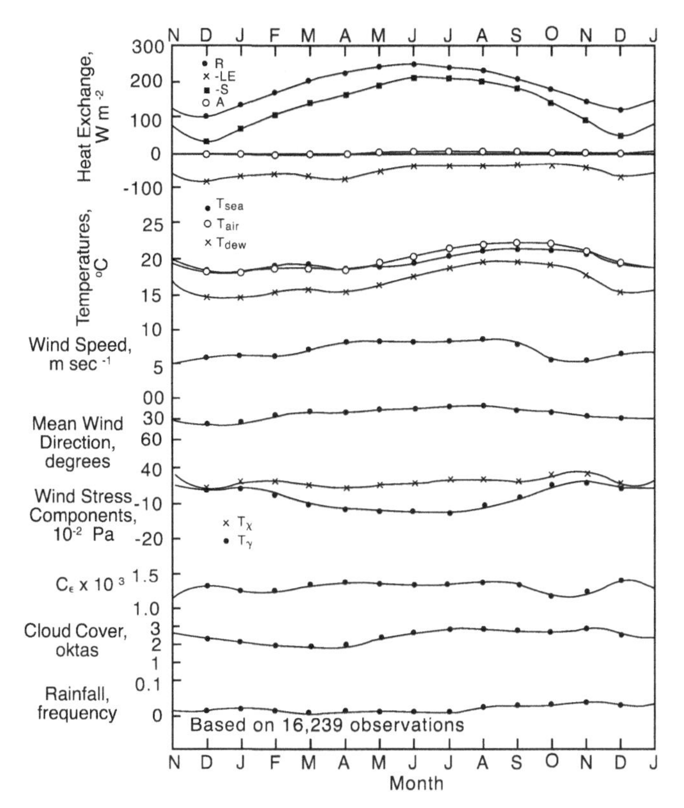

Figure 3.29 As Figure 3.26, but in the Spanish Sahara upwelling region. From Bunker (1976).

to raise the temperature of the water entrained from the oceanic thermocline. Similar conditions prevail in the eastern equatorial Atlantic and Pacific, as we have discussed above.

5. Gulf Stream region, 39°N, 62°W (Figure 3.30). In remarkable contrast to the previous location, the ocean loses heat at extravagant rates through both sensible and latent heat fluxes over the Gulf Stream. Except in midsummer, solar radiation cannot balance the losses, and what we have so far called heat retained by the ocean is large and negative, the heat loss supplied by oceanic advection of warm water. The large heat loss comes from high water to air temperature and humidity contrasts plus strong winter winds. Advection of cold and dry continental air in winter is the main cause of the contrasts; in summer the sea surface temperature-dewpoint difference is smaller, but still enough to sustain a latent heat flux of over 100 W m^{-2}. Bunker (1976) mentions that in summer, "the warm water produces large numbers of cumulus clouds." Apart from diminishing solar radiation, the clouds no doubt actively entrain dry air.
6. Norwegian Sea, 71°N, 17°E (Figure 3.31) . This location has many similarities to the previous one, radiation gain confined to the summer months, fairly large sensible and latent heat transfer in the winter months, combining into

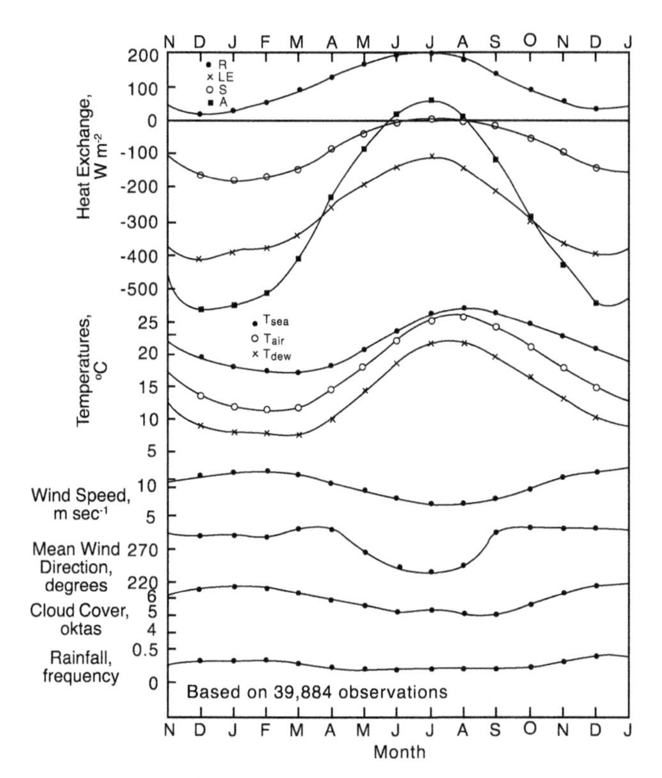

Figure 3.30 As Figure 3.26, but over the Gulf Stream. From Bunker (1976).

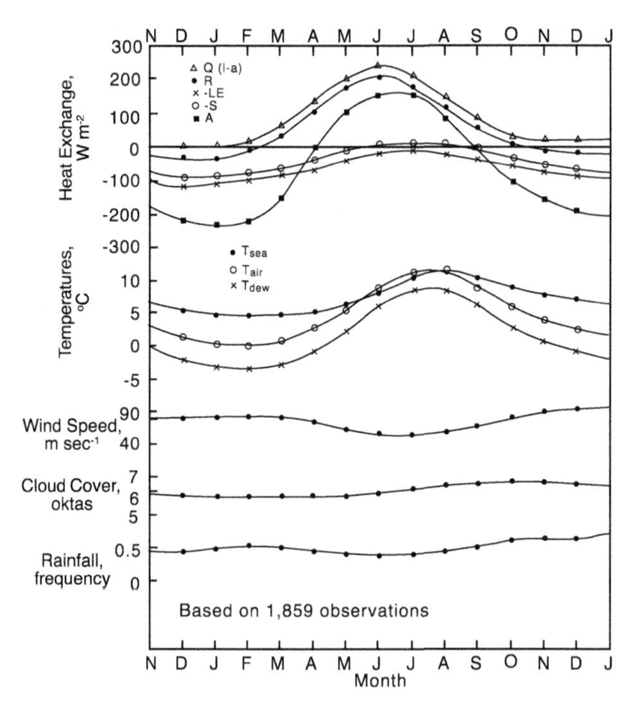

Figure 3.31 As Figure 3.26, but over the Norwegian Sea. From Bunker (1976).

a fairly large oceanic heat loss, supplied by advection from the south via the Norway Current. One difference compared to the Gulf Stream is the substantial oceanic heat gain in the summer, mainly because of very small sensible and latent heat transfer, the latter a consequence of low dewpoint temperature depression, hence presumably of small downward entrainment flux of dry air.

These examples of air-sea interaction cycles put in perspective the terms in the system balance (Equation 3.29). In the Trade Wind region generally the radiant gain R provides the latent heat needed to moisten the entrained air. The latent heat flux is relatively small only in singular locations, such as in upwelling regions. A major entry in several places in the heat budget of the two-mixed layer system, not suspected from out previous discussions of the air-side and water-side mixed layers, is the heat retained or released by the ocean, A_w. Apart from seasonal storage (heating and cooling that tend to balance), this term contains advection of heat from one location to another. The two remaining terms, atmospheric storage and advection plus radiation loss of the atmospheric mixed layer – $A_a + R(Z) - R(0)$ – do not appear to have obvious effects, the radiation loss being small compared to the leading terms in the balances, advection important only near coasts, and over major ocean currents. The approximate equilibrium balance of radiation gain and latent heat transfer that characterizes the Trade Wind region prevails only away from upwelling regions, and then only on a yearly average: Seasonal heating and cooling are important perturbations, as case 3 above demonstrates particularly clearly.

Perhaps the most important conclusion we reach at the end of this chapter is that entrainment of dry air at the top of the atmospheric mixed layer (mainly the work of isolated or stratiform clouds) is one of the key processes in air-sea interaction.

Chapter 4

Hot Towers

The graphic term "hot tower" has originated with Riehl and Malkus (1958), to describe their conceptual model intended to explain the peculiar distribution of "total static energy" or "moist static energy," $c_p T + gz + L_v q$, that they found in the tropical atmosphere (Figure 4.1). Between the energy-rich mixed layer below, and the equally energetic air high in the troposphere, a considerable energy deficit is evident, greatest just above the Trade Inversion. They realized that the high values of total static energy aloft cannot be the result of simple area-wide mixing or advection from below, especially if radiant cooling is taken into account. As Johnson (1969) points out in his review, Riehl and Malkus (1958) proposed that "the ascent [of low level air] .... took place in the embedded central cores of cumulonimbus clouds, protected from mixing with the environment by the large cross-section of the clouds." Release of latent heat turns the central cores of cumulonimbus into hot towers and generates fast ascent in them. In spite of their "large" individual cross section, their aggregate area is small, compared to the area of the global tropical ocean. Mass balance dictates that the total upward mass transport in all the hot towers return to low levels. This takes place in slow subsidence outside hot towers (i.e., over most of the tropical ocean). Hot towers form the upward leg of an overturning atmospheric circulation.

Cumulonimbus clouds spawning showers are familiar features of weather. Their towering height puts them in a different class from the trade cumuli butting against the Trade Inversion. Seen from ground level they have a "cauliflower" look, with many glistening white billows that often grow into an anvil-shaped top, where they spread out their cloud mass horizontally in the upper troposphere, even intruding on the lower stratosphere, above 10 km height. Severe thunderstorms and hurricanes contain hot towers reaching even greater heights, release a great deal of energy and do considerable damage.

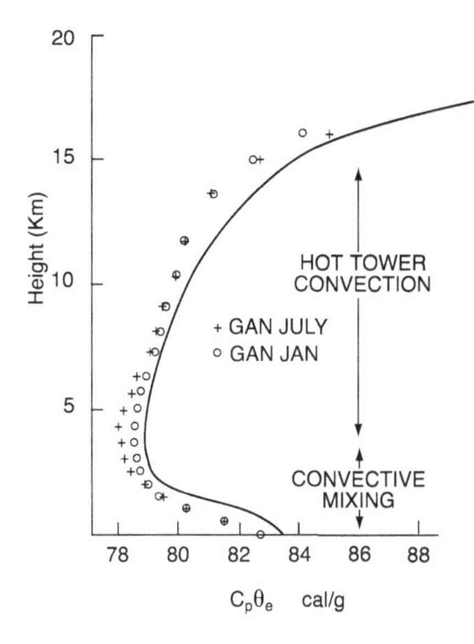

Figure 4.1 Average total static energy versus height in the tropical atmosphere (full line). Points show mean values at a single location, Gan Island (0°41′S, 73°09′E) in the Indian Ocean. From Riehl and Malkus (1958).

From the point of view of air-sea interaction, hot towers perform the important function of drying the air. As the moist air of the mixed layer rises into an environment of dropping atmospheric pressure, most of its moisture condenses and rains out. The latent heat of evaporation remains, leaving the total static energy essentially unchanged, while humidity drops from a typical mixed layer value in the tropics of 20 g/kg to much lower values in the upper troposphere, typically 4 g/kg. As upper tropospheric air descends in the subsidence regions surrounding hot towers, its low humidity remains conserved. As we have discussed in the last chapter, this supply of dry air to the atmospheric mixed layer is the key driving force of sea level latent heat flux.

Oceanic equivalents of hot towers arise from surface cooling, and could be more appropriately called cold fountains, rather than chimneys, as they have lately been christened. They form the descending leg of an oceanic overturning circulation, which differs in many respects from the atmospheric one. There is no analogue of latent heat release in the ocean, nor of drying the air, and all of the cooling as well as heating take place very close to the sea surface. Therefore winds, not heating from below, drive the ascending leg of the overturning circulation, recognizable as upwelling. Nevertheless, the overturning circulation of the ocean is instrumental in large poleward heat transfer, and plays a major role in the global heat balance.

We first discuss atmospheric hot towers and the part they play in air-sea interaction.

4.1 Thermodynamics of Atmospheric Hot Towers

The hot tower conceptual model postulates a pipeline from the atmospheric mixed layer to the upper troposphere, with the air in the pipeline having constant static energy, or constant equivalent potential temperature θ_e, recalling the definition $c_p\theta_e = c_pT + gz + L_vq$. We have already seen such a distribution, observed on a few occasions in the Atlantic Trade Wind Experiment (ATEX), at the research vessel "Meteor," on rainy days (Figures 3.11 and 3.12). The lifting condensation level was low, the transition layer underneath the subcloud mixed layer marked only by a small drop in humidity, the equivalent potential temperature more or less constant above the transition layer, lacking the sharp knee that identifies the Trade Inversion. These are the earmarks of a hot tower.

At constant θ_e, unit mass of the moist air contains constant thermal plus potential energy, so that its entropy does not change in the course of its ascent, as long as any condensed water remains suspended in it. The rainout of the liquid water changes the entropy only by an insignificant amount, however, so that the pressure and temperature changes in the moist air during ascent are nearly adiabatic, or "pseudoadiabatic." The energy balance of this process at any stage of the ascent is then:

$$Tds = c_pdT + gdz + L_vdq = 0 \tag{4.1}$$

condensation of vapor during ascent, $dq < 0$, providing the energy for an increase of the potential temperature, $d\theta = dT + (g/c_p)\,dz > 0$, while $d\theta_e = 0$.

It might surprise the reader that Equation 4.1 implies zero heat gain or loss in a hot tower from an external source, notably from the divergence of radiant heat flux. Clear air certainly loses energy by radiation. That the central core of a cumulonimbus does not, or at least not at an appreciable rate, is because suspended water droplets effectively block any radiation, short wave or long wave. We have seen above how effective stratiform cloud is in this regard, recalling Figure 3.17 of the previous chapter. Another point is the fast upward motion: at the typical convection speed of $5\,\mathrm{m\,s^{-1}}$, moist air rises from the Lifting Condensation Level (LCL) to the top of the troposphere in less than an hour. Even at the typical clear air cooling rate of $2\,\mathrm{K\,day^{-1}}$ the temperature drop in a rising parcel would be less than 0.05 K, making an insignificant change in θ_e.

The blockage of radiation is one consequence of the condensation of water vapor, important for air-sea interaction. Another is that the liquid water content eventually rains out, leaving drier air behind. While meteorological forecasts focus on the rain, what matters for air-sea interaction is the drying out of the air.

4.1.1 The Drying-out Process in Hot Towers

Thermodynamic equilibrium between liquid water and water vapor, or ice and water vapor, limits the partial pressure of vapor that can be present in moist air to a "saturation" pressure that depends on temperature alone. In a rising parcel of originally unsaturated air, both the partial pressure of vapor and the temperature drop, in such a way that the moist air approaches saturation. At the Lifting Condensation Level (LCL), the saturation partial pressure of vapor comes to equal the actual partial pressure: Above this level the moist air would become supersaturated, forcing condensation.

The partial pressure of vapor in moist air, a mixture of the two gases air and water vapor, depends on their relative proportions, and thus on specific humidity q. If the partial densities of dry air and vapor are ρ_d and ρ_v, the definition of specific humidity is $q = \rho_v/(\rho_d + \rho_v)$. An alternative measure of humidity is the mixing ratio $r = \rho_v/\rho_d$. At the small vapor partial pressures of interest these two measures are nearly equal, but r proves easier to work with in thermodynamic argument. The connection is $q = r/(r + 1)$.

According to Dalton's law, the total pressure p in a mixture of gases is the sum of partial pressures, so that if e is the partial pressure of water vapor, $p - e$ is the partial pressure of the dry air. The perfect gas laws connect the partial pressures to the partial densities and the absolute temperature T of the mixture: $\rho_v = e/R_v T$, $\rho_d = (p - e)/R_d T$, with $R_v = 461.5\ \mathrm{J\,kg^{-1}\,K^{-1}}$ and $R_d = 287\ \mathrm{J\,kg^{-1}\,K^{-1}}$, gas constants of water vapor and dry air. Putting $\varepsilon = R_d/R_v = 0.622$ for the ratio of the gas constants, we arrive then at the relationship of the mixing ratio to the partial pressures, $r = \varepsilon e/(p - e)$, or $e/p = r/(r + \varepsilon) \cong r/\varepsilon$. With e/p small, the gas law for the mixture is to a good approximation $\rho = \rho_d + \rho_v = p/R_d T$. Emanuel (1994) lists the exact relationships.

The relationship of pressures to mixing ratio remains true at saturation, so that the saturation mixing ratio is $r^* = \varepsilon e^*/(p - e^*)$, a function of temperature as well as of pressure, because the saturation partial pressure of the water vapor, e^*, is temperature dependent. Its changes with temperature in vapor-liquid equilibrium follow the Clausius-Clapeyron equation:

$$\frac{1}{e^*}\frac{de^*}{dT} = \frac{L_v}{R_v T^2} \tag{4.2}$$

where L_v, the latent heat of vaporization, varies slowly with temperature: $L_v = 2.5 \times 10^6 - 2.3(T - 273\ \mathrm{K})[\mathrm{J\,kg^{-1}}]$. In vapor-ice equilibrium the same equation applies but L_v has to be replaced by L_s, the latent heat of sublimation, $L_s = 2.834 \times 10^6\ \mathrm{J\,kg^{-1}}$.

Integration of Equation 4.2 yields the functional relationship of e^* to temperature. This is useful for calculating saturation partial pressure differences over small ranges of temperature. For the calculation of e^* at a specific temperature a more convenient approximate formula is due to Bolton (1980), valid in the range $-35°\mathrm{C} \leq T \leq 35°\mathrm{C}$:

$$e^* = 6.112 \exp\left(\frac{17.67T}{T + 243.5}\right) \tag{4.3}$$

with e^* the saturation vapor pressure in millibars (= h Pa), T temperature in °C.

The water vapor content of the mixed layer comes from evaporation. In our discussion of the Transfer laws in Chapter 1, we tied sea to air moisture flux to the specific humidity $q(h)$ at a low level h, and the saturation specific humidity at the sea surface temperature, q_s, which we will here denote by q_0^*. The Force driving humidity flux is

$\Delta q = q(h) - q_0^*$, while according to Equation 1.57 the flux is:

$$\overline{w'q'} = C_E U \Delta q \tag{4.4}$$

where C_E is an evaporation coefficient with a typical value of 10^{-3}, and U wind speed at the 10 m level. The latent heat flux carried by the vapor is ρL_v times the humidity flux. At the typical latent heat flux of $100\,\mathrm{W\,m^{-2}}$, and a wind speed of $10\,\mathrm{m\,s^{-1}}$, the humidity difference between the mixed layer air and saturated air at sea surface temperature is about $\Delta q = 3 \times 10^{-3}$.

Consider now changes with height z in the thermodynamic properties of rising moist air. At the "root" of a hot tower in the well-mixed subcloud layer (i.e., below the LCL), the specific humidity q, or the mixing ratio r, is constant with height, and so is therefore the ratio of vapor pressure to total pressure e/p. At sea level, say at a temperature of 20°C, the saturation vapor pressure is $e_0^* = 2340$ Pa, according to Equation 4.3, while the total pressure is $p = 10^5$ Pa. The mixing ratio of saturated air is then $r_0^* = \varepsilon e_0^*/p = 14.5 \times 10^{-3}$. As we just calculated, the mixed layer air is drier by $\Delta r \cong \Delta q = 3 \times 10^{-3}$, so that its mixing ratio is $r = 11.5 \times 10^{-3}$ its vapor pressure $e_0 = rp/\varepsilon = 1850$ Pa.

As the moist air now rises from sea level in a hot tower, its potential temperature, θ, remains constant in the subcloud well-mixed layer, clouds aloft shielding it from radiation loss. The absolute temperature then drops at the adiabatic lapse rate:

$$\frac{dT}{dz} = -\frac{g}{c_p} \tag{4.5}$$

while the total pressure drops following hydrostatic balance:

$$\frac{dp}{dz} = -\rho g = -p\frac{g}{R_d T}. \tag{4.6}$$

Because the e/p ratio is constant in the subcloud layer, vapor pressure changes track total pressure:

$$\frac{de}{dz} = -e\frac{g}{R_d T} \tag{4.7}$$

while the saturation pressure changes with temperature:

$$\frac{de^*}{dz} = \frac{de^*}{dT}\frac{dT}{dz} = -e^*\frac{L_v}{R_v T^2}\frac{g}{c_p}. \tag{4.8}$$

Expecting small proportionate change in absolute temperature, we find upon integrating the last two equations:

$$\ln(e_z^*/e_z) = \ln(e_0^*/e_0) - \frac{gz}{R_d T}\left(\frac{\varepsilon L_v}{c_p T} - 1\right) \tag{4.9}$$

where subscript z designates vapor pressures at height z, subscript 0 those at sea level. At the lifting condensation level $e^* = e$, the left-hand side vanishes, and the height of the LCL can be calculated. Putting $T = 290$ K, with $e_0^* = 2340$ Pa and $e_0 = 1850$ Pa as estimated above, we find for that height 470 m, a typical observed LCL.

Above the LCL the moist air becomes saturated, remaining in a state of thermal equilibrium at first between liquid water and water vapor, then at higher levels between ice and water vapor. In this state, the partial pressure of vapor, e, equals the saturation value e^* at all heights, while the vapor content diminishes. Differentiation of $r = \varepsilon e/(p - e)$, with both r and e now understood to be saturation values, the pressure still hydrostatic, yields the rate of change of mixing ratio in a rising parcel:

$$\frac{dr}{dz} = \frac{r(r+\varepsilon)}{\varepsilon}\left(\frac{g}{R_d T} + \frac{1}{e}\frac{de}{dz}\right). \tag{4.10}$$

Because r is smaller than ε typically by a factor of 30, the factor in front of the bracket equals r (or q) to a fairly good approximation.

The change of temperature with height follows from Equation 4.1:

$$\frac{dT}{dz} = -\frac{g}{c_p} - \frac{L_v}{c_p}\frac{dr}{dz} \tag{4.11}$$

which expresses the balance: rate of change of potential temperature $\theta = T + gz/c_p$ equals rate of latent heat liberation through condensation, divided by heat capacity.

Equation 4.10, with Equation 4.2 substituted, is one relationship between dr/dz and dT/dz; Equation 4.11 another. Eliminating the mixing ratio gradient leads to an expression for the pseudoadiabatic temperature gradient:

$$\frac{dT}{dz} = -\frac{g}{c_p}\frac{1 + rL_v/(R_d T)}{1 + rL_v^2/(c_p R_v T^2)}. \tag{4.12}$$

This is the dry adiabatic atmospheric lapse rate times a factor that depends on the mixing ratio and the three nondimensional parameters $L_v/R_d T$, $L_v/c_p T$, and $L_v/R_v T$, all functions of the temperature.

Eliminating the temperature gradient instead, we find for the mixing ratio gradient:

$$\frac{dr}{dz} = \frac{rg}{R_d T}\frac{1 - \varepsilon L_v/(c_p T)}{1 + rL_v^2/(c_p R_v T^2)} \tag{4.13}$$

showing the mixing ratio gradient to be r divided by a scale height $R_d T/g$ times a factor depending on the same nondimensional parameters as the lapse rate, all containing the temperature. At $T = 273$ K, the scale height is 7835 m. Both the mixing ratio gradient and the pseudoadiabatic lapse rate depend on r as well as T, and their distribution over height requires simultaneous integration of Equations 4.12 and 4.13, a task easily carried out on a personal computer, given initial conditions on the temperature T and the saturation mixing ratio r^*, at the LCL. Figure 4.2 shows the results for a typical starting state. Alternatively, the changes can be read from various widely available graphical representations of moist air properties, along pseudoadiabats, although the drying-out rate is not easily extracted that way.

Equation 4.13 contains the physics of the drying-out process in hot towers: it hinges on the thermodynamic properties of dry air and of the water substance, with gravity playing a controlling role as it determines the scale height of mixing ratio and temperature changes. This tells us then how hot towers deliver the goods, in the form of dry

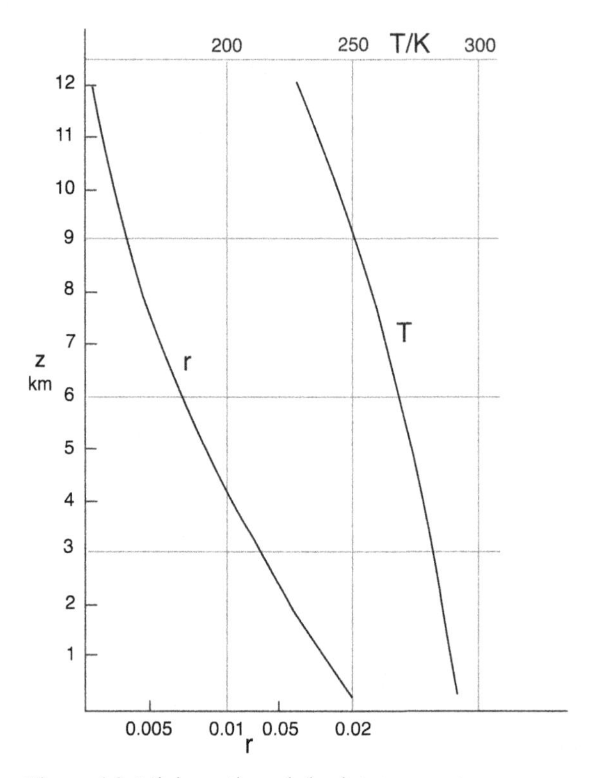

Figure 4.2 Mixing ratio and absolute temperature versus height above the Lifting Condensation Level (LCL) in the saturated ascent of moist air in a hot tower.

air, but not what the driving force is behind the overturning circulation of which hot towers are a part.

4.1.2 The Thermodynamic Cycle of the Overturning Circulation

Various authors (e.g., Kleinschmidt, 1951; Riehl, 1954) have noted that such prime hot towers as hurricanes derive their mechanical energy from a thermodynamic cycle, akin to that of a heat engine. The same holds true for the general circulation of the atmosphere, again an often expressed idea (e.g., Lorenz, 1967). But what kind of thermodynamic cycle? Emanuel (1986) and Rennó and Ingersoll (1996), among others, suggested that the Carnot cycle is a suitable idealized model of thermodynamic processes in convectively driven circulation. An examination of those processes suggests a different cycle, however, with a thermal efficiency roughly half of a Carnot cycle's, operating between the same temperature limits.

A well-known theorem of thermodynamics states that for a heat engine operating between two temperature limits T_1 and T_2, $T_1 > T_2$, the maximum attainable efficiency (greatest fraction of heat input converted into mechanical energy) is the Carnot cycle efficiency, $\eta = (T_1 - T_2)/T_1$. A simple demonstration of this rests on the second law of thermodynamics, and the introduction of entropy S as a state variable.

Let the working medium of a heat engine go through an arbitrary series of processes, returning to its original state at the end, a combination known as a "closed" thermodynamic cycle. Expressing heat input and output during the cycle in terms of entropy changes we have:

$$Q_1 = \int T dS \ldots\ldots (dS > 0)$$

$$Q_2 = \int T dS \ldots\ldots (dS < 0)$$

the integrations to extend over all parts of the cycle with heat addition or rejection, respectively. The temperature T must remain between the limits, $T_2 < T < T_1$, but is not necessarily equal to either limit in the course of heat input or output. The cycle being closed, entropy returns to its original value at the end. The aggregate entropy change during heat input, ΔS, is then equal and opposite to entropy change during heat rejection. We may write the inputs and outputs of heat then as $Q_1 = T_i \Delta S$ and $Q_2 = -T_o \Delta S$, where T_i is the weighted average temperature associated with heat gain, T_o with heat loss. The net heat added in a cycle (and converted to mechanical energy according to the first law of thermodynamics) is $Q_1 + Q_2$. The efficiency is then $\eta = (Q_1 + Q_2)/Q_1 = 1 - T_o/T_i$. Maximum efficiency requires a cycle in which all heat input takes place at $T_i = T_1$, and all output at $T_o = T_2$. The simplest such cycle is the Carnot cycle, consisting of single heating and cooling legs at constant temperature, connected by isentropic expansion and compression.

The classical representation of the Carnot cycle is its temperature-entropy, *TS*, diagram (Figure 4.3d). Between points 1 and 2, heat is added at constant temperature T_1, while between points 3 and 4 heat is rejected at T_2. In a gas, isentropic expansion from 2 to 3, and isentropic compression from 4 to 1 complete the closed cycle. The entropy change in the course of heat input is the same as during heat output, while no heat input or output occurs between points 2 and 3, or 4 and 1, as the working fluid first expands, then returns to its original temperature at constant entropy. The area enclosed by the diagram equals $(T_1 - T_2)\Delta S$, and is proportional to the heat converted to mechanical energy in a Carnot cycle.

Early steam engines operated on something close to a Carnot cycle, the heating and cooling legs evaporating water and condensing steam. There are other thermodynamic cycles: in an internal combustion engine heat addition occurs with the piston in its extreme position compressing the mixture of air and fuel. As a spark ignites the mixture, heat is added at essentially constant volume, both temperature and entropy rising steeply. Heat rejection takes place at roughly constant pressure, as the exhaust is released to the atmosphere. Compression and expansion take place ideally at constant entropy, as in a Carnot cycle. The efficiency of such an idealized cycle is, however, much less than the Carnot cycle ideal.

How about the sequence of thermodynamic processes in the overturning circulation of hot towers? Various authors have identified these. As we have seen in the last chapter,

Betts and Ridgeway (1988) described and schematically illustrated the pathway of the "working fluid": moist air (Figure 3.25). Starting near sea level in the tropical and subtropical ocean, the air streams to the hot towers of the ITCZ, rises there to great heights, and returns to subsidence regions outside hot towers, where it descends and yields up its heat gain via long wave radiation to space.

Betts and Albrecht (1987) portrayed the component thermodynamic processes in a "conserved variable diagram," specific humidity q against equivalent potential temperature θ_e, the latter a proxy for total static energy (Figure 4.3a). Starting at what Betts and Albrecht (1987) call "CBL top" (CBL = Convective Boundary Layer), meaning just above the atmospheric thermocline, unit mass of low moisture air descends to sea level (identified as "mixed layer"), picking up moisture and hence latent heat, both its humidity and total static energy increasing. The moist air then rises in a hot tower, and precipitates its moisture at constant θ_e. The now dry air moves away from the hot tower and descends to the level where it started, losing heat by radiation so that its total static energy returns to its initial value. The entire cycle consists of three "legs": (1) "CBL mixing" = moistening; (2) precipitation = ascent in a hot tower, and (3) radiation = subsidence. The important point is that the three processes form a closed cycle, returning the working fluid to its original state at the end, and forming a closed loop in the conserved variable diagram.

Another representation of the same cycle consisting of three legs is implicit in Riehl and Malkus' (1958) moist static energy diagram (Figure 4.1). We can replace the abscissa $c_p\theta_e$ by θ_e to make it easier to compare with Betts and Albrecht's diagram, and assume that the observed values represent the state of the working fluid in the descent from large height to the Trade Inversion (points 3-1 in Figure 4.3b), followed by descent to sea level while gaining heat and vapor (points 1-2). The cycle closes by isentropic expansion, represented by a straight vertical line connecting the sea level value of θ_e to the height where the observed value of θ_e is the same, points 2-3 in Figure 4.3b. This is now a $\theta_e - z$ diagram, the three points 1-3 separating the sea level heat gain, hot tower rise and radiation cooling legs and forming a closed loop.

The same three processes represented in a *TS* diagram also form a closed loop consisting of three legs (Figure 4.3c). The working fluid is moist air, unit mass of which starts well outside a hot tower, just above the atmospheric thermocline, where its moist entropy $c_p\theta_e$ is lowest, owing mainly to low humidity, point 1 in the diagram. The reader may find helpful to look back at Figures 3.11 and 3.12 of the preceeding chapter, which show typical distributions of moist air properties over height. As the dry air subsides into the mixed layer, evaporation from the ocean moistens it, sensible heat transfer warms it, while radiation cools it. The rate of change of its total static energy content per unit height is as follows: input to the mixed layer from the ocean, less radiation cooling, divided by the subsidence mass flux $-\rho w$:

$$T ds = c_p dT + g dz + L_v dq = \left(-\frac{dQ_r}{dz} + \frac{L_v E}{Z} + \frac{Q_s}{Z} \right) \frac{dz}{(-\rho w)} \tag{4.14}$$

where Q_r is radiant heat flux, Q_s sensible heat input from the ocean, E is rate of

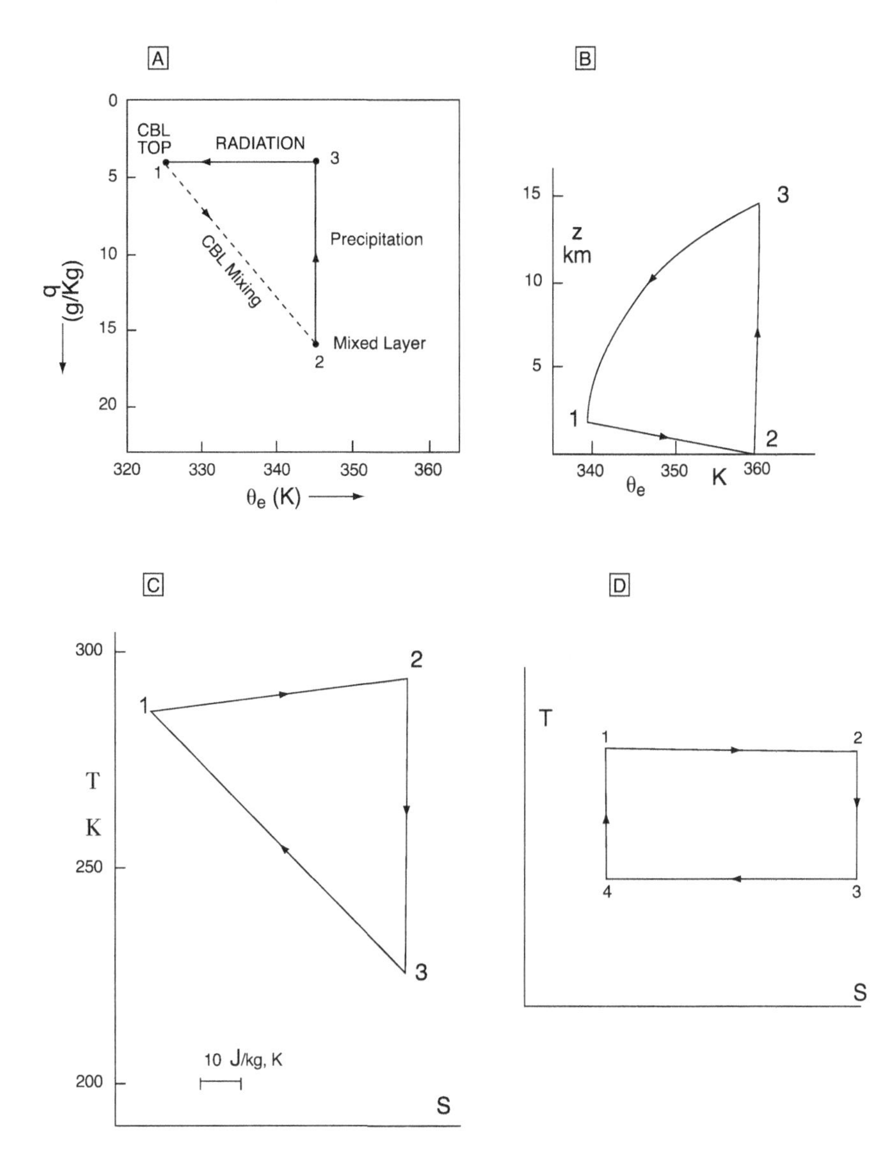

Figure 4.3 Different representations of the "gentle" hot tower thermodynamic cycle: a. Schematic $\theta_e - q$ diagram, showing descent through the mixed layer, points 1 2, labeled "CBL mixing," hot tower ascent, points 2-3, labeled "Precipitation," and descent through troposphere, points 3-1, labeled "Radiation," from Betts and Albrecht (1987); b. $\theta_e - z$ diagram showing the same three legs adapted from Figure 4.1, connecting the sea level temperature by a constant θ_e leg to the presumed hot tower top; c. *TS* diagram of the same three processes between realistic temperature-entropy limits; d. The classical Carnot cycle in a *TS* diagram for comparison.

evaporation, Z mixed layer height, and w is subsidence velocity at the top of the mixed layer, a negative quantity. A convenient way to carry the radiant heat flux divergence is as a cooling rate, $D = -(dQ_r/dz)/(\rho c_p)$, in K s^{-1}. The evaporation rate distributed over the mixed layer translates into specific humidity change: $dq = (-E/\rho w)(dz/Z)$, matching the dq term in the middle expression for moist entropy.

In the same expression, the other two terms combine into $c_p d\theta$, with θ the potential temperature, which then matches the net heating term on the right, sensible heat input less radiant cooling.

Leg 1-2 in the *TS* diagram represents this descent of moist air through the mixed layer to sea level, from height $z = Z$ to $z = 0$. Integration of Equation 4.14 yields the net energy input in this leg:

$$\int_1^2 T ds = c_p(\theta_2 - \theta_1) + L_v(q_2 - q_1) = -c_p \frac{D}{w} Z - \frac{Q_s}{\rho w} - \frac{L_v E}{\rho w}. \tag{4.15}$$

Here D/w is positive, ratio of two negative quantities, so that the potential temperature difference across the atmospheric thermocline is negative if radiation loss exceeds sensible heat gain:

$$\Delta\theta = \theta_2 - \theta_1 = -\frac{D}{w} Z - \frac{Q_s}{\rho c_p w}. \tag{4.16}$$

In the observations of Augstein et al. (1974), discussed in the last chapter, $\Delta\theta$ was some -7 K, so that even if Q_s was negligible, the cooling rate in the mixed layer was nearly 2 K/day (putting $w = -4 \times 10^{-3}\,\mathrm{m\,s^{-1}}$ following Betts and Ridgeway, 1988; and Z $= 1500$ m). The total heat input to the mixed layer, including latent heat, was moderate, because the latent heat gain of the moist air, with $\Delta q = q_2 - q_1 = 0.01$, was on the low side. Hence, $L_v \Delta q = 2.5 \times 10^4\,\mathrm{J\,kg^{-1}}$, for a net heat input of around $1.8 \times 10^4\,\mathrm{J\,kg^{-1}}$. Writing the net heat input as $T_i \Delta s$, with T_i the average temperature in the mixed layer, where all the heat input takes place, and putting $T_i = 290$ K, the entropy increase was 62 $\mathrm{J\,kg^{-1}\,K^{-1}}$. The absolute temperature difference $T_2 - T_1$ was fairly small, about 8 K, so that this leg, in the Augstein et al. (1974) observations, was not very different from the constant temperature heat input leg of a Carnot cycle.

The moist air now streams to the base of a hot tower with no further entropy or temperature change, its state still represented by point 2 (this is realistic for "gentle" hot towers without a significant pressure deficit at their center, as most of them are). In the hot tower, the air rises and expands isentropically to a low pressure and temperature, shedding much of its moisture content in the process. Point 3 in the *TS* diagram represents its state at the end of the expansion. The expansion leg 2-3 is similar to the expansion leg of the Carnot cycle, or of other heat engine cycles. The endpoint, at temperature T_3, is typically reached at a height $Z_3 = 12$ km or so, where a typical temperature is $T_3 = 225$ K.

From the high level of hot tower top the now dry air moves some distance away and descends to the top of the atmospheric thermocline, while losing heat through radiation. This leg, the cooling leg, point 3 to 1, closes the cycle. The heat balance in the course of descent is between potential temperature advection and rate of radiant cooling, as we have pointed out in the previous chapter. Here we write this in terms of entropy change:

$$T ds = c_p dT + g dz = -\frac{dQ_r}{dz}\frac{dz}{\rho w} \equiv c_p \frac{D}{w} dz \tag{4.17}$$

where again both w and D are negative, their ratio positive and presumably different from what it is in the mixed layer, as well as variable over the great height of the descent. Integration over the descent yields:

$$\int_3^1 T\,ds = c_p(T_1 - T_3) + g(Z - Z_3) = c_p\frac{D}{w}(Z - Z_3) \tag{4.18}$$

with D/w a weighted average value. From this we find for the average temperature gradient or lapse rate between the top of the thermocline and the level where the hot tower discharges the dry air:

$$\frac{T_1 - T_3}{Z_3 - Z} = \frac{g}{c_p} - \frac{D}{w}. \tag{4.19}$$

Observation shows this lapse rate to be of the order of half the dry adiabatic lapse rate g/c_p. This means $D/w = 0.5\,g/c_p$ or so, implying a cooling rate of about $2 \times 10^{-5}\ \mathrm{K\,s^{-1}}$.

Because the temperature differences involved in the legs 1-2 and 3-1 are considerably less than the absolute temperatures, the lines connecting these points, found by integrating the entropy Equations 4.15 and 4.17, are nearly straight, as indicated in the diagram. To the straight-line approximation, the entropy change Δs in leg 1-2 is:

$$\Delta s = \frac{2}{T_1 + T_2}\int_1^2 T\,ds \tag{4.20}$$

and similarly for leg 3-1, with somewhat less justification. The area enclosed by the cycle is then a triangle, equal to $(T_2 - T_3)\Delta s/2$. This area represents the energy converted by the thermodynamic cycle into mechanical energy, work done on the environment. The ratio of the mechanical energy gained to the total heat input in leg 1-2 is the thermodynamic efficiency of the cycle, η. To the accuracy of the straight-line approximation, the efficiency is then:

$$\eta = \frac{T_2 - T_3}{T_1 + T_2}. \tag{4.21}$$

Recalling now that T_1 differs little from T_2, this is very nearly $\eta = (T_2 - T_3)/(2T_2)$, or half the efficiency of a Carnot cycle operating between the temperature limits T_2 and T_3.

To take a typical example, let the three temperatures separating the legs be $T_1 = 300$ K, $T_2 = 289$ K, and $T_3 = 225$ K. The efficiency is then $\eta = 0.127$. For heat input in leg 1-2 we suppose $q_2 - q_1 = 0.01$ and $-Q_s/\rho w = 4000\ \mathrm{J\,kg^{-1}}$ (about $Q_s = 20\ \mathrm{W\,m^{-2}}$), $Z = 1800$ m, and $D/w = 5 \times 10^{-3}\ \mathrm{K\,m^{-1}}$. The net heat input in leg 1-2 is then $20000\ \mathrm{J\,kg^{-1}}$, and the energy converted into mechanical form is $2540\ \mathrm{J\,kg^{-1}}$.

Summing up this thermodynamic discussion, hot towers constitute the expansion leg of a thermodynamic cycle that energizes an atmospheric overturning circulation. They also dry out the air, a prerequisite for heat input into the cycle from the ocean. High efficiency of the cycle rests upon tall hot towers with cold tops. We now turn to the experimental evidence on hot towers, to see how they actually operate, and perform their important role in large-scale air-sea interaction.

4.2 Ascent of Moist Air in Hot Towers

A simple view is that moist air in hot towers rises because it is warmer and therefore lighter than the air surrounding the hot tower, owing to the release of latent heat from condensing vapor. This is true as far as it goes, but it is not the whole truth: heating of a column of air has dynamical effects that go well beyond making the air buoyant. For a start, continuity requires that upward motions be accompanied by downward ones, of comparable vigor, adding up to much more complex flow fields than the simple view would suggest.

Latent heat release in a rising parcel of moist air starts at the Lifting Condensation Level, LCL, and continues as long as the air remains saturated. The buoyancy force per unit mass driving an ascending parcel of air is $g(T_{vp} - T_{ve})/T_{ve}$, where T_{vp} is the virtual temperature of the parcel, T_{ve} that of the environment (we recall the definition, $T_v = T + 0.61\, qT$). If a parcel originating at the LCL did not mix with the ambient air, nor lose heat some other way, its temperature would change with height pseudoadiabatically. In that case, its total energy gain in the course of its ascent would be the integral of buoyancy, what meteorologists call Convective Available Potential Energy, CAPE:

$$CAPE = \int_{LCL}^{EL} g \frac{T_{vp} - T_{ve}}{T_{ve}} dz \tag{4.22}$$

where the upper limit EL is the equilibrium level, the level of vanishing virtual temperature difference between parcel and environment.

CAPE may be calculated from the distribution of temperature over height observed in the air surrounding the hot tower, known as a "sounding." Figure 4.4 shows a sample sounding, taken over the so-called western Pacific warm pool near Australia, from Lucas et al. (1994). The representation follows current practice in meteorology: The ordinate is the logarithm of pressure (in mb = h Pa), but instead of an abscissa, diagonal lines rising toward the right mark constant temperatures in °C. Contours of constant potential temperature are dotted, concave, and rise to the left. Another set of broken contours, convex and curving to the left up to the 200 mb level, denote pseudoadiabats. The smooth thick full line is the pseudoadiabat from the LCL, the irregular full lines the observed temperature and dewpoint distributions.

The kink in the temperature distribution at 950 mb is the LCL: The pseudoadiabat starting there almost parallels the sounding, virtual temperature differences between the two being nearly constant and close to 5 K in the middle troposphere (from 700 mb to 200 mb). According to Lucas et al., total CAPE calculated from Equation 4.22, is 1750 $\mathrm{J\,kg^{-1}}$.

The parcel theory supposes that the rising parcel does not mix with ambient air, and that condensed water falls out of it. Actually, mixing is pretty vigorous between adjacent updrafts and downdrafts. An admixture of drier air from downdrafts causes re-evaporation and cooling in the updraft, tending to reduce its buoyancy. Furthermore, water droplets remain suspended, so that their weight has to be supported by the moist

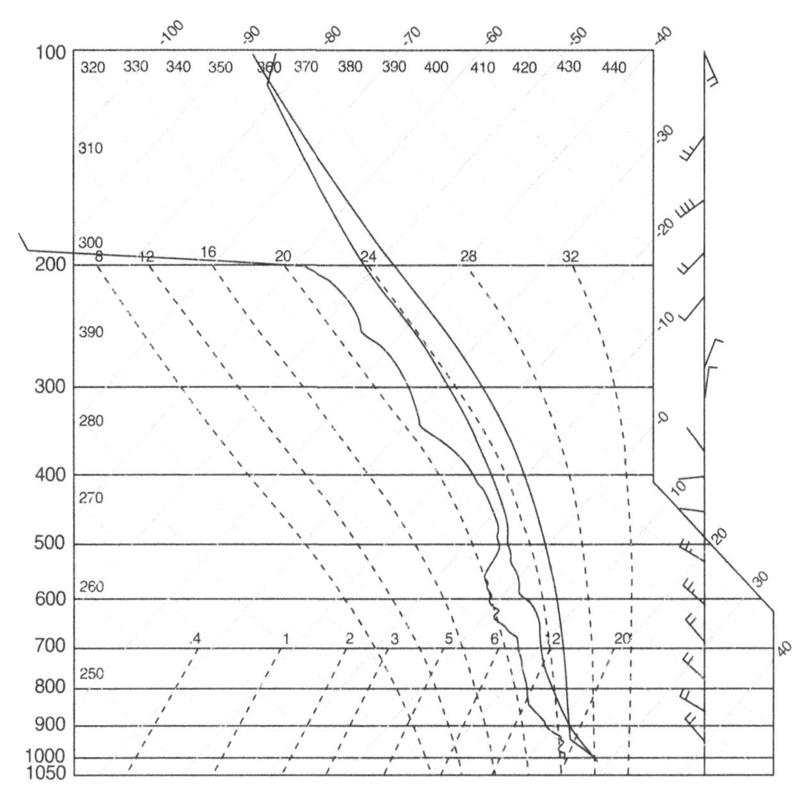

Figure 4.4 Temperature sounding in standard meteorological representation, from Darwin, Australia. The two irregular full lines show observed temperatures and dewpoints in a skewed coordinate system marked on the right side and on top of the figure, see the text. From Lucas et al. (1994).

air, again diminishing buoyancy. Observed virtual temperature deviations are therefore much smaller than parcel theory indicates.

Jorgensen and LeMone (1989) analyzed aircraft observations in cumulonimbus clouds over the seas around Taiwan, taken during the Taiwan Area Mesoscale Experiment (TAMEX). Research aircraft traversing the clouds recorded the vertical velocity, temperature, and humidity while flying through many updrafts and downdrafts. The analysis distinguished "cores" of updrafts and downdrafts from weaker motions by the criterion that the velocities, positive or negative, exceeded 1 m s^{-1} over a distance of over 0.5 km. The observed average virtual temperature excesses were of the order of $\Delta T_v = 0.5°\mathrm{C}$ in updrafts in the middle troposphere, where an undiluted parcel would have shown an excess of about 4°C. In downdrafts the virtual temperature was still higher by a fraction of a degree than surrounding the hot tower. From their observations over the western Pacific warm pool, discussed above, Lucas et al. (1994), showed histograms of the frequency of virtual temperature deviation in updrafts and downdrafts. Deviations occurring with significant frequency were again an order of magnitude lower than the parcel-theory values of 5 K or so.

A somewhat puzzling finding was the positive, if small, temperature anomaly in many *down*drafts: if they are buoyant, what propels them downward? Jorgensen and LeMone (1989) discuss this at length and end up concluding that the weight of suspended liquid water is principally responsible, reducing the buoyancy by $q_l g$. Given high condensation rates at low altitude, q_l can be as high as 0.01 by the time a parcel reaches the mid-troposphere. This is equivalent to a negative virtual temperature anomaly of about 2.7 K, an amount that easily overwhelms the parcel's observed positive ΔT_v of order 0.2°C.

Jorgensen and LeMone's (1989) observations in TAMEX extended over 359 updrafts and 466 downdrafts over a total flight distance of nearly 13000 km. They recorded the diameters and the velocities in all observed updrafts and downdrafts, as well as in the subset of cores. The strong updraft cores moved with a velocity of nearly $9\,\mathrm{m\,s^{-1}}$ and had diameters around 2 km in size, more or less constant over much of the troposphere above the inversion. The frequency distribution of diameters and highest velocities was approximately lognormal (Figure 4.5). This type of distribution characterizes nonnegative properties of turbulent flow (such as the concentration of an admixture). Therefore, while individual updrafts may perhaps be classed as hot towers, the aggregate of updrafts and downdrafts certainly constitutes convective turbulence. The near constancy of velocities and diameters with height above the mixed layer supports this point of view. In any case, if we view the updrafts as hot towers, we have to conclude that they come in quantity, and bundled with downdrafts.

4.2.1 Hot Tower Clusters

That hot towers congregate into assemblies of updrafts and downdrafts was first discovered in the course of the Thunderstorm Project. Reporting on that project, Byers and Braham (1948, quoted by Emanuel, 1994) illustrated the observed arrangement of updrafts and downdrafts in a thunderstorm (Figure 4.6). Later research showed the multicellular structure to be universal: as Mapes (1993) puts it, tropical convection is "gregarious." Mapes also makes the case that this is no accident, but arises from the interplay of a heated column of air with the stratified troposphere. One feature of that interplay is compensating subsidence adjacent to the convective ascent, the subsidence in turn giving rise to a further adiabatic ascent, which favors additional convection. In other words, the interaction generates new updrafts near an old one, which eventually decays.

Cooperative field experiments beginning with GATE (GATE = GARP Atmospheric Tropical Experiment, where GARP = Global Atmospheric Research Program; GATE was carried out in the eastern tropical Atlantic in 1974) showed tropical convection to be organized into "cloud clusters" with diameters of 200 km and more. They have typical lifetimes of 12 to 24 hours and contain either few or many centers or lines of deep convection. Stratiform clouds spreading out from these centers produce copious amounts of rain. Overall effects of the cluster include heat release distributed over height in a characteristic profile, peaking in the mid-troposphere, and corresponding

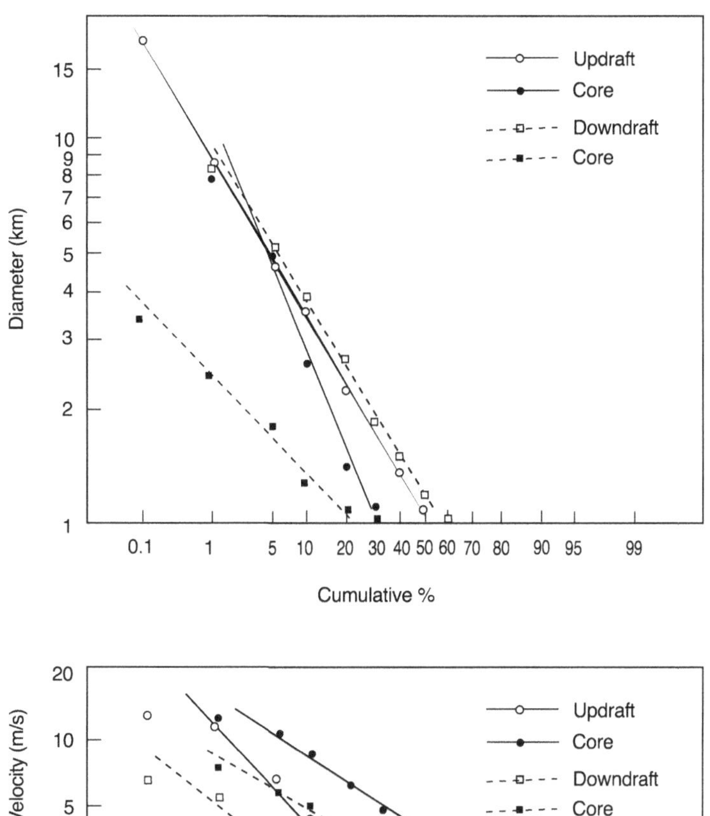

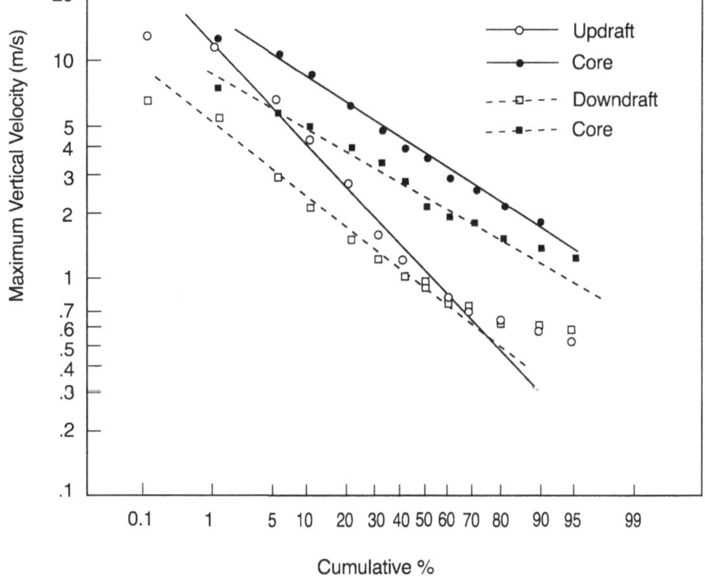

Figure 4.5 Lognormal plots of the cumulative distributions of updraft-downdraft diameters and peak velocities. From Jorgensen and LeMone (1989).

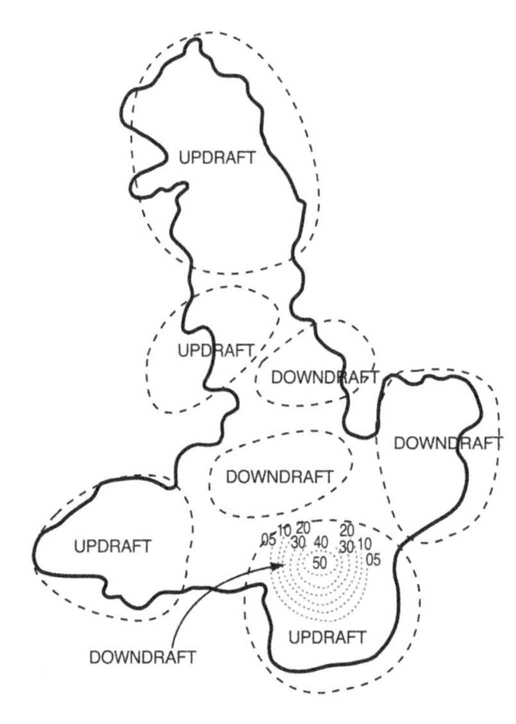

Figure 4.6
Updraft-downdraft distributions in a thunderstorm: the typical diameter of the cells outlined by broken lines is 5 km. The full line is the limit of the radar echo of cloud. From Byers and Braham (1948).

average upward motion, as well as the drying out of the rising air. Not only do these effects add up to the expansion leg of the hot tower thermodynamic cycle, but they also govern the nature of the dynamic interaction of the cluster with the stratified atmosphere.

The work of Frank and McBride (1989) gives a good summary of area-average properties of cloud clusters in GATE and AMEX (Australian Monsoon Experiment). Both experiments covered an area with horizontal dimensions of a few hundred kilometers within which active hot towers with cloud tops colder than $-70°\mathrm{C} = 203\,\mathrm{K}$ were centers of deep convection. Frank and McBride analyzed budgets of heat, mass, and humidity and determined rainfall rates as well as the distribution of area-average vertical velocity and diabatic heating rate versus height, in five stages of the life-cycle of four cloud clusters. Vertical velocities, expressed as pressure tendency, varied with the stage of a cluster's life cycle, reaching $-300\ \mathrm{mb\,day}^{-1}$ in the most active stage of AMEX (see Figure 4.7a) (pressure tendency $dp/dt = w dp/dz = -\rho g w$, so that $-300\ \mathrm{mb\,day}^{-1}$ corresponds to about $w = 0.05\ \mathrm{m\,s}^{-1}$ in the mid-troposphere). In the GATE clusters, the average velocity was more or less constant between the trade inversion and about 300 mb, the typical atmospheric pressure at 9 km height. By inference, the boundary layer under the inversion was convergent, the high troposphere divergent, conforming to the hot tower conceptual model. The area-average ascent velocity was, however, only one order of magnitude higher than the subsidence velocity outside hot towers. The AMEX results show convergence below 600 mb and divergence above, inflow and outflow distributed fairly evenly. The diabatic heating rates mirror this

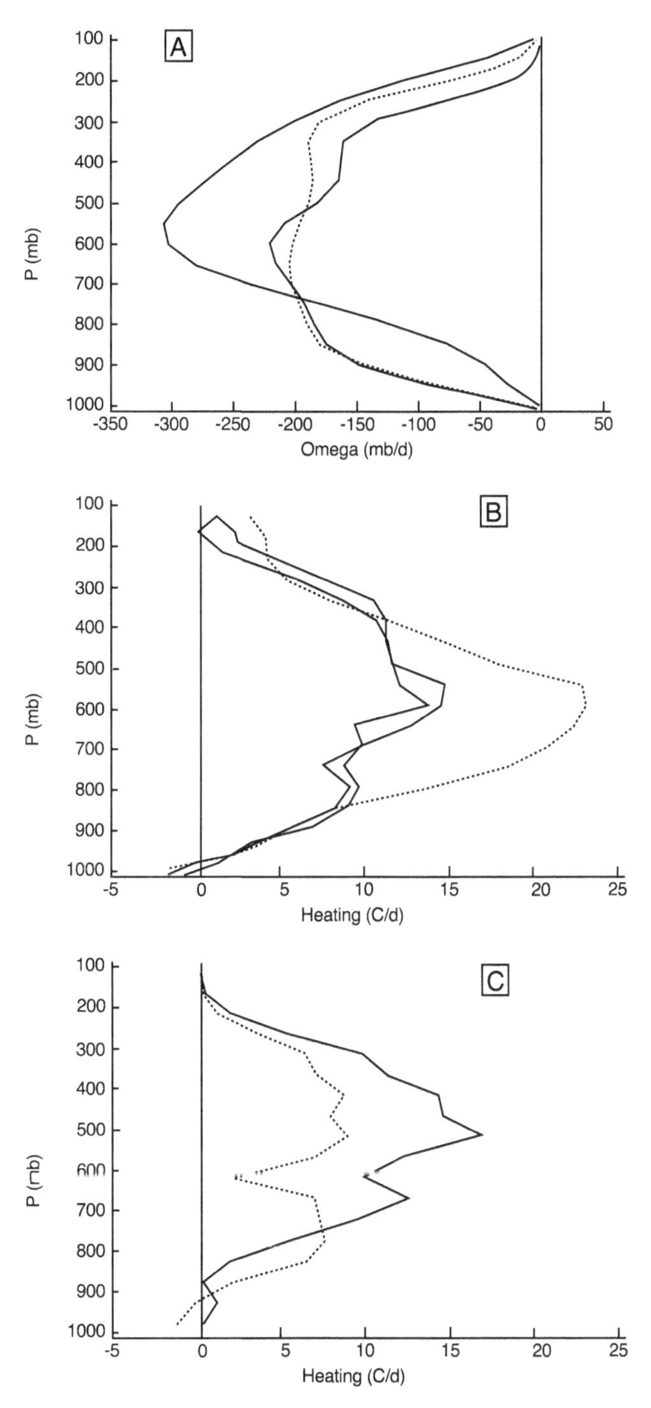

Figure 4.7 Vertical profiles of: a. Vertical "velocity" in h Pa/day, b. heating rate in °C/day and c. drying rate in units of equivalent warming in °C/day, in active stages of cloud cluster convection in the AMEX and GATE cooperative experiments. From Frank and McBride (1989).

difference (see Figure 4.7b), the peak heating rate in AMEX being double the nearly constant mid-tropospheric heating rate in GATE. Evidently, hot tower clusters exhibit more complex behavior than the hot tower conceptual model would suggest.

How hot tower clusters accomplish the drying out of the air was also determined by Frank and McBride (1989). They reported it in terms of drying-equivalent heating, latent heat times rate of decrease of specific humidity, $-L_v dq/dt$, in °C day^{-1}, a conventional but inconvenient method. Figure 4.7c shows AMEX drying rates in these terms, in the later stages of a cluster's life cycle. Converting to $dq/dz = (dq/dt)/w$, the peak drying rate at 600 mb corresponds to about $dq/dz = -1.73 \times 10^{-6}\,\mathrm{m}^{-1}$, or a decrease by 17.3 g/kg in 10 km, as one would expect for ascent along a pseudoadiabat. The drying-equivalent heating and the ascent velocity vary with height in a similar way, so that their ratio, proportional to dq/dz, remains much the same over the middle troposphere. Lin and Johnson (1996) calculate much lower (by about a factor of three) drying-equivalent heating rates in GATE and at some other locations, a puzzling discrepancy. What is certain is that specific humidity at tropopause level is lower than in the mixed layer by 10 to 20 g/kg, so that dq/dz in air ascending through the troposphere has to be much as deduced from Frank and McBride's (1989) results on drying-equivalent heating.

4.2.2 Squall Lines

Are there any organized flow structures within cloud clusters? Observations of local ascent velocities or buoyancy on the one hand, area average properties of clusters on the other hand, certainly make one wish for a pattern into which to fit all the information. A relatively simple pattern is the quasi-twodimensional motion in "squall lines," hot towers arranged along a line or front, propagating perpendicularly to the front. In a remarkable study, carried out as part of TOGA-COARE (Tropical Ocean Global Atmosphere Coupled Ocean Atmosphere Response Experiment, a new record for acronym length) over the western Pacific warm pool, Jorgensen et al. (1997) documented the flow pattern of a squall-line system. They described the squall line motion as follows: "At the system's leading edge (the eastern side), a line of deep convective storms was organized in a multicellular fashion that resulted from discrete propagation; that is, new cells developed ahead of the line of mature cells because of low-level convergence from the outflowing gust front." They illustrated the flow pattern by the distribution of average velocities, relative to the moving squall line, in a cross section (Figure 4.8). The flow (and the radar echo of cloud generated by suspended droplets or ice particles) reaches a height of 16 km, where a sounding and an infrared satellite photograph show the cloud top temperature to be 189 K. The highest point of the flow, and coldest cloud top, lies some 30 km to the rear of the squall line.

Further to the rear of the squall line the air flow turns downward. The descending air is relatively dry and cool, marked by a minimum in equivalent temperature θ_e. Dryness comes from an admixture of upper air, cooling from the re-evaporation of condensed vapor. Where the descent reaches the sea surface, the flow re-curves toward the squall

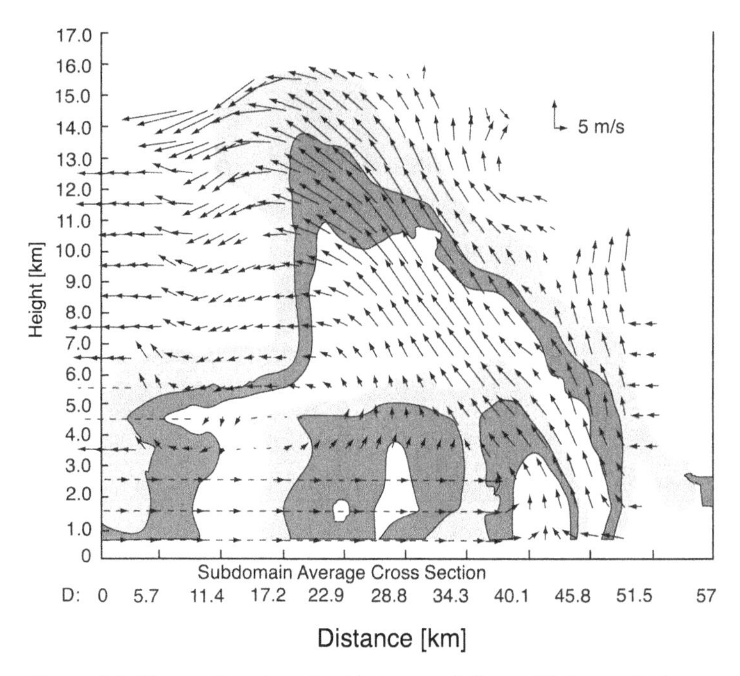

Figure 4.8 Flow pattern in and behind a squall line, with the vertical motion exaggerated. The shaded area is the radar echo of cloud. From Jorgensen et al. (1997).

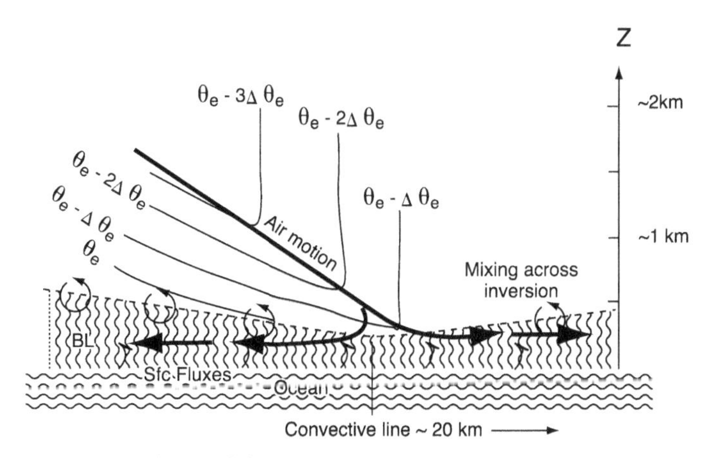

Figure 4.9 Schema of the air motion 20 km behind the squall line interacting with the mixed layer. Descending air cooled by evaporation enhances surface fluxes. From Jorgensen et al. (1997).

line. Here the relatively dry and cool air and strong wind combine to generate intense fluxes of latent and sensible heat from the underlying warm sea. This constitutes positive feedback to the energy supply of the squall-line circulation. Figure 4.9 is Jorgensen et al.'s (1997) schema of the low level air flow impinging on the subcloud layer, some 20 km behind the squall line.

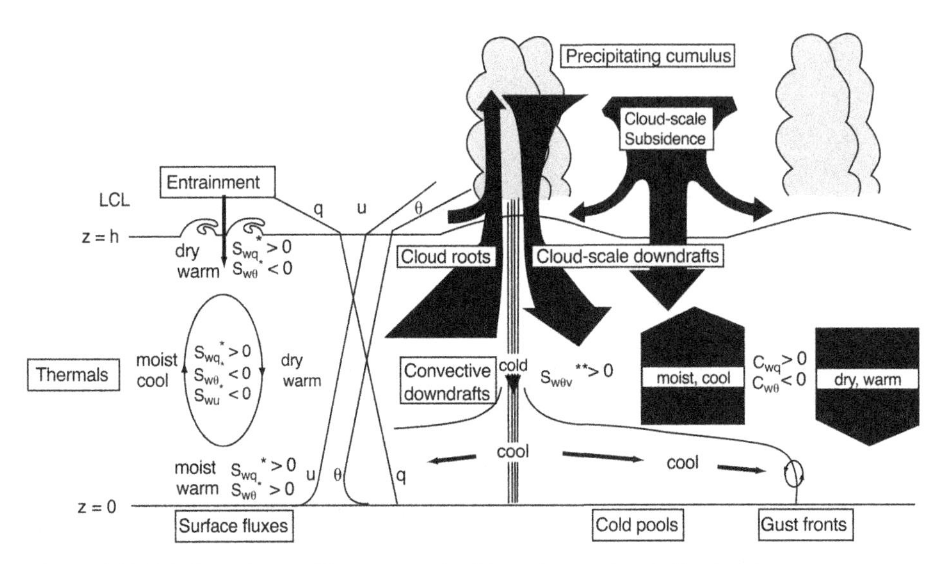

Figure 4.10 Another schema of hot tower-mixed layer interaction, indicating the many complexities of moist air behavior. From Williams et al. (1997).

This kind of boundary layer interaction with hot towers was the subject of an aircraft study by Williams et al. (1997), also part of TOGA-COARE. Figure 4.10 summarizes their findings in a cartoon, distinguishing between small-scale and cloud-scale processes. On the left-hand side are the familiar mixed layer processes, various Reynolds fluxes designated by capital S, subscripts identifying the fluxes. The larger scale processes to the right include the "cloud root" updrafts, and cold downdrafts. The latter create cold air pools on the sea surface that greatly accelerate evaporation and sensible heat transfer. Of particular interest is the variation of Turbulent Kinetic Energy (TKE) near the sea surface with the intensity of convection. As an index of the latter, Williams et al. use the large-scale mean vertical velocity w at cloud base, inferred from the convergence of horizontal motion. TKE rises by an order of magnitude under strong precipitating convection, from 0.3 $m^2 s^{-2}$ with weak convection, that is at vanishing vertical velocity, to 3.5 $m^2 s^{-2}$ in strong convection at the relatively high $w = 0.08$ m s^{-1}.

The structure of flow in squall line convection revealed by Jorgensen et al.'s study closely parallels Houze et al.'s (1989) conceptual model, as quoted by Pandya and Durran (1996) (Figure 4.11). Two remarkable properties of the squall line phenomenon are what Jorgensen et al. (1997) called "discrete propagation," the spontaneous appearance of new cells ahead of old ones, and the rise of the anvil cloud to great height well behind the convective cells, where the coldest cloud tops are seen by the satellites. Pandya and Durran (1996) have analyzed the underlying dynamics in some detail, and were able to show that both these properties are due to the interaction of vertically distributed heating in the convective cell with the stratified atmosphere: The heating disturbs the stratification and evokes a response that induces convergence at low levels ahead of the cell, as we have already seen in connection with "gregarious convection" (Mapes, 1993). An additional effect is upward motion and adiabatic expansion above

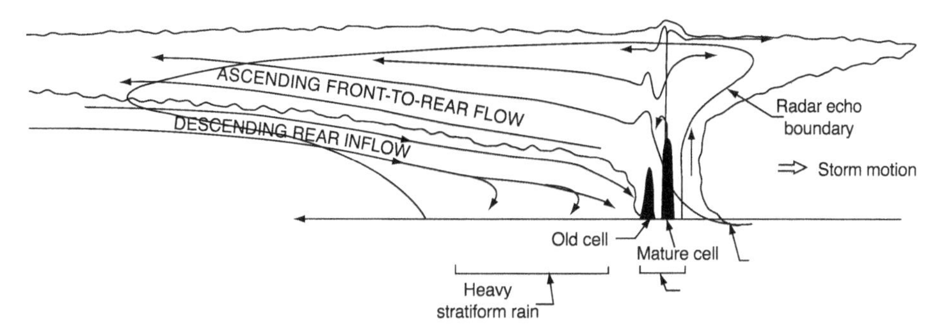

Figure 4.11 A larger scale schema of a multicell squall line with trailing stratiform cloud, from Houze et al. (1989), as quoted by Pandya and Durran (1996).

the heated cell, in the trailing anvil cloud behind the cell. This extends the expansion leg of the thermodynamic cycle to lower temperatures, beyond where the buoyancy of the air would take it. The coldest temperature in the ascending air, that is the low temperature limit of the thermodynamic cycle, thus depends on the heated column-stratified atmosphere interaction, and not only on the buoyancy of ascending parcels.

Further information on the low temperature limit comes from the study of Yutter and Houze (1998), who analyzed a large data set from the Pacific warm pool taken in the course of TOGA-COARE. They correlated the horizontal extent of precipitation with satellite determined cloud top temperature. They found that when 75% or more of a "coarse area" (240 km diameter) is covered by precipitation, area-average cloud top temperatures are uniformly low at about 200 K. Presumably, these low temperatures come from anvil clouds over individual "convective cells dispersed over a large area within the precipitation region."

4.3 Hurricanes

Hurricanes and typhoons – or tropical cyclones as they are also called – are the prominent representatives of hot towers over the ocean, and their properties and mechanics are of particular interest. They have been the subject of many observational and theoretical studies, documented in a very large meteorological literature. The authoritative volume of Palmén and Newton (1969) contains a thorough discussion of tropical cyclones, their structure, dynamics, and thermodynamics. In the past three decades, as instruments and methods of observation have become more and more sophisticated, including especially instruments carried on aircraft and satellites, a wealth of further information on hurricanes has come to light. Flying into hurricanes to observe their properties must take steady nerves, but it seems to be taken for granted nowadays: Much of our present knowledge comes from aircraft studies.

Just as squall lines, hurricanes possess a well-defined structure, seen in a cartoon in the very first illustration in Chapter 1, Figure 1.1. An "eyewall" where the heaviest rain and strongest winds are found, surrounds a calm and rain-free eye, which is also distinguished by a major depression of atmospheric pressure, by 100 h Pa and more. The eyewall is the prototype hot tower: It reaches into the tropopause and above, and conveys air of high static energy upward. Profiles of horizontal wind and equivalent potential temperature at tropopause heights (at the low pressure levels of 190–100 h Pa) due to Holland (1997) clearly show this. The data were collected by aircraft along a radius from the center of "Supertyphoon" Flo in the Pacific in 1990. The horizontal wind peaked in the eyewall at about 55 m s^{-1}, while the equivalent potential temperature droped by 15 K or so across the eyewall. The eyewall had an annulus-shaped cross section, its diameter increasing with height and thus sloping outward, to end in a high stratiform cloud deck.

The high-speed horizontal wind circling the core of the hurricane is of course its most destructive feature. Marks and Houze (1987) refer to the azimuthal wind as the primary circulation, to the radial inflow and upward motion as the secondary circulation. Their schematic illustration shows the streamlines of the latter with arrows, contours of azimuthal wind by dashed lines, in Hurricane Alicia off the Texas coast, August, 1983 (Figure 4.12). Shading marks the air volume sending back radar echo because of rain. Figure 4.13 from the same work is a schematic plan view of the radar echo, indicating the eyewall and different "rainbands" outside the eyewall. The distribution of the azimuthal wind in a radial-vertical section is shown in Figure 4.14, peaking at 50 m s^{-1} in this hurricane. The radial wind speed peaks at 6 m s^{-1} in the inflow at the sea surface, at 8 m s^{-1} in the outflow at 13 km height. The illustrations reveal a radial

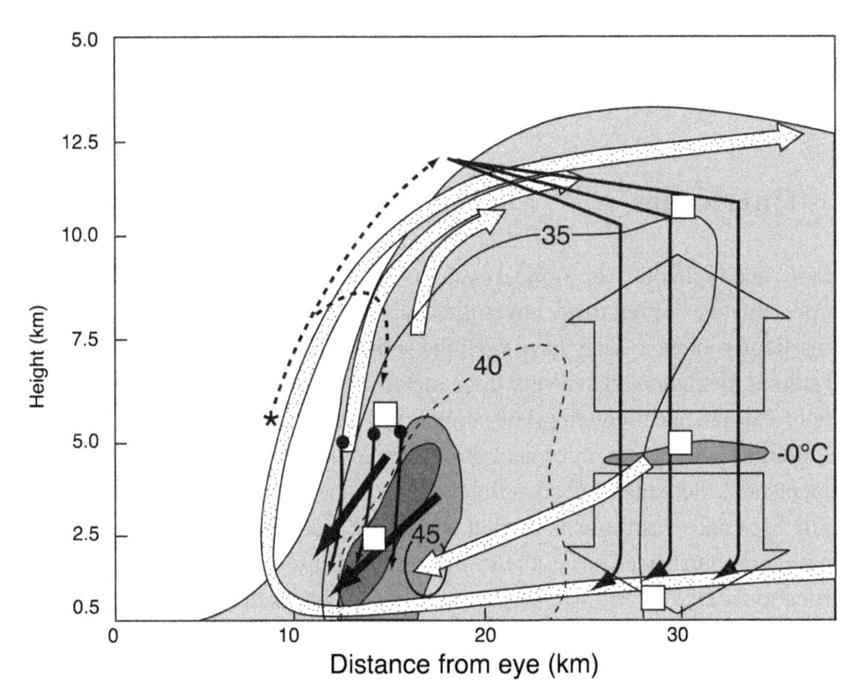

Figure 4.12 Schema of hurricane circulation in a radial section, showing also observed paths of "hydrometeors" (droplets or icicles). From Marks and Houze (1987).

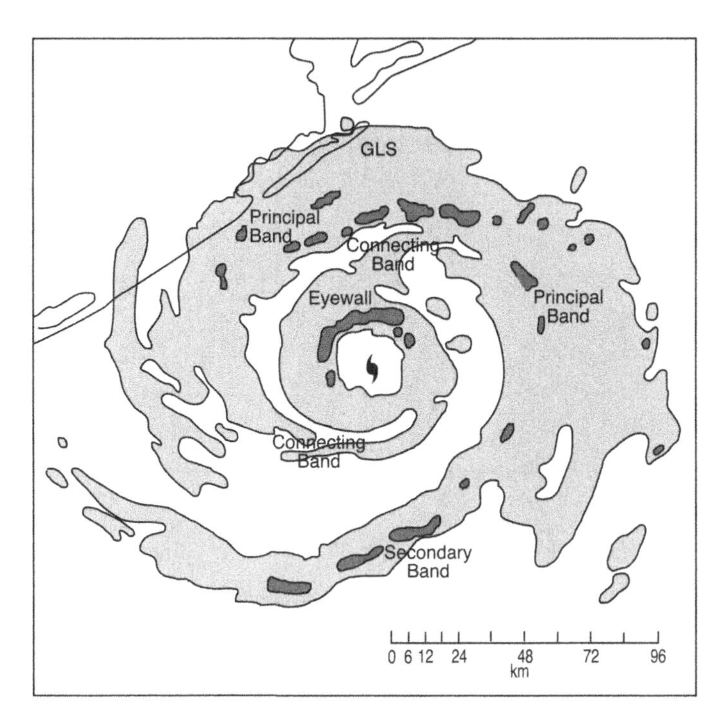

Figure 4.13 Radar echo of Hurricane Alicia off the Texas coast, showing the eyewall as well as the different rainbands, "principal," "secondary," and "connecting." From Marks and Houze (1987).

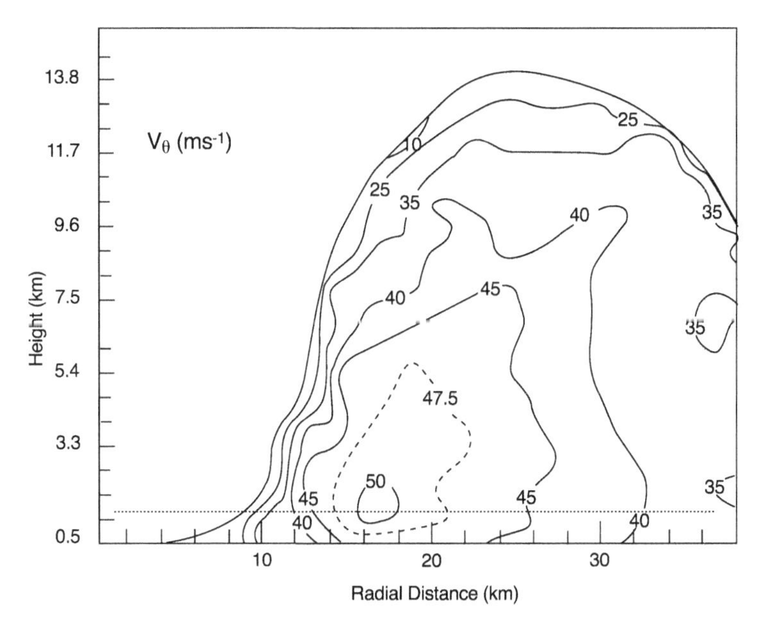

Figure 4.14 Distribution of the tangential wind in Hurricane Alicia, from Marks and Houze (1987).

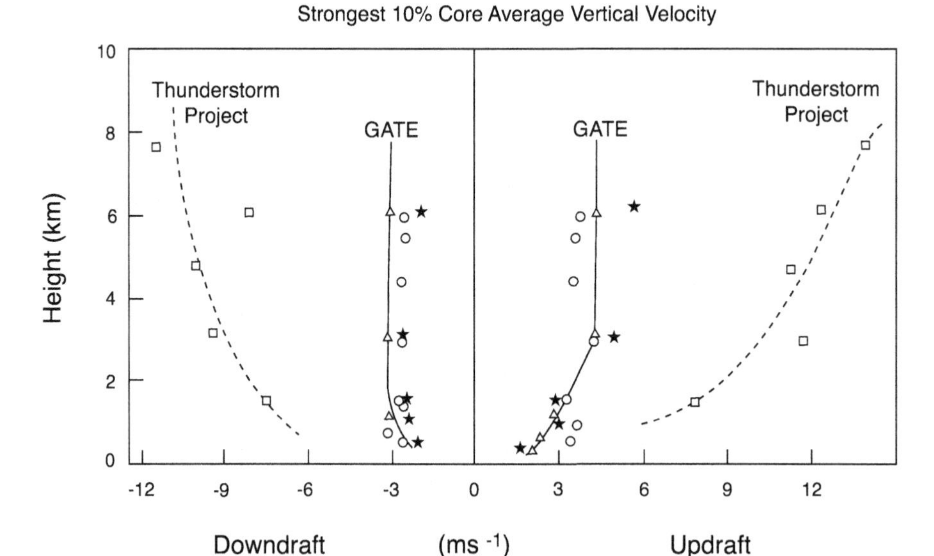

Figure 4.15 Vertical velocity in strongest updrafts and downdrafts versus height in oceanic convection in TAMEX (stars) and GATE (triangles) cloud clusters and several hurricanes (circles), all falling on the same line, in sharp contrast to updrafts and downdrafts in thunderstorms over land (squares). From Jorgensen and LeMone (1989).

inflow-outflow circulation extending over the entire troposphere: substantial radial inflow to 3 km height, well above the subcloud layer, outflow at tropopause level. As Willoughby et al. (1984) point out, these and some other features of the hurricane flow-field result from an interaction of the hot tower with the troposphere, including the already described adiabatic expansion at cloud top into the stratosphere, and the generation of new hot towers adjacent to old ones.

Jorgensen et al. (1985) reported research aircraft investigations on four "mature" hurricanes, observing updrafts and downdrafts at several levels, and classifying them in the same way as they did for cloud clusters in TAMEX, GATE, and in other tropical experiments. In spite of the well organized structure of hurricanes, very different from the extensive cloud clusters with their updrafts and downdrafts distributed over a much wider area, the peak vertical velocities in hurricanes were much the same. Figure 4.15 shows Jorgensen and LeMone's (1989) comparison of strongest 10% core velocities in the four hurricanes, with cloud cluster velocities. Surprisingly, updraft velocities in thunderstorms over land are much higher, as data from the Thunderstorm Project (carried out in the years 1946–47) have shown (Figure 4.15).

Again, as in cloud clusters, Jorgensen et al. (1985) found downdrafts as well as updrafts in hurricanes, both in the rainbands and in the eyewall. The diameters of the updraft cores were greater in hurricanes than in cloud clusters, however, as were the vertical mass transports (velocity integrated along the flight legs). The net upward mass transport (updrafts minus downdrafts) at these levels in the eyewalls was about three times higher than the net transport in the 10% strongest GATE updrafts minus downdrafts. High mass transport in the eyewalls is therefore one feature that sets hurricanes apart from cloud clusters.

The sea level signatures of hurricanes differ much more dramatically from cloud cluster properties: Minimum Sea Level Pressure (MSLP) at the eye of the hurricane and Maximum Sustained (low level) Wind (MSW) in the eyewall remain the properties by which hurricanes are classified as to strength. MSW is the principal practical measure of hurricane strength. Observation shows it to be closely related to the minimum sea level pressure. Given the MSW, an empirical relationship as in Dvorak (1984) yields an estimate of the MSLP, and vice versa. The lowest central pressures, 900 h Pa and less, or pressure deficits of 100 h Pa and more, occur in hurricanes with winds up to $80\,\mathrm{m\,s^{-1}}$.

While the strength of individual hurricanes varies within wide limits, statistics of strength depend critically on Sea Surface Temperature (SST). Very strong hurricanes only occur over SST of 29°C or so, and no hurricanes (defined as cyclones with MSW $> 33\,\mathrm{m\,s^{-1}}$) develop over water colder than 25°C. Figure 4.16 from Merrill (1988) shows the frequency of hurricanes of given MSW against sea surface temperature, with the approximate central pressure, according to the empirical formula, shown on the right. From these statistics Merrill deduces an empirical "potential maximum intensity" of hurricanes, as the limit toward lower temperature of observed instances

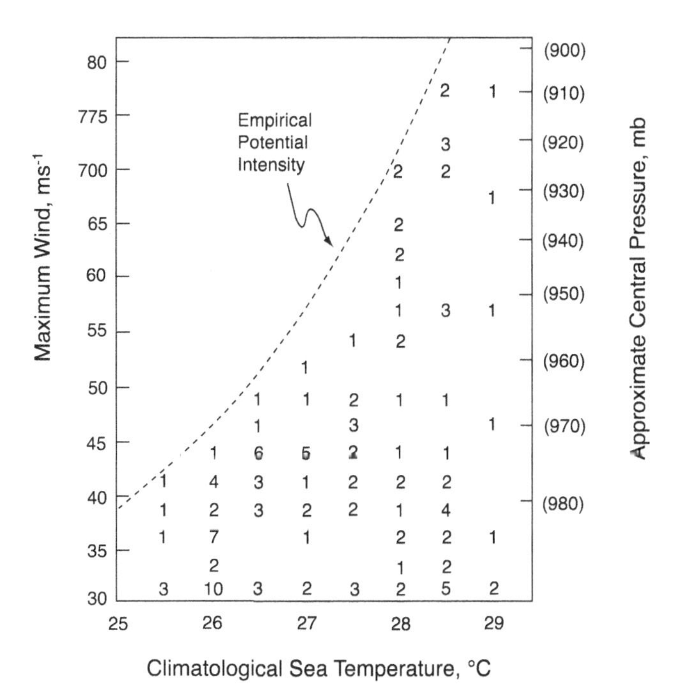

Figure 4.16 Empirical relationship between long-term mean sea surface temperature and Maximum Sustained Wind (MSW) in hurricanes, showing number of hurricanes observed in bins of size 0.5°C, $4\,\mathrm{m\,s^{-1}}$. Dashed line is an apparent upper bound on hurricane intensity. From Merrill (1988).

(the dashed line in the figure). A similar diagram, of the highest MSLP observed at given SST, is seen in Holland (1997). Holland also connects MSLP to cloud top temperature, remarking that "the maximum height of the warm core and vigor of the convection is a major factor for determining cyclone intensity." The logic of this is that the warm core creates pressure deficit at hurricane center, the higher the core the greater the deficit, while the the pressure deficit determines the strength of the cyclonic circulation. Hurricanes with central pressures of around 900 h Pa have cloud top temperatures of $-80°$C, while those with 950 h Pa sport cloud tops of $-55°$C. This is only part of the story, however. The pressure deficit divided by density is actually energy available to support the circulation. The strength of that circulation depends on the balance of the energy available, and energy dissipation involving an irreversible increase of entropy.

4.3.1 Entropy Sources in Hurricanes

The hot tower thermodynamic cycle discussed earlier is an idealization of slow circulation of moist air, consisting of ascent, outflow, and subsidence, in the course of which the moist air picks up or rejects heat, expands or contracts. The yield of the thermodynamic cycle ηQ_i, heat converted into mechanical energy, where Q_i is the heat input per unit mass of the moist air, causes pressure differences and drives horizontal and vertical motions. All of this mechanical energy is ultimately dissipated (i.e., irreversibly reconverted into heat), increasing the entropy of the system. Entropy "sources," processes that irreversibly increase entropy, include such Force-Flux combinations as heat flow down a temperature gradient, vapor flow down a concentration gradient, and most importantly, momentum transfer down a velocity gradient.

Actual entropy increases occur at the smallest scales of turbulence via the molecular processes of viscosity, heat conduction, and vapor diffusion, as the final step in energy degradation that begins with the production of TKE, TTV, and THV, by downgradient Reynolds fluxes, refer to our discussion of mixed layer turbulence in the preceeding chapter. Budgets of TKE, TTV, and THV have sink terms labeled dissipation; they are proportional to entropy increases. An apparent bypass of the direct route from TKE to dissipation is through the force of gravity, conversion of TKE to gravitational potential energy by negative buoyancy flux. This comes, however, at the expense of increasing TTV or THV production, which is eventually dissipated, leaving the ultimate entropy increase unchanged, as discussed in Chapter 1.

In a hurricane, the path of the moist air through the hot tower cycle includes the inflow in the subcloud layer from a large radius to the center, plus ascent in the hot tower, the eyewall. Low air density within the eyewall causes the sea level pressure at the bottom of the tower to be substantially lower than at some distance from the hot tower. Earth rotation combined with the radial pressure gradient turns the flow approaching the hot tower into a high-speed cyclone, so that the subcloud air spirals in from all directions, accelerating as it approaches the hot tower, the perimeter shrinking with the radius. Very high azimuthal (tangential) wind speeds in the inward spiraling moist air in contact with the sea surface make this a prime dissipation zone. The resulting high azimuthal shear stress sustains radial inflow (as "Ekman transport") toward the eyewall, that ends up transporting much more sensible and latent heat than the approach flow of a "gentle" hot tower, owing to higher air-sea transfer rates. In a hurricane's thermodynamic cycle, we then have to include the extra heat input, to realistically estimate the mechanical energy gain. We first discuss how that energy gain is dissipated.

An idealization underlying simple models of hurricanes is that the approach flow in the subcloud layer is steady and axially symmetric (e.g., Emanuel, 1991). In such steady mean flow, the Bernoulli equation with friction quantifies changes in mechanical energy along a streamline, as Emanuel (1991) and Rennó and Ingersoll (1996) pointed out:

$$d\left(\frac{u^2+v^2}{2}\right)+d(gz)+\frac{dp}{\rho}+\widetilde{F}\cdot d\widetilde{\ell}=0 \tag{4.23}$$

where $\widetilde{F}$ is friction force per unit mass and $d\widetilde{\ell}$ is streamline element, both vectors in a scalar product; u is mean radial velocity outward from the hot tower; v mean azimuthal (tangential) velocity counterclockwise, in a polar coordinate system with the origin at the center of the hot tower. The friction force in the subcloud layer is the divergence of Reynolds momentum flux, varying in some manner with height:

$$\widetilde{F}=\frac{\partial\overline{u'w'}}{\partial z}\widetilde{i}+\frac{\partial\overline{v'w'}}{\partial z}\widetilde{j} \tag{4.24}$$

where $\widetilde{i}$ and $\widetilde{j}$ are unit vectors in the radial and azimuthal direction, respectively. Axial symmetry implies that streamlines at a given level are similar. A line element on any such streamline covered in a short time δt, is:

$$\delta\widetilde{\ell}=u\delta t\widetilde{i}+v\delta t\widetilde{j}. \tag{4.25}$$

The inner product of force and line element is then:

$$\widetilde{F}\cdot\delta\widetilde{\ell}=\delta t\left(\frac{\partial\overline{u'w'}}{\partial z}u+\frac{\partial\overline{v'w'}}{\partial z}v\right). \tag{4.26}$$

The inner product can also be expressed as the projection of the friction force on the tangent to the streamline, F_b for "braking" force, say, times the length of the streamline element $|\delta\widetilde{\ell}|$. That length furthermore equals the radial component of $\delta\widetilde{\ell}$, δr, divided by $\cos\vartheta$, where ϑ is the angle between radius and the tangent to the streamline. The inner product on the left of Equation 4.26 may then be written as:

$$\widetilde{F}\cdot\delta\widetilde{\ell}=F_b\frac{\delta r}{\cos\vartheta}. \tag{4.27}$$

Integrating the shear force-line element product over the height of the subcloud layer, we have:

$$\int_0^h(\widetilde{F}\cdot\delta\widetilde{l})\,dz=\delta t\int_0^h\left(\frac{\partial\overline{u'w'}}{\partial z}u+\frac{\partial\overline{v'w'}}{\partial z}v\right)dz. \tag{4.28}$$

Although the mean velocity components u, v are not strictly zero at the sea surface, they are small. Neglecting them, and supposing vanishing shear stress at the top of the subcloud layer, partial integration of Equation 4.28 leads to:

$$\int_0^h (\widetilde{F} \cdot \delta \widetilde{l})\, dz = -\delta t \int_0^h \left(\frac{\partial u}{\partial z} \overline{u'w'} + \frac{\partial v}{\partial z} \overline{v'w'} \right) dz. \tag{4.29}$$

Here we recognize the integrand (with the negative sign included) as the local shear production of TKE at height z in the mixed layer, the integral the total production. In the absence of buoyant production, and supposing that none of the TKE is exported upward or downward, total production equals total dissipation, hence:

$$\int_0^h (\widetilde{F} \cdot \delta \widetilde{l})\, dz = \delta t \int_0^h \varepsilon\, dz. \tag{4.30}$$

All of the work done by the friction force thus appears as TKE production, which then equals total dissipation in the subcloud layer.

The calculation so far has yielded the energy dissipated in time increment δt, along streamline elements of lengths varying with height in the subcloud layer. In the same time increment, moist air moves closer to the hot tower at the depth-average radial velocity u_a:

$$\delta r = u_a \delta t = \frac{\delta t}{h} \int_0^h u\, dz. \tag{4.31}$$

Vertical mixing is efficient in the subcloud mixed layer and eradicates any differences in particle displacement between streamlines. It is then legitimate to put $\delta r / u_a$ for δt on the right of Equation 4.30. Integrating from a large distance $r \to \infty$, to $r = r_0$, the outer radius of the eyewall we arrive at:

$$\int_0^h \int_\infty^{r_0} \frac{F_b}{\cos \vartheta}\, dr\, dz = \int_0^h \int_\infty^{r_0} \frac{\varepsilon}{u_a}\, dr\, dz \tag{4.32}$$

having substituted from Equation 4.27.

The integrand on the left of this equation is still the inner product of friction force and line element of a streamline, the quantity $\widetilde{F} \cdot \delta \widetilde{\ell}$ that appears in the Bernoulli Equation, 4.23. The integral with respect to the radius is the same as the line-integral of the friction force along the streamline from a distant point to the bottom of the eyewall.

Integrating the other terms in Equation 4.23, the gz term makes no contribution, but the kinetic energy change is significant, from a low value at a large radius, to the eyewall, where the air leaves spiraling upward with high azimuthal velocity. In strong hurricanes, at least, the pressure term provides most of the balance for the friction work. Its integral is:

$$\int_\infty^{r_0} \frac{1}{\rho} \frac{dp}{dr}\, dr = -R_d T \ln \left(\frac{p_\infty}{p_0} \right) \tag{4.33}$$

the pressure varying only with the radius, independently of azimuth, in the axisymmetric approximation. For the small proportionate changes in pressure, we may expand the logarithm, and express the balance of friction work and pressure work as:

$$\int_0^h \int_\infty^{r_0} \frac{F_b}{\cos\vartheta}\, dr\, dz = \frac{h(p_\infty - p_0)}{\rho} - e_t h \tag{4.34}$$

having put e_t for the energy of the air leaving in the eyewall. Substituting into Equation 4.32, we have finally:

$$\frac{(p_\infty - p_0)}{\rho} = \frac{1}{h}\int_0^h \int_\infty^{r_0} \frac{\varepsilon}{u_a}\, dr\, dz + e_t \tag{4.35}$$

a balance of pressure work and energy dissipation, per unit mass of the moist air passing through the hot tower cycle. The right-hand side contains TKE dissipation in the subcloud layer divided by the volume rate of inflow $u_a h$, that is mechanical energy irreversibly converted into heat per unit mass of the inflowing air, and summed over the approach, plus e_t.

Significant entropy sources other than the subcloud layer, include the eyewall and the rainbands, with their many updrafts and downdrafts, as well as the cold downdrafts interacting with the "root" of the hot tower, and with whatever tropospheric circulation is induced by the hot tower. We have no good way of quantifying dissipation in all these processes, other than writing down an integral of ε over a large volume of the troposphere above the subcloud layer for a second component of energy loss, and divide it over the inflow to yield dissipated energy per unit mass in J kg^{-1}:

$$e_t = \frac{1}{u_0 h_0}\int_h^H \int_\infty^0 \varepsilon\, dr\, dz \tag{4.36}$$

where H is the height or the troposphere or the upper limit of hot tower influence, and $u_0 h_0$ is the volume flow rate from the subcloud layer into the ascending part of the hot tower, at the outer radius of the eyewall. The sum of the two integrals in Equations 4.35 and 4.36 then yields the total loss of mechanical energy per unit mass of the "working fluid," that should equal the total gain from the hot tower thermodynamic cycle.

4.3.2 Thermodynamic Cycle of Hurricanes

The "gentle" hot tower thermodynamic cycle we discussed earlier (applying to slow overturning circulations, such as found in cloud clusters) did not take into account the center pressure deficit in the calculation of the energy input, nor the extreme conditions caused by wind speeds in excess of 50 m s^{-1}. Various observers noted that the spray from churning waves under such extreme winds turns the mixed layer air into something close to a two-phase fluid. The two phases being then necessarily in thermodynamic equilibrium, both are at the same temperature, the air as humid as it can get (i.e., it is saturated), and both water droplets and air are at the surface temperature of the water. Similar conditions should prevail within some 30 km of a strong hurricane's center, inside the radius of maximum wind.

As the air spirals toward the center, the pressure drops, but the temperature remains at the sea surface temperature, so that the two-phase fluid picks up additional sensible

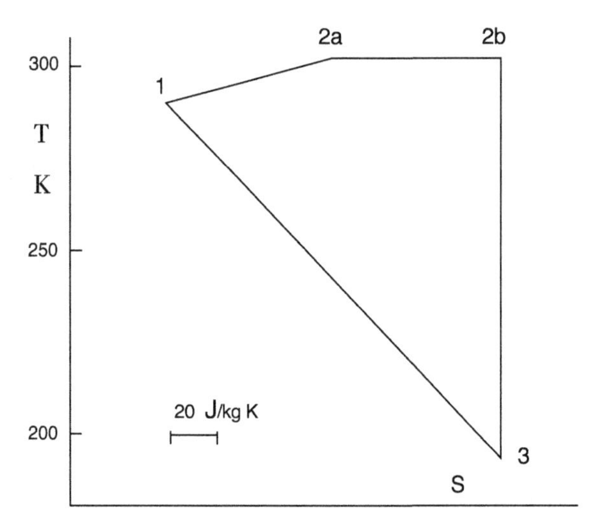

Figure 4.17 *TS* diagram of the hurricane thermodynamic cycle, showing constant temperature heating along the approach to the eyewall, points $2a$–$2b$. In a comparison with Figure 4.3c the different entropy scale should be noted, as well as the colder endpoint of expansion, point 3.

heat from the sea surface to keep its temperature constant while expanding (actually dropping somewhat owing to heat loss in the water-side mixed layer). Specific humidity also increases as more water evaporates at the lower pressure. In such cases, thermodynamic equilibrium, not the transfer laws of Chapter 1, determines the final state of the moist air, prior to ascending the hot tower (or else C_h and C_q can be regarded as very large, their ratio determined by thermodynamic properties of moist air, not by hurricane dynamics, contrary to arguments of Emanuel, 1995).

In the gentle hot tower thermodynamic cycle, the air was supposed to stream in from the surrounding atmosphere toward the hot tower center at constant temperature, with no change of entropy (Figure 4.3c). To realistically represent conditions in a hurricane, we have to modify this cycle, extending leg 1-2 by a constant temperature expansion from pressure p_∞ to p_0. Increase of entropy in this sub-leg turns the *TS* diagram of the cycle from a triangle into a trapezoid (as illustrated in Figure 4.17), the endpoints of the new sub-leg labeled $2a$ and $2b$.

Reversible entropy changes in the constant temperature sub-leg are due to latent heat addition and pressure change:

$$T ds = L_v dq - \frac{dp}{\rho}. \tag{4.37}$$

We express latent heat addition by first raising humidity close to saturation at constant pressure, then adding the effect of the pressure drop, calculated from $q = 0.622e/p$, with p dropping, but the vapor pressure e remaining unchanged at constant temperature. Integration of Equation 4.37 from p_∞ to p_0 thus yields the total heat input in the new sub-leg, calculated to first order in the pressure ratio $\Delta p/p_\infty$:

$$\int_{p_\infty}^{p_c} T ds = \frac{\Delta p}{\rho_\infty} + L_v q_{2b} \frac{\Delta p}{p_\infty} + L_v(q_{2b} - q_{2a}) \tag{4.38}$$

where $\Delta p = p_\infty - p_0$ as before and index ∞ indicates conditions far from the hurricane center, at sea level. With the aid of the gas laws, the two Δp terms can be combined and the total heat input in the new sub-leg expressed as:

$$I_2 = L_v(q_{2b} - q_{2a}) + \frac{\Delta p}{\rho_\infty}\left(1 + \frac{L_v q_{2b}}{R_d T_2}\right) = T_2 \Delta s_2 \tag{4.39}$$

where Δs_2 is the entropy increment, T_2 the temperature in this sub-leg. The heat input in leg 1-2a remains what we wrote down in Equation 4.15 above:

$$I_1 = L_v(q_{2a} - q_1) + c_p(\theta_2 - \theta_1) = \frac{T_1 + T_2}{2} \Delta s_1 \tag{4.40}$$

where we approximated the weighted average temperature by the arithmetic mean of the endpoints, and Δs_1 denotes the entropy increment in leg 1-2a.

For the heat loss in the final leg we similarly write $(T_1 + T_3)(\Delta s_1 + \Delta s_2)/2$. The net heat input is the mechanical energy gain:

$$G = \frac{T_1 + T_2}{2}\Delta s_1 + T_2 \Delta s_2 - \frac{T_1 + T_3}{2}(\Delta s_1 + \Delta s_2). \tag{4.41}$$

Expressing the entropy increments in terms of the heat inputs, this can be written as:

$$G = \frac{T_2 - T_3}{T_1 + T_2} I_1 + \left(1 - \frac{T_1 + T_3}{2T_2}\right) I_2. \tag{4.42}$$

The factors multiplying the heat inputs are partial efficiencies, η_1 and η_2. The thermal efficiency of the whole cycle, η, is the gain G divided by the total heat input $I_1 + I_2$ and is the weighted average of the two partial efficiencies.

As we discussed in the preceding subsection, the mechanical energy gain is dissipated in the subcloud layer and in the hurricane circulation above. Putting gain equal to loss, the loss from Equations 4.35 and 4.36, we have:

$$G = e_t + \frac{\Delta p}{\rho_\infty}. \tag{4.43}$$

Writing out the heat inputs by the middle expressions in Equations 4.39 and 4.40, and substituting into Equation 4.42, also putting for G the last result, a fairly long equation emerges, containing $\Delta p/\rho_\infty$ on both sides. It can be reduced to:

$$e_t + \frac{\Delta p}{\rho_\infty}(1 - m\eta_2) = \eta_1 I_1 + \eta_2(I_2 - I_p) \tag{4.44}$$

where $I_p = m(\Delta p/\rho_\infty)$ and $m = 1 + L_v q_{2b}/(R_d T_2)$, I_p being the pressure drop-related part of the heat input in the new sub-leg. With humidities and temperatures given, and e_t somehow estimated, this equation could be solved for $\Delta p/\rho_\infty$. There is no good basis for independently estimating e_t, however, while there is good information on the pressure deficit of strong hurricanes, which then makes an estimate of e_t possible.

To take a strong hurricane example, let the temperature limits of the hot tower cycle be: the SST of 29°C = 302 K the high temperature limit, $T_2 = 302$ K. A realistic low temperature limit is $T_3 = -80°C = 193$ K. Well outside the hot tower, we take the temperature just above the inversion to be $T_1 = 292$ K, the humidity a typical $q_1 = 4 \times 10^{-3}$. With inversion height at 1800 m this temperature implies a potential temperature difference of $\theta_2 - \theta_1 = -8$ K between a neutral subcloud layer and the top of the inversion. For the specific humidity in the subcloud layer far from the hot tower we take $q_{2a} = 18 \times 10^{-3}$, at the base of the hot tower $q_{2b} = 24 \times 10^{-3}$, close to saturation.

The two partial efficiencies calculated for this cycle are $\eta_1 = 0.183$ and $\eta_2 = 0.197$, resulting in an overall efficiency of $\eta = 0.190$. The heat inputs in the two sub-legs are very similar at $I_1 = 27000$ J kg^{-1} and $I_2 = 28000$ J kg^{-1}. The net energy gain is 10450 J kg^{-1}. If we take the pressure drop to be $\Delta p = 100$ h Pa, the estimated energy loss above the subcloud layer is $e_t = 2750$ J kg^{-1}.

Perhaps the most important point about this example is that slightly more than half of the total heat input comes during the approach of the moist air to the hot tower's base. Without that extra, the net energy gain would have been just under 5000 J kg^{-1}. With e_t presumably not very different (judging by the similarity of convection in hurricanes and cloud clusters), there would not have been enough energy to generate more than a center pressure deficit of about 30 h Pa. The intensity of strong hurricanes thus depends on the extra heat input induced by their own circulation, which is a positive feedback identified by Emanuel (1986) in his "air-sea interaction theory for tropical cyclones." Or as we put it at the very beginning of Chapter 1, "the lifeblood of a hurricane is intense sea to air transfer of heat and water vapor," that is partly self-generated.

4.4 Oceanic Deep Convection

In Chapter 3 we have discussed the important role of the oceanic mixed layer in insulating deeper waters from surface processes. While insulation operates over much of the world's oceans most of the time, sustained strong surface cooling brings about buoyant convection intense enough to break through the thermocline under the mixed layer, and mix surface water down to a much greater depth. Such deep convection occurs, however, only in certain locations and at certain times, usually in late winter near the edges of continents unleashing strong cold and dry winds over the adjacent seas.

Just how far "deep" convection reaches depends on circumstances. In midlatitudes, near the poleward edge of subtropical gyres, a very strong "permanent" thermocline sets a limit to convective motions at depths of 300–500 m. At higher latitudes, where the ocean is weakly stratified, deep convection mixes the water down to a thousand meters or more, in some locations to the seafloor.

This already makes clear that oceanic deep convection differs from its atmospheric counterpart in its geographic location and depth range. A more consequential contrast is that the ocean lacks the analogue of latent heat release, which in the ocean would be cooling distributed over depth. All of the cooling occurs right at the sea surface. Nor is there any significant heating below the optical depth of surface waters, at most a few tens of meters. All heat gain and loss thus take place at essentially constant atmospheric pressure. The associated thermodynamic cycle converting heat input to mechanical energy is then very inefficient, the gain of mechanical energy from the cycle low.

Also, the negative buoyancy $b' = \alpha g \theta'$arising from a small temperature depression θ' is about an order of magnitude smaller than those driving atmospheric convection, owing to the low thermal expansion coefficient of water, α (particularly at near-freezing temperatures, where much of deep convection takes place). Salinity excess adds very little to buoyancy fluxes $\overline{w'b'}$. As a result, "CAPE," Convective Available Potential Energy to support convection, is much smaller in the ocean than in atmospheric hot towers.

In spite of the differences, the details of oceanic deep convection resemble the atmospheric prototype, with a novel nomenclature, however. Jones and Marshall (1993) define "plumes" as convective elements generated by surface cooling, 1 km in typical diameter, sinking analogues of atmospheric updrafts. They appear in groups (clearly gregarious again), flanked by rising fluid. The aggregate of plumes and risers in a group is called a "chimney" of well-mixed, convecting fluid, the analogue of a cloud cluster. The terminology originates from the notion that the average density within the convecting column being greater than in its neighborhood, the column is an inverted analog of the flue on a fireplace. When surface cooling stops, the chimney "collapses," spreading out at depth, ambient fluid converging at the surface and "capping" the chimney. The mixed fluid within the chimney then sinks to its level of neutral buoyancy, where it is accommodated in the thickening of isopycnic layers (= layers between two constant density surfaces). Thickened isopycnic layers are known as "pycnostads," containing nearly homogenous water, and are found in many high-latitude seas (McCartney, 1982). Stretching a point a little, they are the analogues of anvils and stratiform clouds at the top of atmospheric hot towers.

While the terminology is new, the physics of convection is the same as in the atmospheric mixed layer: Surface buoyancy flux B_0 and convecting column depth h are the key independent variables, and vertical velocities are proportional to $w^* = (B_0 h)^{1/3}$. With typical oceanic values of buoyancy flux $B_0 = 10^{-7}\ \mathrm{m^2\,s^{-3}}$, similar to what we encountered in night-time cooling of the oceanic mixed layer, convecting column depth $h = 1000$ m, the velocity scale w^* is about 0.05 m s^{-1}, two orders of magnitude less than in atmospheric convection. The same surface buoyancy flux sustained for a period t generates a density excess of $\Delta\rho/\rho = B_0 t/gh$; a 10-day long episode of cooling then results in $\Delta\rho/\rho = 10^{-5}$, a change in the second decimal of density expressed as $\sigma_t = \rho - 1000\ \mathrm{kg\,m^{-3}}$, the conventional oceanographic measure of density (or rather $\sigma_\theta = \rho_\theta - 1000\ \mathrm{kg\,m^{-3}}$, with σ_θ potential density, analogous to potential temperature, and equal to the density of a parcel adiabatically brought to the surface).

The buoyancy flux $\overline{w'b'}$ decreases linearly with depth from its surface value B_0 in a convecting column, just as it decreases with height in atmospheric convection.

Where $\overline{w'b'}$ is positive, it acts as a source of TKE. At the bottom of a convecting column it becomes negative and entrains cooler and heavier fluid from below, at the rate governed by Carson's law, the peak negative buoyancy flux equal in magnitude to about 20% of B_0. This adds to the potential energy of the convecting column, and also increases its depth.

The change of depth is very slow when a typical seasonal thermocline underlies the convecting column. We have seen this in Chapter 3, in nighttime cooling of the mixed layer. To entrain all of the thermocline fluid and finally break through such a thermocline, the slow surface cooling-induced thermocline erosion must continue for several fall and winter months. Oceanographers have christened this period the "preconditioning" phase of deep convection. It is shallow not deep convection, but it does most of the work, continuously adding potential energy to the entrained fluid. That it takes so long is a direct consequence of weak buoyancy flux.

The term "chimney" also serves as a reminder that the pressure difference from chimney top to bottom being greater, owing to higher density, than in the surrounding slightly warmer fluid, a deficiency of pressure tends to appear at the surface, excess pressure at the bottom. This is what induces capping at the surface, spreading at the bottom. The inward and outward motions induced by the small pressure differences are slow, however, slow enough for earth rotation to deflect them into a "rim current," in which the Coriolis force balances the pressure gradient. Subsequent breakdown because of hydrodynamic instability of the rim current into warm core eddies and turbulence eventually induces more inflow and allows the capping to be completed. Laboratory simulations of the process by Maxworthy and Narimousa (1994), and numerical experiments of Jones and Marshall (1993) have given a vivid picture of the processes at work in the capping of chimneys.

Spreading out at the bottom of a convecting column is a similar process in principle, and results in the formation of a well-mixed "bolus," a lens-shaped pycnostad between constant potential density surfaces. These are stores of potential energy, as the following simple calculation shows.

In pycnostad formation, neighboring isopycnal surfaces move apart, to accommodate a bolus of nearly well-mixed fluid. An idealization is shown in Figure 4.18, an originally linear increase of potential density with depth, $-d\rho/dz = const.$ replaced by a constant-density pycnostad of depth h. In the lower half of the bolus, the density is less than outside, and the hydrostatic pressure inside the bolus decreases less rapidly with height. In the upper half, the opposite is the case. A simple calculation shows that the maximum excess pressure, at the center of the bolus, is:

$$\Delta p = -g \frac{d\rho}{dz} \frac{h^2}{8}. \tag{4.45}$$

Divided by density, the excess pressure is excess potential energy per unit mass in $\mathrm{J\,kg^{-1}}$. The depth-average potential energy excess per unit mass of the bolus is:

$$\frac{\Delta p}{\rho} = \frac{gh}{6} \frac{\Delta \rho}{\rho} \tag{4.46}$$

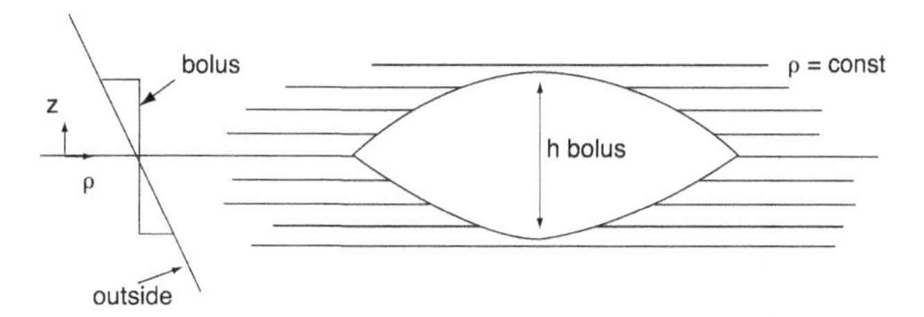

Figure 4.18 Idealized density distribution versus depth inside and outside of a pycnostad (left), also showing an idealized arrangement of the isopycnal surfaces.

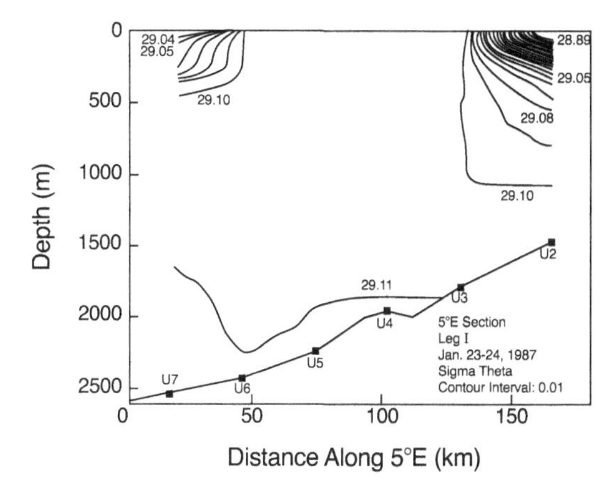

Figure 4.19 Cross section of potential density σ_θ across a chimney observed in the Gulf of Lions in the Mediterranean in late January 1987. From Leaman and Schott (1991).

where ρ is the density at the center of the bolus, and $\Delta\rho = -(h/2)(d\rho/dz)$ is the density jump at the top and bottom of the bolus. In reality, the bolus is less sharply defined of course, and these formulae are overestimates of pressure and energy gain. As we will see, pycnostads are a significant energy source for ocean circulation. First, we have look at the evidence on oceanic deep convection.

4.4.1 Observations of Oceanic Deep Convection

One location where oceanic convection is deep and frequent is the Gulf of Lions in the Mediterranean Sea. This is where the first major cooperative field experiment on deep convection took place in 1969 (MEDOC Group, 1970). Strong surface cooling and evaporation episodes occur here under winter outbreaks of cold and dry air, the "Mistral" blowing down from the Rhone valley, the "Tramontane" wind from the north side of the Pyranees. Schott and Leaman (1991) and Leaman and Schott (1991) reported field observations carried out on a chimney in the Gulf of Lions in January 1987. Schott et al. (1996) described another such experiment carried out in 1992. The diameters of the three chimneys observed in the three experiments ranged from 50 to 100 km, according to Schott et al (1996). A hydrographic section of potential density σ_θ across the chimney of 1987 may be seen in Figure 4.19 (Leaman

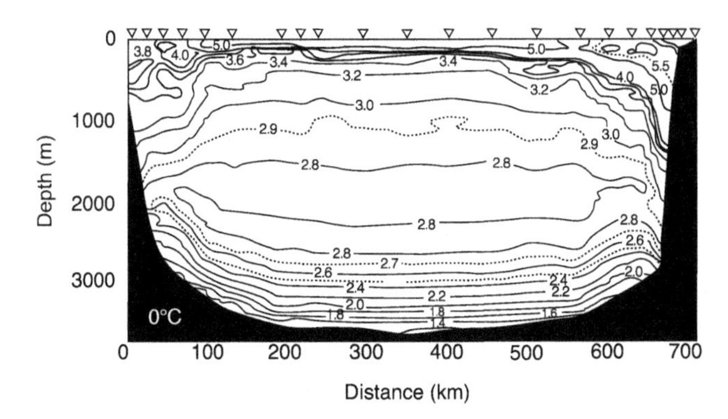

Figure 4.20 Autumn hydrographic section across the Labrador Sea showing potential temperature in the lens-shaped pycnostad centered at about 2 km depth, formed by convection in past winters. From Marshall et al. (1998).

and Schott, 1991). The chimney fluid was homogenous to within 0.01 in σ_θ, or to 10^{-5} kg m^{-3}. A surface mixed layer poised to intrude over the top of the chimney is lighter than the chimney fluid by only 2×10^{-4} kg m^{-3}. Observations showed sinking motions in the plumes with a vertical velocity of up to 0.1 m s^{-1}, rising ones half as fast. The variations of vertical motion arose from plumes being convected past the instrument, and showed the plumes to be of order 1 km in diameter. Current meter records also revealed eddies detached from the mixed layer front at the rim of the chimney, of order 5 km in diameter, attributable to instability of the rim current. The geostrophic rim current velocity, calculated from the density distribution, ranged from 0.06 to 0.16 m s^{-1}.

Another well-explored site of deep convection is the Labrador Sea, the source of Labrador Sea Water, a large volume of cold waters (near 3°C) spread over the north-south extent of the deep Atlantic Ocean. Clarke and Gascard (1983) and Gascard and Clarke (1983) gave a detailed account of large- and small-scale processes involved in the formation of this water mass. More recently, Marshall et al. (1998) described early results of a major cooperative experiment in the Labrador Sea, begun in the autumn of 1996 and aimed at elucidating the details of oceanic deep convection. A hydrographic cross section of the Labrador Sea, taken at the start of the experiment, showed the distribution of potential temperature (Figure 4.20). The lens-shaped bolus of nearly constant potential temperature extending down to about 2000 m is the Labrador Sea water formed in one or more previous winters. It is covered by a seasonal thermocline and a mixed layer of about 5.2°C temperature. Figure 4.21 shows the progressive cooling of this mixed layer, along with a very slow erosion of the thermocline, the "preconditioning" phase of convection in November 1996 (compare this with Figure 3.20, nighttime cooling of a mixed layer, with no visible effect on the seasonal thermocline). By the end of January 1997, little stratification was left. Floats following the vertical motions of water parcels revealed continuous convection in late February to early March, some parcels penetrating to 1000 m (Figure 4.22).

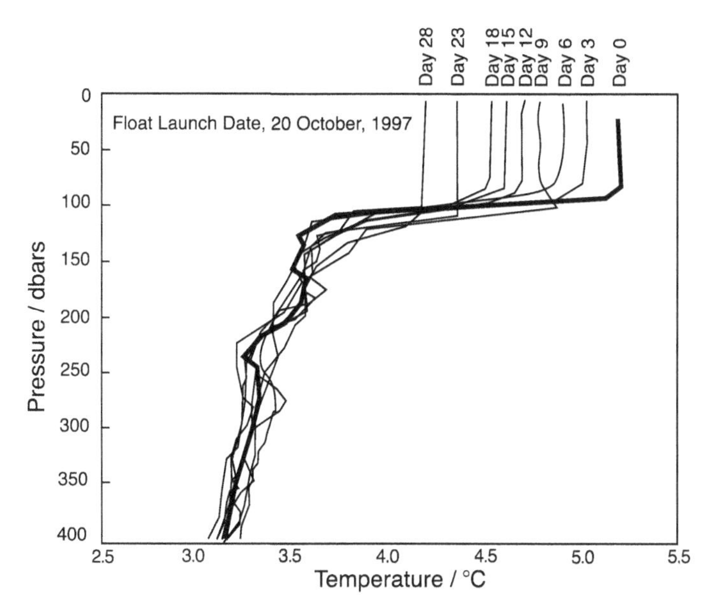

Figure 4.21 Temperature profiles at 3-day intervals in the Labrador Sea documenting the preconditioning phase of convection: the cooling of the mixed layer proceeds without significant entrainment. From Marshall et al. (1998).

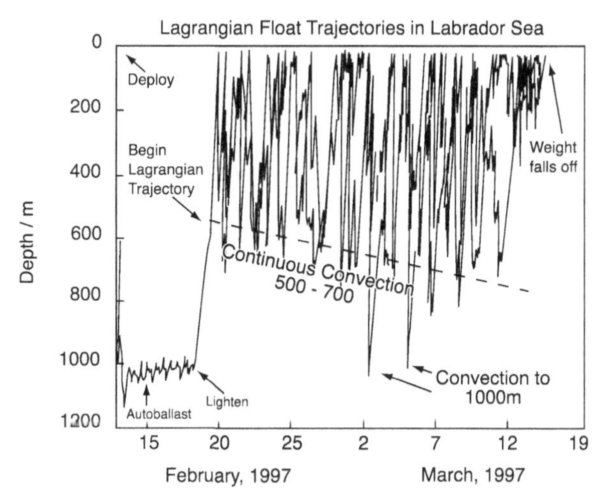

Figure 4.22 Continuous convection in the Labrador Sea in late winter, sensed by a free-floating instrument. From Marshall et al. (1998).

Much eddy activity and other complexity accompanied the convection, the deep portion of which (mixed layer depth > 700 m) was confined to a horizontal area of about 300 km × 200 km.

The final phase of oceanic deep convection, pycnostad formation, is perhaps most interesting. As we have just seen in Figure 4.20, this results in thick isopycnic layers.

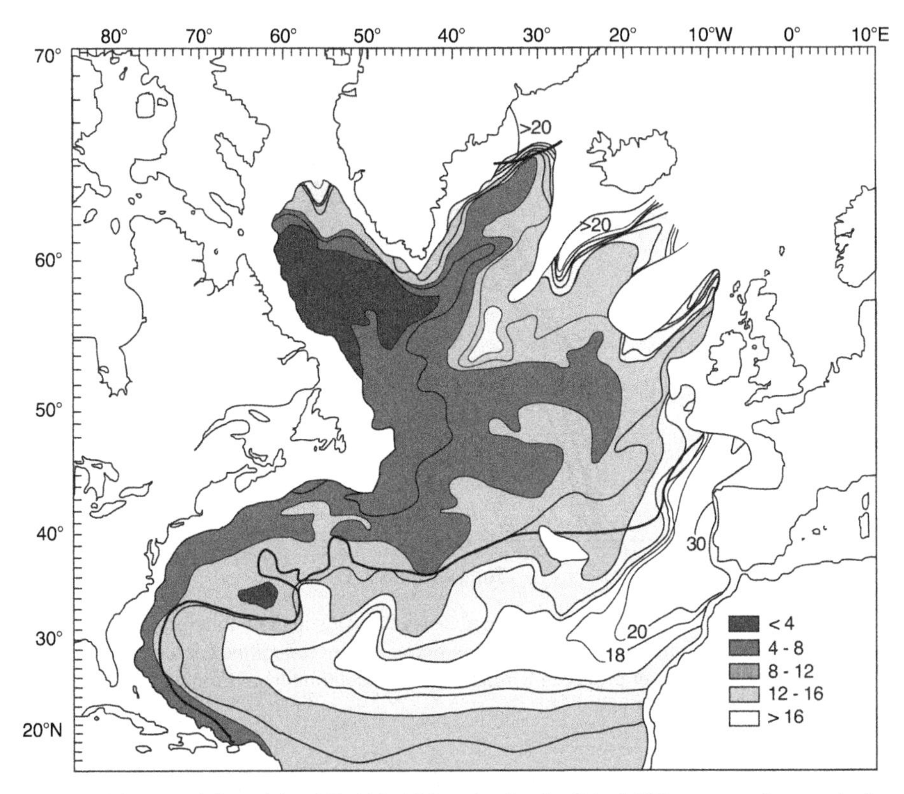

Figure 4.23 Potential vorticity $(f/\rho)(d\rho/dz)$ at the depth of the LSW pycnostad center in the North Atlantic. This quantity is inversely proportional to pycnostad thickness and its minimum in the Labrador Sea is marked by the deepest shading. From Talley and McCartney (1982).

The first instance of a pycnostad to be so recognized was Worthington's (1959) "eighteen degree water" in the North Atlantic, just south of the Gulf Stream. This water of nearly constant temperature and salinity forms each winter and reaches its greatest depth of about 500 m in early March. It remains a persistent feature; in Worthington's words: "During the spring and summer a seasonal thermocline develops but the identity of the winter surface layer is never entirely destroyed: it persists throughout the year as a nearly isothermal/isohaline layer between the bottom of the seasonal thermocline and the top of the main thermocline until it is renewed at the end of the following winter." Worthington has also pointed out that eighteen degree water spreads far southward from its formation region, at the 300 m level to southerly locations where "there is no possibility of its having been formed locally."

Another well-explored pycnostad is the one containing Labrador Sea Water (LSW), a picture of which at its formation region we have seen in Figure 4.20. Energy stored in the pycnostad allows it to spread far from its formation region, much as the eighteen degree water does. Talley and McCartney (1982) produced a striking illustration of this (Figure 4.23). It shows potential vorticity, proportional to pycnostad thickness, at the center of the level where the LSW spreads out (average depth 1500 m). The darkest shading marks the thickest pycnostad in the formation region of LSW, and along

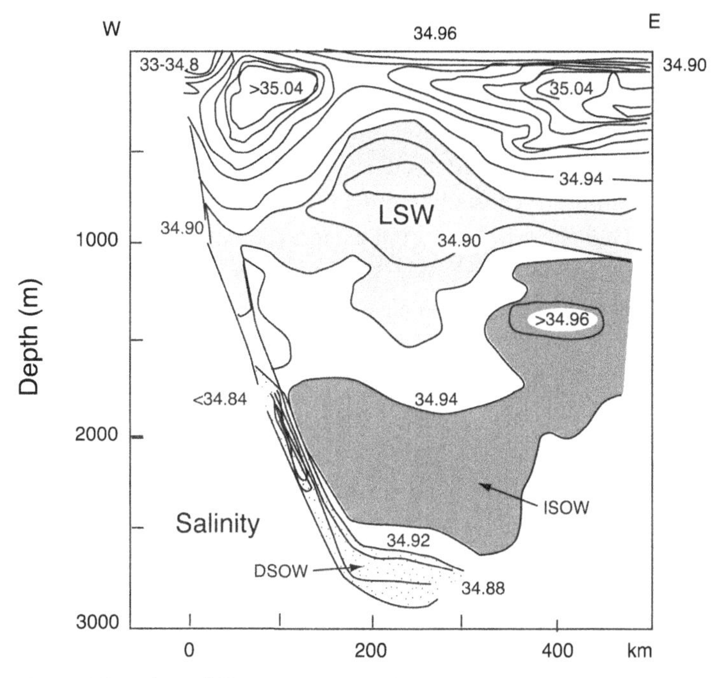

Figure 4.24 Three different pycnostads in the Irminger sea, LSW coming from the Labrador Sea, ISOW from the Iceland-Scotland Ridge east of Iceland, DSOW from the Denmark Strait Overflow, west of Iceland. On their passage to the Southern Hemisphere these waters combine to form NADW, North Atlantic Deep Water.

the continental slope of North America, where it forms a "deep western boundary current." The energy stored in the initial pycnostad is apparently enough to energize this boundary current for a trip of several thousand kilometers.

The northern North Atlantic is a prolific producer of cold pycnostads. Small differences in salinity allow oceanographers to identify the products of different chimneys in diverse locations, in the Norwegian Sea, Irminger Sea, among others. A striking illustration of this is the disposition of pycnostads in the Irminger Sea east of Greenland (Figure 4.24) from Dickson and Brown (1994). It shows a northward extension of the LSW pycnostad, and two others, coming from further east and north, across the Denmark Strait and the Iceland-Scotland overflow. All these pycnostad waters make their way southward, their slight salinity differences eroding, and arriving as North Atlantic Deep Waters, as far south as the Antarctic Seas. Their long journey begs the question of where the energy comes from, since pycnostad energy runs out by the time these water cross the equators, as Figure 4.23 shows.

Two well-known pycnostads are the Labrador Sea Water source region, and the eighteen degree water in the North Atlantic subtropical gyre. Putting $\Delta\rho/\rho = 10^{-4}$, $h = 1000$ m in the formula for the Labrador Sea, we find the potential energy stored in the pycnostad to be $0.16\ \mathrm{J\,kg^{-1}}$. Remember that this is an overestimate, so that the motions energized by it must be very slow. The energy gain has to support the rim current and the subsequent dissipation of its energy by eddies. The observed rim current velocity of order 0.3 $\mathrm{m\,s^{-1}}$ uses up 0.045 $\mathrm{J\,kg^{-1}}$ for its generation, leaving the rest for the eddies. The fluid spreading out at depth has a different fate, and leaves via a Deep Western Boundary Current, as we discuss in the next chapter.

In the subtropical gyre, realistic numbers are $\Delta\rho/\rho = 10^{-3}$, $h = 300\,\mathrm{m}$, with the energy gain from the pycnostad according to the formula being 0.5 $\mathrm{J\,kg^{-1}}$. This serves to enhance Gulf Stream transport in winter. To put the numbers on energy gain in perspective, other energy sources for ocean circulation have to be examined. We discuss these points further in the next chapter.

Chapter 5

The Ocean's WarmWaterSphere

Among Georg Wüst's many contributions to oceanography are his almost pole-to-pole meridional sections of the Atlantic Ocean (Wüst, 1935). Figure 5.1 here shows his "main" or western section, forcefully making the point that most of the ocean waters are cold, excepting only a shallow surface pool from about latitude 52°N to 43°S. The bottom of the warm pool is the 8°C isotherm according to Wüst's choice, but the 10°C, or 5°C, one serves just as well. The depth of the warm water varies from 200 m at the equator, to 1000 m at the center of the "subtropical gyre" in the North Atlantic. Wüst called this the "Warmwassersphäre," or alternatively the Troposphere, versus the deeper Stratosphere. The latter terminology is less apt, the troposphere analogy being somewhat forced. In English, WarmWaterSphere is gratifyingly descriptive, and the acronym WWS fits both saxon tongues.

The influence of the WWS on the climate of continents, especially coastal regions, is pronounced. Trondheim, Norway is a pleasant place to visit in summer, with 24 hours of daylight, and no need for a coat while walking home from a party. At the same latitude South, at the northern tip of the Antarctic Peninsula, nobody lives. The difference is that the annual mean sea surface temperature off Trondheim is 8°C, at the northern limit of the WWS, versus −1°C off Antarctica, the 8° isotherm being there 1500 km away equatorward.

A more dramatic example of WWS control of climate are the monsoons of India. Massive quantities of water vapor, collected from a huge area of the tropical Indian Ocean, invade the land mass of Asia, to fuel there superclusters of hot towers that drop rain in some places at extravagant rates. How important this is to the 20% or so of humanity that lives there is made clear by the catastrophic effects of a failing monsoon. A less extreme manifestation of the same vapor invasion of a continent are the sticky summers and tornadoes of eastern North America, orchestrated by the Gulf of Mexico.

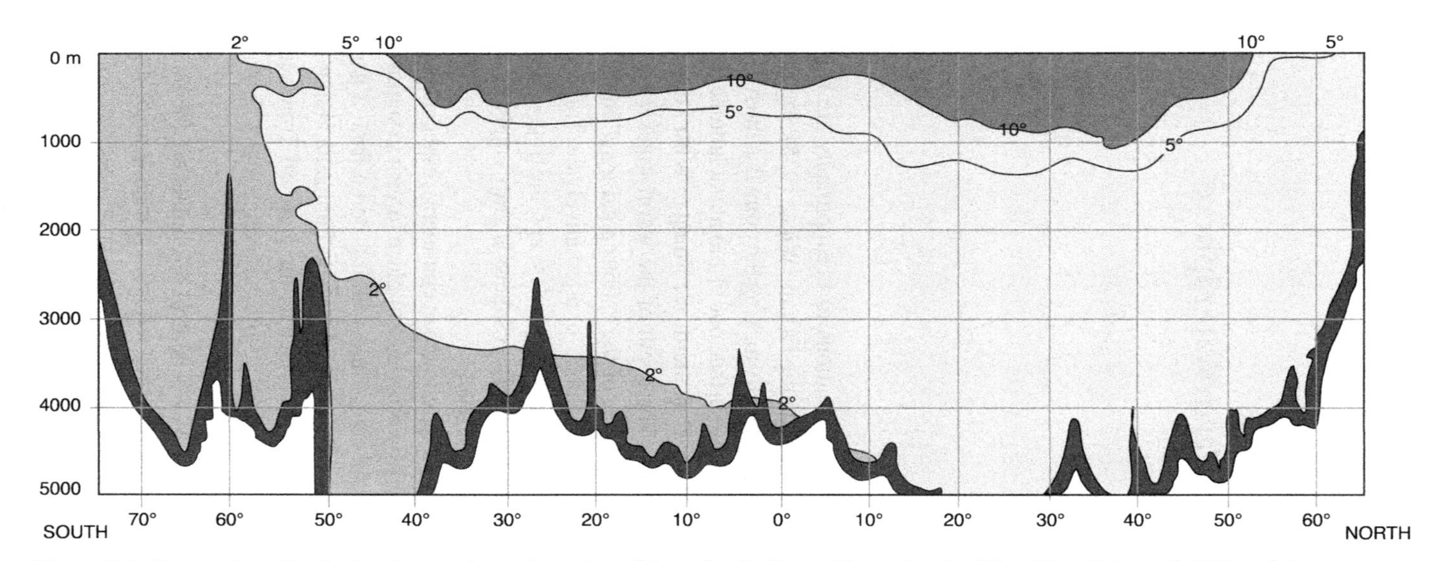

Figure 5.1 Temperature distribution in a north-south section of the Atlantic Ocean illustrating the WarmWaterSphere-ColdWaterSphere division. Any isotherm between 5 and 10°C will serve as a boundary. From Wüst (1935).

Behind these local effects lies a global-scale interplay of ocean and atmosphere, in which the WWS plays a critical role. Because evaporation rates are high only over warm surface waters, the WWS supplies most of the water vapor crucial to atmospheric circulation and dynamics, and of course to the hydrologic cycle. The WWS collects the requisite latent heat from solar radiation, mainly over vast stretches of the tropical and subtropical ocean, and redistributes this heat income to higher latitudes in the manner of a welfare state. Major WWS currents, such as the Gulf Stream, transport massive amounts of heat poleward and play an important role in this process. In return, winds energize WWS circulation, directly through wind drag and indirectly through promoting differential heating and cooling of the ocean. Laws and mechanisms of this interplay, the subjects of this last chapter, are in effect a synthesis of laws governing constituent processes that we discussed in preceding chapters.

5.1 Oceanic Heat Gain and Loss

As we have seen in Chapter 3 on mixed layers in contact, the net gain or loss of heat by the ocean's WWS is an outcome of a variety of processes acting in concert, radiation, wind stress, advection, turbulence, and such large-scale phenomena as subsidence or upwelling. Different processes dominate the heat budget of the oceanic mixed layer in different locations and in the changing seasons, as we have seen in Bunker's (1976) analysis of air-sea interaction cycles (Section 3.5.3). They bring about a complex geographical distribution of annual mean oceanic heat loss and heat gain.

Observations from "ships of opportunity" as well as from research vessels have built up over the years a massive data base on air-sea transfer processes worldwide. Data coverage of the North Atlantic is particularly good – the southern seas, however, are sparsely documented. Meteorologists and oceanographers have painstakingly put together this evidence over the different oceans and portrayed the annual average net heat flow into the ocean in worldwide maps.

Figure 5.2 shows such a map for the Atlantic Ocean (Bunker, posthumously published in 1988; the North Atlantic portion was published by Bunker and Worthington in 1976). The contours connect points of equal heat gain in W m^{-2} (negative if loss). The zero-gain contour cuts through this ocean along a diagonal roughly from Spain to the island of Hispaniola in the Caribbean. North of this contour the ocean loses heat, at spectacularly high rates over the warm waters of the Gulf Stream. Here the annual average rate of loss exceeds 200 W m^{-2}. On the other side of the ocean, off the Norway coast, a northward tongue of the WWS is still responsible for heat losses between 50 and 100 W m^{-2}, and even higher off Lapland.

South of the zero-gain contour, over most of the subtropical gyre, the ocean gains heat as colder waters flow southward and absorb solar heat, the energy gain through this "cold water advection" process being, however, moderate, typically 25 W m^{-2}. In this region, evaporation is also high, raising the salinity of surface waters.

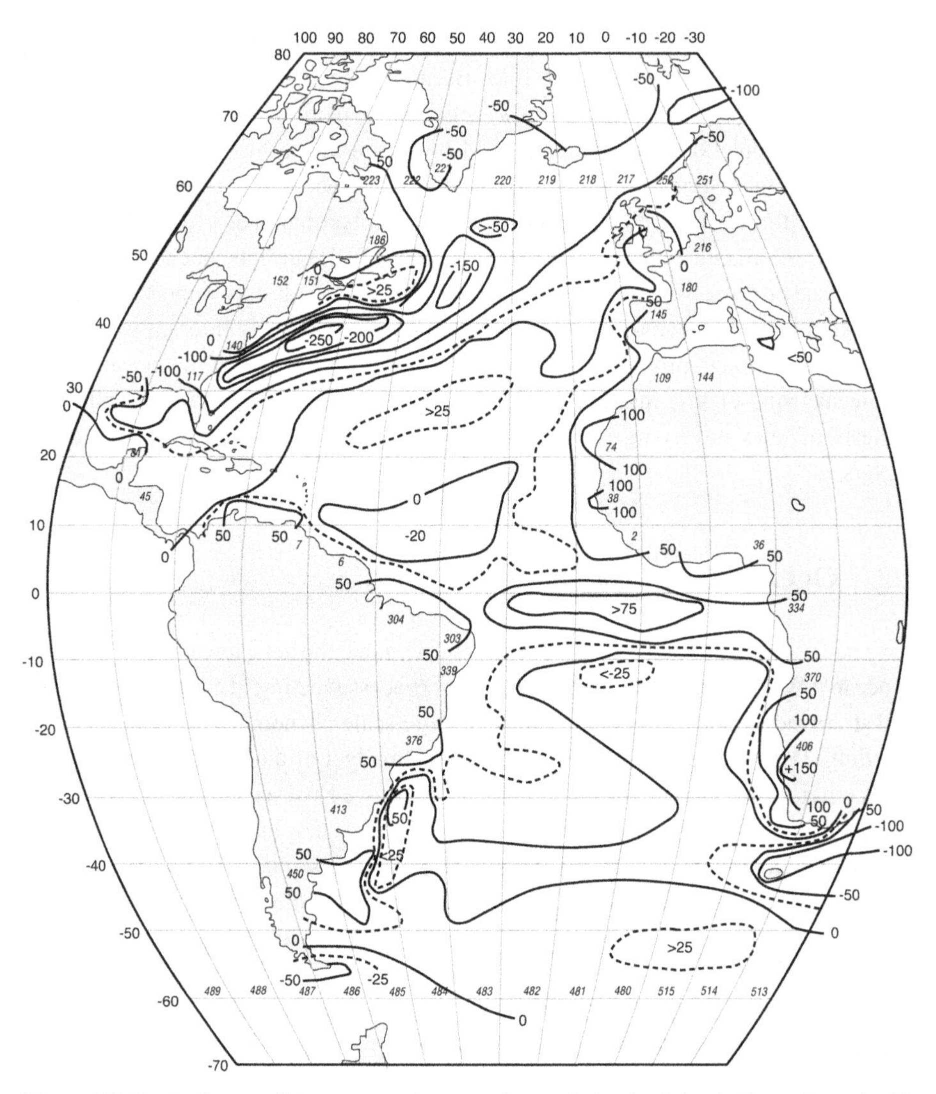

Figure 5.2 Bunker's map of the net annual average heat gain by the Atlantic Ocean (negative if loss). Full lines are at contour intervals of 50 W m^{-2}, broken lines at half that.

Other locations of relatively high heat gain lie along coasts where upwelling is vigorous because of alongshore winds, blowing with the coast to their left, and driving mixed layer "Ekman transport" (see below) offshore. These are the Saharan coast, the Venezuelan coast, and portions of the eastern coast of North America.

The equatorial high heat gain region was not conspicuous on the early (1976) version of this chart, but it did show a heat gain in excess of 50 W m^{-2}, attributable to equatorial upwelling. The full version of this illustration, including the South Atlantic (available in manuscript form circa 1981), shows oceanic heat gain over that entire ocean down to 60°S, missing only the Antarctic seas. The equatorial high heat gain tongue now stands out, as does the upwelling zone of the Benguela Current off South

Africa. The moderate heat gain region over the subtropical gyre is also present in the Southern Hemisphere.

What differs dramatically from the Northern Hemisphere is the large moderate heat gain region lying south of about 40°S, and extending to the Antarctic seas. In this region, the westerlies drive cold surface waters northward. In southern summer, these waters rapidly gain heat from the sun, air-sea transfer of sensible and latent heat being suppressed by the cold temperatures. At similar latitudes in the North Atlantic, high heat loss is universal; winter cooling dominates the annual mean heat budget. Bunker points out that the cold South Atlantic gains heat at the total (area-integrated) rate of about 0.5 PW (≡ petawatt, 10^{15} W), while the warm Northern Hemisphere loses it at the rate of some 0.7 PW.

The underlying difference in surface temperatures between the hemispheres is indeed considerable: Figure 5.3 shows annual mean surface temperatures worldwide (from Levitus, 1982). The Atlantic is unfortunately cut up in this projection of the world ocean, but the map still shows that the WWS extends in this ocean by ten to twenty degrees latitude farther poleward in the north than in the south.

On the heat losses in the high latitude northern seas of the North Atlantic, Häkkinen and Cavalieri (1989) have published fascinating charts of monthly average heat loss. In mid-winter, off Lapland, local heat losses reach 600 $\mathrm{W\,m^{-2}}$, while in August they range from 20 to 40 $\mathrm{W\,m^{-2}}$. Here the edge of the ice sheet terminates the range of air-sea heat exchange.

As the Atlantic, the Pacific Ocean also extends to high northern latitudes, but its northern waters are relatively cold, and there is no temperature contrast between south and north similar to the Atlantic's. A net oceanic heat gain map for the world ocean in Hsiung (1985) shows neither much heat loss nor heat gain in the northern Pacific (Figure 5.4). This map shows large heat losses over the Kuroshio off Japan and the adjacent portions of the subtropical gyre, similar to those over the Gulf Stream and adjacent region. The other conspicuous feature is the large heat gain over the eastern equatorial Pacific, where upwelling is intense.

For the Atlantic, a comparison of Hsiung's heat gain map with Bunker's shows many differences in detail, revealing the magnitude of the uncertainties in such calculations. Sparsity of meteorological data in some regions is the main cause, but different formulations of the transfer laws also contribute. Steering clear of the poorly explored southern seas, Hsiung's map only extends southward to about 40°S, northward to 60°N, covering thus about twenty degrees of latitude less than Bunker in both hemispheres, and missing the high heat losses in the North Atlantic, as well as the large area of moderate heat gain in the South Atlantic.

A recent compilation of net heat gain in the Northern Pacific by Moisan and Niiler (1998), from a much larger database, yielded the map in Figure 5.5. This differs from Hsiung's map in Figure 5.4 by showing moderate heat gain over much of the northern North Pacific, as well as in the coastal upwelling band along the coast of North America. The zero heat gain contour cuts through the Pacific from Japan to California almost along a latitude circle, with a heat gain region in the north, in dramatic contrast

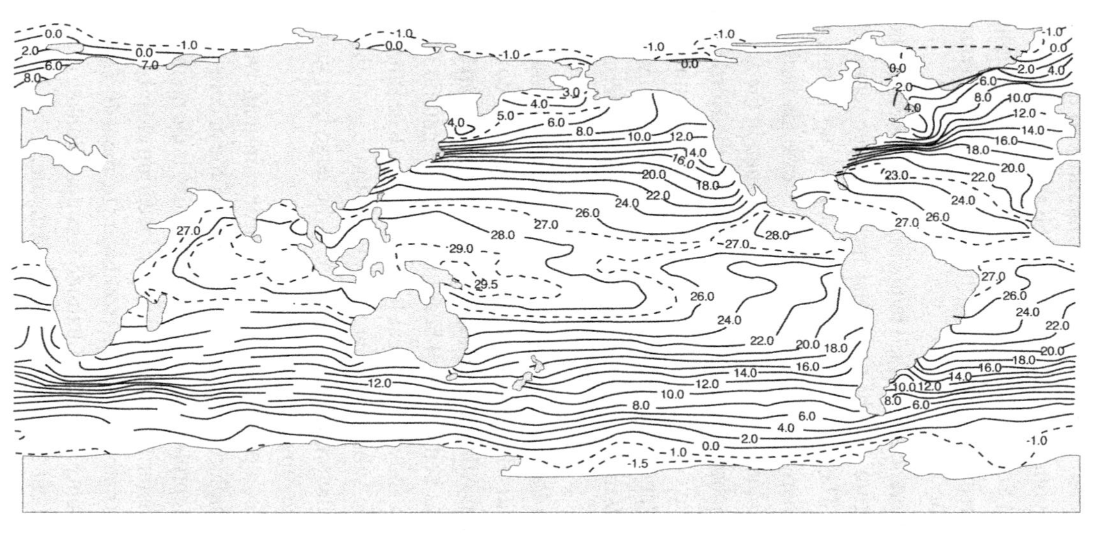

Figure 5.3 Annual average surface temperature of the world ocean, °C, from the Atlas of Levitus (1982).

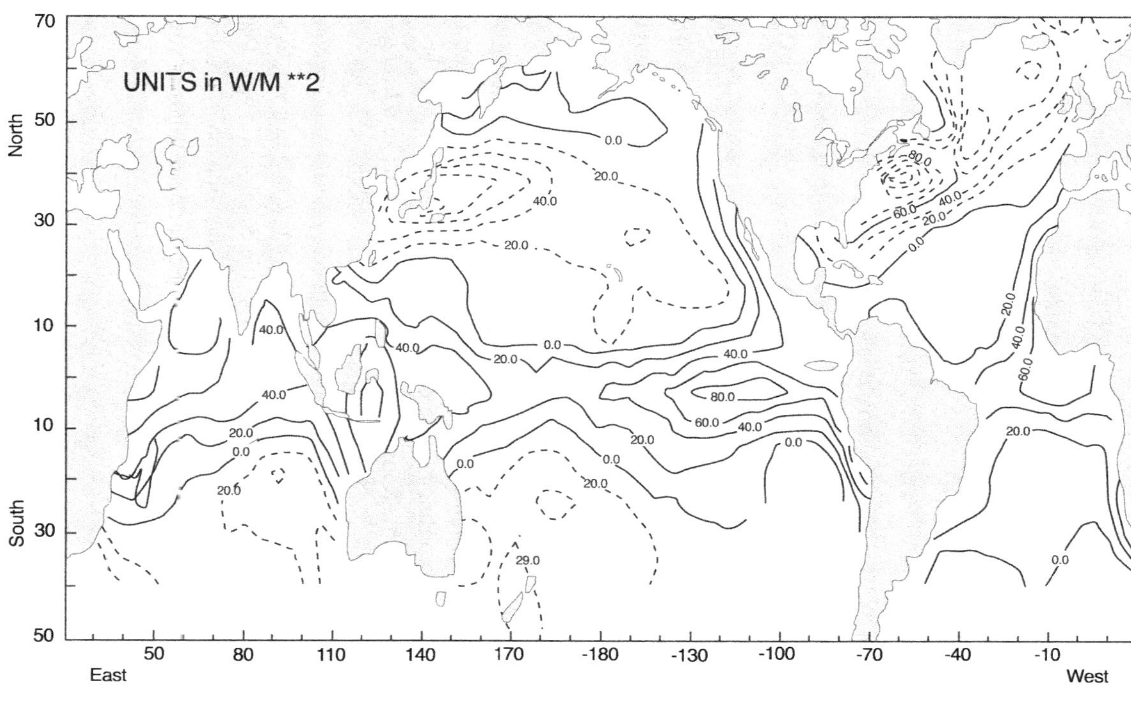

Figure 5.4 Annual average heat gain by the world ocean in 20°C contour intervals, full lines indicating heat gain, broken lines heat loss. From Hsiung (1985).

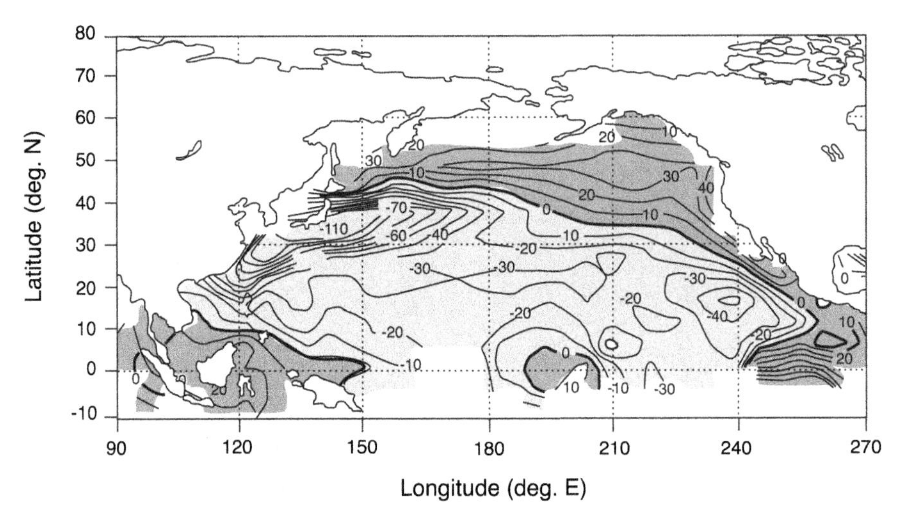

Figure 5.5 Annual average heat gain of the North Pacific Ocean in 10°C contour intervals. Darker shading indicates heat gain, lighter shading heat loss. From Moisan and Niiler (1998).

to the Atlantic. The high latitude heat gain region roughly coincides with the subpolar gyre, where cyclonic winds drive low intensity upwelling. In respect to heat gain, the North Pacific thus behaves somewhat similarly to the high latitude South Atlantic.

In the Indian Ocean Hsiung's map shows fairly high heat gain in the Arabian Sea and along the east coast of Africa, associated with the boundary currents there, moderate heat losses south of about 20° S, with higher southern latitudes unexplored. A more detailed map from Hastenrath and Lamb (1980) of the Indian Ocean shows that the high heat gain regions are confined to the coastal upwelling zones along the African and Arabian coasts (Figure 5.6). Remarkable is the location of the zero heat gain contour, roughly coincident with the 10° S latitude circle. The subtropical gyre south of that circle loses heat at a moderate rate owing to high evaporation rates, according to Hastenrath and Lamb (1980). In the heat budget of the oceanic mixed layer, this loss is balanced presumably by warm water advection from the equatorial region that gains heat. On the air side, this region is "the major source of the atmospheric water vapor carried across the coastline of southern Asia during the northern summer southwest monsoon."

5.1.1 Mechanisms of Heat Gain

Some features of the above heat gain maps are certainly unexpected and call for further discussion. In Chapter 3 we have seen that oceanic heat loss or gain is one entry in the heat balance of the two mixed layers in contact. We return to that heat balance in order to understand how the interplay of different processes produces the observed oceanic heat loss or gain.

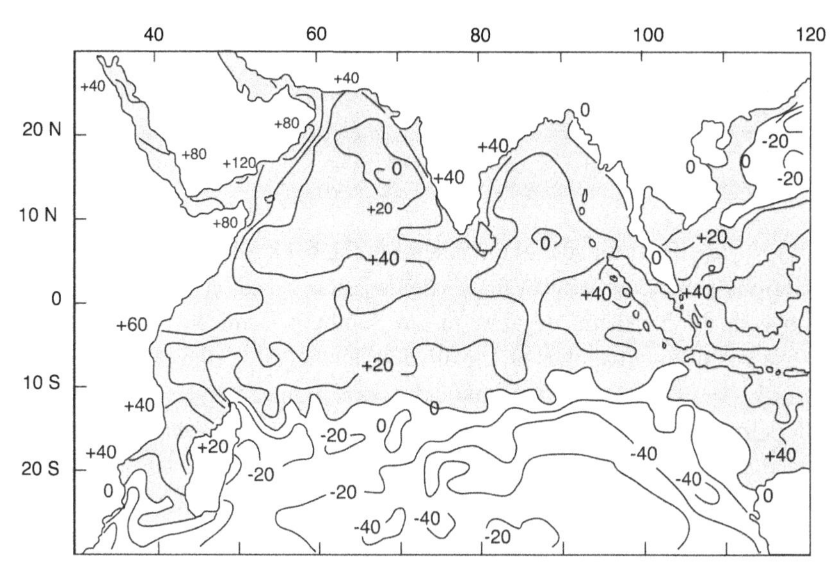

Figure 5.6 Annual average heat gain of the Indian Ocean in 20°C contour intervals. Stippling indicates heat gain. From Hastenrath and Lamb (1980).

We defined the net heat gain of the oceanic mixed layer, A_w, in Chapter 3 as the algebraic sum of gains from horizontal heat advection, upwelling of colder water, and radiant heat loss at the bottom of the mixed layer. The latter is only significant in very shallow mixed layers and can be neglected in the global context. That leaves:

$$A_w = \rho_w c_{pw}(hd\theta/dt - \overline{w'\theta'}(h)) \quad (5.1)$$

where the total derivative $d\theta/dt$ includes local heat storage, $\rho_w c_{pw} h\partial\theta/\partial t$, as well as temperature advection by the depth-average velocity components u, v in the mixed layer, $u\partial\theta/\partial x + v\partial\theta/\partial y$. The Reynolds flux of temperature at mixed layer bottom, $-\overline{w'\theta'}(h)$, may also be expressed as an entrainment or upwelling velocity times the temperature defect of the upwelling water, $w_e\Delta\theta$.

The heat balance of the mixed layer relates the heat gain to surface transfers:

$$A_w = -R(0) - SH - LE \quad (5.2)$$

the terms on the right being in order downward irradiance, sensible, and latent heat transfer to the atmosphere (Equation 3.31). The last two equations tell us, first, how the oceanic mixed layer absorbs any heat gain, by heating advected or upwelling water; second, where the heat gain comes from, or rather where it does not go, in places of low evaporation.

According to Equation 5.1, heat advection is proportional to mass transport. In the two mixed layers in contact, an important part of the mass transport is "Ekman transport," flow induced by the shear stress on the air-sea interface. When no other forces act on the air or water, the Coriolis force from earth rotation, acting on the induced flow, balances the shear stress. In this case, the mass transports along- and across-latitude (zonal and meridional transports, along x and y) in the two mixed layers are:

$$\begin{aligned} \rho V_E &= \pm\frac{\tau_{x0}}{f} \\ -\rho U_E &= \pm\frac{\tau_{y0}}{f} \end{aligned} \tag{5.3}$$

where τ_{x0}, τ_{y0} are the shear stress components at the sea surface, and $f = 2\,\Omega\sin(\phi)$ is the Coriolis parameter, with Ω the earth's angular speed, ϕ the latitude, the latter positive in the Northern, negative in the Southern Hemisphere (see e.g., Gill, 1982). The positive sign applies on the air side, the negative sign on the water side. $U_E = uh$, $V_E = vh$ are Ekman (volume) transports in mixed layers of depth h. Density differs from air side to water side and so does the Ekman volume transport, but the surface shear stress and the Coriolis parameter are common, and with them the magnitude (but not the sign) of the mass transports.

Zonal and meridional heat advection in the water-side mixed layer are now $c_{pw}\rho_w U_w \partial\theta_w/\partial x$ and $c_{pw}\rho_w V_w \partial\theta_w/\partial y$, with $\rho_w U_w$, $\rho_w V_w$ the sum of Ekman mass transports, and mass transports due to other causes. Most important of the latter are mass transports sustained by a balance of Coriolis force and pressure gradients, known as geostrophic transports. In many important cases, Ekman transports are (nearly) meridional while geostrophic transports are (nearly) zonal. In these cases, they do not interfere much with each other. Examples are the low latitude easterly winds that exert a nearly zonal shear stress westward on the oceanic mixed layer. In the northern hemisphere, the associated Ekman transport is northward, in the southern hemisphere southward. The midlatitude westerlies drive opposite oceanic Ekman transports. Because the ocean exerts an equal and opposite shear force on the air, atmospheric Ekman transports are the opposites of the oceanic ones.

To see how winds and their Ekman transports shed light on the global heat gain map, consider the Southern Ocean at the latitude of the strongest westerly winds, between 40°S to 60°S latitude. The Coriolis parameter is here negative, close to $f = -1.0 \times 10^{-4}$ s^{-1}, the wind stress 0.1 Pa or higher. The Ekman volume transport on the water side, V_{wE}, is then positive or northward and close to 1 m^2 s^{-1} or higher. Taking this to be the total volume transport, and estimating the northward temperature gradient in the South Atlantic from Figure 5.3 at about 5×10^{-6} K m^{-1}, we find the meridional heat advection $c_{pw}\rho_w V_w \partial\theta_w/\partial y$ to be about 20 W m^{-2}, pretty much what Bunker's map shows (Figure 5.2). Upwelling and zonal heat advection are apparently minor influences. Note that our argument here rests on Equation 5.1, on how the ocean disposes of its heat income, while Bunker's estimate of net heat gain came from Equation 5.2, from the balance of irradiation and heat transfer to the atmosphere.

It is also interesting to note that, in the same location in the South Atlantic, the atmospheric Ekman volume transport is southward at the fairly substantial rate of some 800 m^2 s^{-1}. South of the strongest westerlies this southward volume transport diminishes, so that its convergence has to ascend. Over a presumed effective ascent region 1000 km wide, the ascent velocity would be about 0.8×10^{-3} m s^{-1}, high enough to significantly reduce entrainment of dry air in the atmospheric mixed layer. As we have seen in Chapter 3, dry air supply is critical for evaporation, in this location already limited by low temperatures. Low LE in Equation 5.2 goes a long way toward explaining moderate oceanic heat gain at such high latitudes, with the large radiant heat gain in summer not used up to support evaporation.

The high latitude North Pacific is another location with counterintuitive heat gain (Figure 5.5). In this location, the positive heat gain region roughly coincides with the subpolar gyre, where surface currents circulate in the cyclonic (anticlockwise) direction, driven by the polar easterlies and lower latitude westerlies. Oceanic Ekman transports are in such regions divergent, southward under the westerlies, northward under the easterlies. Divergent oceanic Ekman transport, unless canceled by opposite geostrophic transport, implies upwelling into the mixed layer, in the case of the subpolar gyre at a typical velocity of $w_e = 10^{-6}$ m s^{-1}. The corresponding upward temperature advection would be $w_e \Delta\theta$, with $\Delta\theta$ the excess temperature of the mixed layer over the substratum. The observed value of $\Delta\theta$ in the subpolar gyre is about 4 K, so that $-\overline{w'\theta'}(h) = w_e \Delta\theta = 4 \times 10^{-6}$ K m s^{-1} would balance a net heat gain of 16 W m^{-2}, close to the heat gain inferred from the surface heat balance (Figure 5.5). The oceanic mixed layer apparently uses most of the heat gain to heat upwelling cold water.

In the overlying atmospheric mixed layer, Ekman transports are opposite to those in the ocean, and are therefore convergent. Again, as in the Southern Ocean, the associated mean ascent of air suppresses entrainment of dry air, presumably reducing evaporation, and allowing the oceanic mixed layer to retain much of the heat gained from irradiance.

A further location of counterintuitive heat loss is the tropical South Indian Ocean, at south latitudes higher than 10°S (see Figure 5.6). The 10°S latitude circle roughly coincides with the strongest easterlies and corresponding southward oceanic Ekman transport in this region. South of that zero heat gain contour, low intensity heat loss, of order 20 W m^{-2}, arises from the difference between fairly high irradiance, about 100 W m^{-2}, and somewhat higher latent heat loss of around 120 W m^{-2}. Hastenrath and Lamb (1980) point this out and fill in some details: The high evaporation rate brings about the cloudiness that reduces irradiance, while subsidence of dry air reduces low level humidity and supports evaporation. Oceanic Ekman transport brings warm mixed layer water south across the zero heat gain contour, while the corresponding northward atmospheric Ekman transport takes away the subsiding air, enriched in vapor by evaporation, and supplies the water vapor for the monsoon, as already mentioned.

5.2 Oceanic Heat Transports

The large annual mean heat losses in the high latitude North Atlantic totaling some 0.7 PW, or comparable losses in the northwest Pacific, do not result in steadily dropping surface temperatures, and thus must be countered by oceanic heat transport from regions of heat gain. The required transports are large, far beyond the capacity of mixed layer oceanic Ekman transports. Instead, major boundary currents such as the Gulf Stream are the conveyors of warm water from low to high latitude, while other currents return the mass transports of the boundary currents (the ocean cannot pile up water anywhere) as they convey colder waters back. The mass transport exchange involves also deep waters of three oceans, not only mixed layers, the return transports in particular typically taking place outside the WWS. The large oceanic heat transports are key links in the global climate, and much effort has gone in recent years into determining their magnitude and global distribution.

5.2.1 Direct Estimates of Heat Transports

A "direct" method of gauging oceanic heat transport, by observing velocities and temperatures with the aid of sufficiently dense instrumentation in an oceanic cross-section between two coasts, is not a practical possibility today. A combination of observation and simple theory has, however, yielded fairly good estimates in data-rich regions.

Hall and Bryden (1982) adopted this approach to estimate oceanic heat transport across latitude circle 25°N in the North Atlantic. Integrated temperature advection across the section of an ocean along a latitude circle yields the net meridional heat transport:

$$Q = \iint \rho c_p \theta v \, dz \, dx \tag{5.4}$$

where Q is heat transport in W, positive northward, θ is potential temperature, v meridional velocity, and the integration extends over the depth and width of the ocean between two continental boundaries. The mass transport in the same section is:

$$M = \iint \rho v \, dz \, dx \tag{5.5}$$

where M is nonzero, $Q = \overline{c_p \theta} M$, weighted average heat capacity times the mass transport. With M vanishing, we can write $M_N + M_S = 0$, where M_N is the integral in Equation 5.5 evaluated over the portion of the section where $v > 0$, M_S over the rest of the section. If the corresponding weighted average temperatures are $\overline{\theta_N}$ and $\overline{\theta_S}$, then the heat transport is $Q = c_p(\overline{\theta_N} - \overline{\theta_S})M_N$. In the Atlantic, which is almost closed in the north (only about 1.5×10^6 m^3 s^{-1} enters from the Arctic Ocean, as Hall and Bryden 1982, note), supposing M zero causes little error, while the northward mass transport and the weighted average temperatures may be estimated fairly accurately.

Hall and Bryden (1982) carried out this scheme by splitting the heat transport into three components: (1) the well-explored transport across the Florida Straits; (2) the wind-driven Ekman transport; and (3) the geostrophic transport, the latter two over the rest of the section east of the Straits. The geostrophic transport again consists of two parts: (1) the "baroclinic" part, with the meridional velocity calculated from the density distribution, and (2) the barotropic part, calculated from mass balance and observed temperatures. The division of the mass transport in this manner facilitates estimation of the component mean temperatures. Hall and Bryden arrived at a heat transport estimate of $Q = 1.2 \times 10^{15}$ W, or 1.2 PW, with a possible error of 0.3 PW. They also showed that the Reynolds flux of heat, $\overline{\theta' v'}$, makes a negligible contribution; this should be true in the ocean in most locations, except where eddy activity is as vigorous as it is in the Drake Passage of the Southern Ocean (Bryden, 1979).

A later calculation by Fillenbaum et al. (1997) of oceanic heat transport across essentially the same latitude circle took advantage of further direct observations, of boundary currents over the Bahama Escarpment that forms the western boundary of the Atlantic east of the Florida Straits. There are two well-defined boundary currents here – one a warm surface current flowing northward, and one deep current centered at 2000 m or so, both within well defined temperature limits. Their result was $Q = 1.44 \pm 0.33$ PW, higher than Hall and Bryden's (1982), although within the error bars of either estimate. Recent studies of the eastern boundary current near the latitude of this section may further improve this estimate.

Also from observation aided by theory, at the higher latitude of 55°N, between Greenland and Ireland in the Atlantic, Bacon (1997) found a northward heat transport of 0.28 ± 0.06 PW.

Similar estimates based on observation cum theory are available at a few locations also in the North Pacific. They are all essentially spot readings from which a global picture of oceanic transports would be difficult to piece together.

5.2.2 Syntheses of Meteorological Data

A global view of oceanic heat transports first emerged from the integration of the annual mean net oceanic heat gain. Although as we have seen the data on heat gain are reasonably reliable only at certain latitudes and not in all oceans, Hastenrath (1980, 1982) has nevertheless succeeded in putting together from them a remarkable global schema of oceanic heat transports. He supposed all three oceans – Atlantic, Indian, Pacific – closed in the north, so that the meridional heat transport across latitude y is $Q(y)$, as calculated from Equation 5.4. Restricting attention to the long-term mean value of this transport, heat storage $\partial\theta/\partial t$ integrated over the ocean mass north of latitude y is legitimately neglected. The cross-latitude oceanic heat transport, $Q(y)$, then has to balance the heat loss integrated over the area of the ocean lying to the north of latitude y:

$$Q(y) + H(y) = 0 \tag{5.6}$$

where:

$$H(y) = \iint A_w \, dx \, dy$$

with A_w the net oceanic heat gain from Equation 5.2 (i.e., radiant heat gain less latent and sensible heat transfer to the atmosphere), on a long-term annual average, in W m^{-2}. Integrating the heat loss or gain shown by maps similar to Figure 5.2 at different latitudes, a global schema of oceanic heat transport emerges. Figure 5.7 shows the schema so obtained by Hastenrath (1982). This differs from the original version in Hastenrath (1980), making use of later data, with northward heat transport in the Atlantic reduced by 0.4 PW at latitudes south of 25°N. The magnitude of the correction is similar to the uncertainty of direct determinations above.

Other workers followed in Hastenrath's footsteps in calculations of oceanic heat transports from data on net heat gain, using different, improved, or extended databases. These later estimates of oceanic heat transport are nevertheless subject to similarly large uncertainty. Hsiung et al. (1989) give detailed monthly estimates of heat transport, adding up to an Atlantic yearly total at 25°N of about 1.0 PW, and a global total (Atlantic and Pacific) of 1.75 PW, versus 2.2 PW according to Hastenrath (1982). Talley (1984) concluded, however, that "confident determination of even the sign of the heat transport in the North Pacific is not possible." Helped by considerably more data, Moisan and Niiler (1998) put the northward heat transport in the Pacific at 25°N at 0.3 ± 0.15 PW, for a global total (with Fillenbaum et al.'s 1997 estimate for the Atlantic) identical to Hsiung et al.'s (1989). According to Moisan and Niiler (1998) weak southward transport at 40°N conveys equatorward the moderate heat gain of the high latitude Pacific cyclonic gyre. Between this latitude circle and 5°N the North Pacific loses heat. Their northward cross-equatorial transport estimate of 0.4 PW is remarkable in its similarity to the Atlantic's 0.6 PW, given the very different transport distribution in the two oceans further north.

Heat transport estimates for the South Pacific scatter particularly widely. The cross-equatorial northward transport estimate of 0.4 PW from Moisan and Niiler (1998) agrees with Hsiung's (1985), but it conflicts with Hastenrath's (1982) (Figure 5.7). Hsiung (1985) puts the peak heat transport in the South Pacific at 0.5 PW, southward, at latitude 50°S. Hastenrath's (1982) map shows a peak southward transport of 2.07 PW at 20°S, and as much as 1.19 PW southward, going into the Circumpolar Current, at 60°S.

The Circumpolar Current girdles the globe and transports about 130×10^6 m^3 s^{-1} of fairly cold water, connecting all three oceans: Atlantic, Indian and Pacific. Georgi and Toole (1982) put together the available information on mass transport and temperature of the waters in this current, and calculated the changes in the heat transport as the Circumpolar Current passes each ocean. The changes in heat transport may be taken as estimates of southward heat transport in each ocean into the Circumpolar Current, roughly at 60°S. The results were:

Atlantic	Indian	Pacific
−0.335	0.648	−0.317

all southward transports in petawatts, with error bars about as large as the transports, except in the Indian Ocean, where at least the sign of the transport is reliable. Hastenrath's (1982) value of 1.19 PW at this latitude in the South Pacific certainly seems incompatible with Circumpolar Current heat transports.

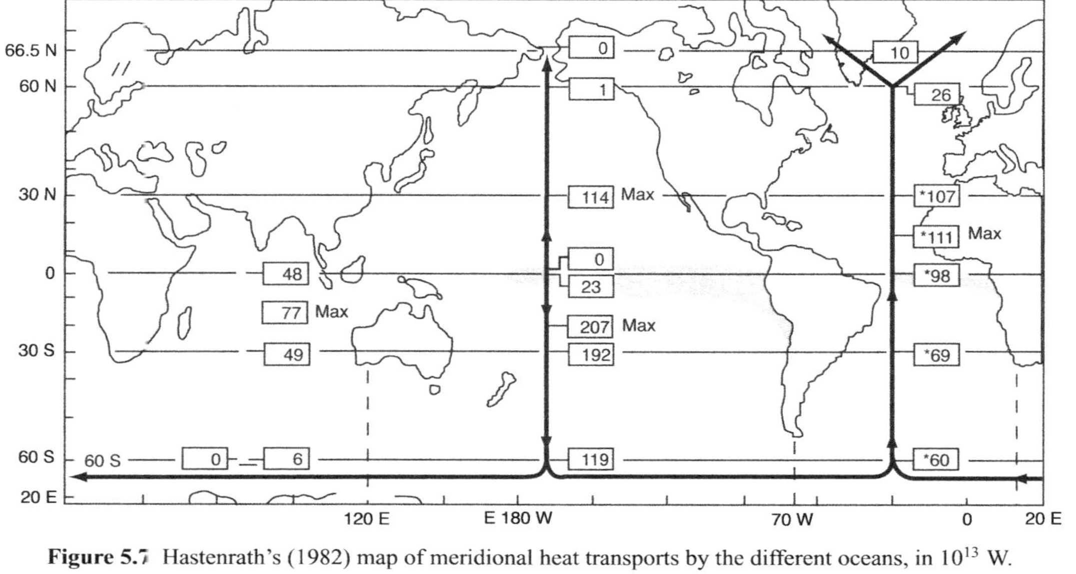

Figure 5.7 Hastenrath's (1982) map of meridional heat transports by the different oceans, in 10^{13} W.

Even though the various estimates scatter in all oceans, they agree on one point: that the Atlantic transports substantial quantities of heat northward, from the Southern Ocean to the northern North Atlantic Seas. Transports at the southern end of this conduit particularly require confirmation, because of the scarcity of data in the Southern Hemisphere, and because the estimated transports have the cold southern seas exporting heat. A thorough study of circulation and heat transport in the South Atlantic led Rintoul (1991) to conclude that at 32°S the Atlantic transports 0.25 ± 0.12 PW heat northward. On the other hand, Saunders and King (1995) gave the result of a recent direct determination of heat transport across a section from 45°S in the west to 30°S in the east as 0.5 ± 0.1 PW. A dense network of hydrographic stations plus current measurements with acoustic doppler meters underlie this estimate. At the lower end of the error bars, at 0.4 PW, this just avoids conflicting with Rintoul's result, and easily fits within the wide error bars of Georgi and Tool's (1982) estimate. The same reduced heat transport is also compatible with estimates further north: at the Equator, the northward transport is 0.6 PW, according to Hsiung (1985) and Hsiung et al. (1989) and 0.54 PW according to Bunker (1981). Hastenrath's figure of 0.6 PW at 60°S is on the high end of Georgi and Tool's (1982) error bars. Adjusting it to Georgi and Tool's mean estimate of 0.35 PW , and leaving the change from 60°S to the Equator the same as in Hastenrath's map, we get a result similar to Hsiung's (1985) at the Equator. At 25°N , as we have already discussed, direct oceanographic estimates of heat flux by Bryden and Hall (1980, 1982) yield a northward heat transport of about 1.1 PW, while a recent calculation of the same kind by Fillenbaum et al. (1997) shows 1.4 PW. These do not differ much from Hastenrath's or Hsiung's estimates and are compatible with the Southern Hemisphere estimates.

A welcome check on the order of magnitude of oceanic heat transport estimates comes from satellite observations and meteorological data, through a procedure pioneered by Oort and Vonder Haar (1976). The satellite yields the net radiant heat gain or loss of the globe north of latitude y, $H(y)$ in a heat balance similar to Equation 5.6. $Q(y)$ in that heat balance is the sum of global meridional heat transports by the ocean and the atmosphere put together. Subtracting meridional heat transport by the atmosphere then yields an estimate of the global oceanic heat transport.

Because the earth does not gain or lose heat on average, pole-to-pole integration of the satellite estimate of heat gain should yield a vanishingly small residue. Actually, residues are of the order of 15% of maximum global heat transports, supplying an estimate of errors in $H(y)$. The global meteorological network provides the database for calculating atmospheric heat transports, clearly again subject to some error. The errors of the two estimates compound by the differencing, so that the results are good only as to order of magnitude. The poleward oceanic transports so determined turn out to be higher by some 50% than the results of the oceanic heat gain integrations, but they also track the latitudinal distribution of the latter, seen in Figure 5.8, from Hsiung (1985). The global total of oceanic heat transports computed as a residual of satellite derived total and atmospheric transports by Carissimo et al. (1985) varies with latitude very similarly to Hsiung's global total. While the actual magnitude of the global oceanic transports remains somewhat

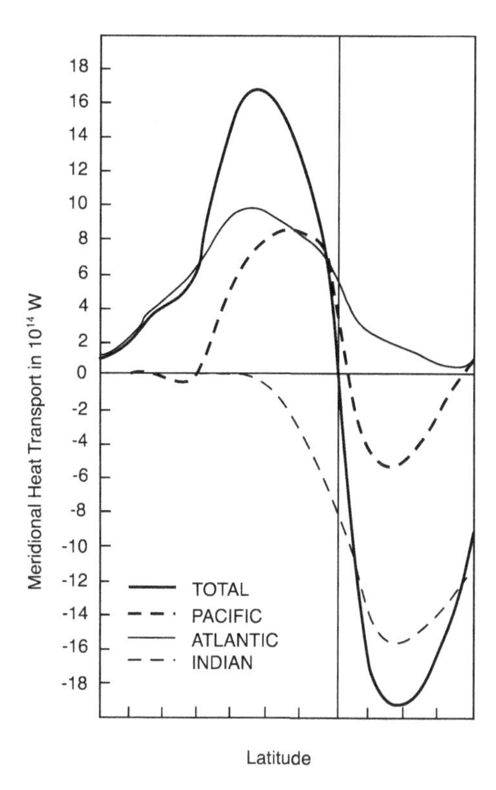

Figure 5.8 Hsiung's (1985) diagram of meridional heat transports by individual oceans, and the global total.

uncertain, satellite data confirm that these transports are well in the several petawatt range.

Large error bars notwithstanding, oceanic heat transports in the petawatt range are comparable to poleward atmospheric heat transports and are thus major players in the heat balance of higher latitudes. A dramatic confirmation of their role is the global distribution of the atmosphere's net heating rate in January, as a mass-weighted vertical average (see e.g., Johnson, 1984). Over the Kuroshio in the Pacific, and the Gulf Stream in the Atlantic, the heating rates are of the same magnitude, about 2 K day^{-1}, as over the major tropical heat sources (because of superclusters of hot towers) in the western Pacific warm pool and equatorial South America. In striking contrast, no comparable midlatitude concentrations of heating exist in southern winter in the Southern Hemisphere.

To sum up, the WWS of the ocean is responsible for large poleward heat transports and plays an important role in the global climate by redistributing the solar heat income. An intriguing aspect of this redistribution is that, as Bunker pointed out, the cold South Atlantic gains heat from solar radiation, while the warm North Atlantic loses it to the atmosphere. What is peculiar is that the Atlantic transfers heat from a cold body to a warm one, much like a heat pump. The second law of thermodynamics says that this requires input of mechanical energy. The details of this large-scale heat transfer process, its mechanism and energy sources, have been the subject of much recent observational work, speculation, and modeling effort. The next few sections examine the process in detail and analyze its thermodynamics.

5.3 Warm to Cold Water Conversion in the North Atlantic

The high net heat loss of the ocean in the northern North Atlantic Seas, looked at from a Lagrangian point of view, implies that individual surface water parcels cool as they make their way northward by a circuitous route, ultimately to escape from the WWS. Surface cooling reaches only the water-side mixed layer, in the heat budget of which cooling appears as negative heat gain $A_w < 0$. At a fixed location, this is made up from a combination of stored heat loss, $c_p\rho\partial\theta/\partial t$, advection of warm water to a colder location, $c_p\rho(u\partial\theta/\partial x + v\partial\theta/\partial y)$, and upwelling, $-c_p\rho\overline{w'\theta'}(h)$ (Equation 5.1 above). The former two together give the rate of change of heat content following the horizontal motion of a parcel in the mixed layer. Upwelling of colder water is a cause of temperature drop, acting much as surface cooling, and is in fact often a consequence of surface cooling.

Surface cooling generates convective turbulence, which entrains cooler and heavier fluid from below. According to Carson's Law (see Chapter 3), the rate of cooling (tantamount to buoyancy loss) on account of entrainment is proportional to the surface buoyancy flux, and hence to the rate of surface cooling. The rate of heat loss following the horizontal motion of a parcel in the mixed layer therefore equals the rate of surface cooling, plus a percentage from upwelling, according to empirical data approximately 25%.

Sustained upwelling under the high winter cooling rates in the northern seas eventually erodes the thermocline. Deep convection then sets in at a few locations and entrains still lower strata into the now very deep mixed layer. The net result is conversion of waters of the WWS into a cold water mass of very different temperature and salinity characteristics, which eventually sinks to a deep level, sometimes to the bottom, there to spread out horizontally in a pycnostad and flow southward.

In a series of important contributions, McCartney and Talley (1982, 1984; see also Talley and McCartney 1982) identified the pathways of water parcels to sites of deep convection in the North Atlantic, and quantified the magnitude of WWS mass and heat loss. The principal poleward conduit is the North Atlantic Current, passing west of Ireland (Figure 5.9) before it splits into an eastern branch ending in the Norwegian Sea, and a western branch crossing the Mid-Atlantic Ridge south of Iceland. It continues on to the Irminger Sea east of Greenland, and then on to the Labrador Sea, which is a major deep convection site discussed in Chapter 4. The western branch is part of the cyclonic circulation of the North Atlantic. Late winter surface temperatures along this branch reflect the heat loss of the mixed layer following the circulation (Figure 5.10), while the depth of the mixed layer increases from 200 m in the mid-Atlantic to 400 m at the eastern edge of the Labrador Sea, suddenly diving there to 1000 m, (Figure 5.11). Downward transfer of waters via "chimneys" (oceanic hot towers) takes place at several sites in the northern seas, not only in the Labrador Sea. They are indicated in Figure 5.9 by curly ends of the warmwater paths. Cold water currents taking away the sinking water masses are seen as white arrows in this figure, surface current full lines, deep currents dashed, all eventually crossing the 50°N latitude circle southward, off of Newfoundland.

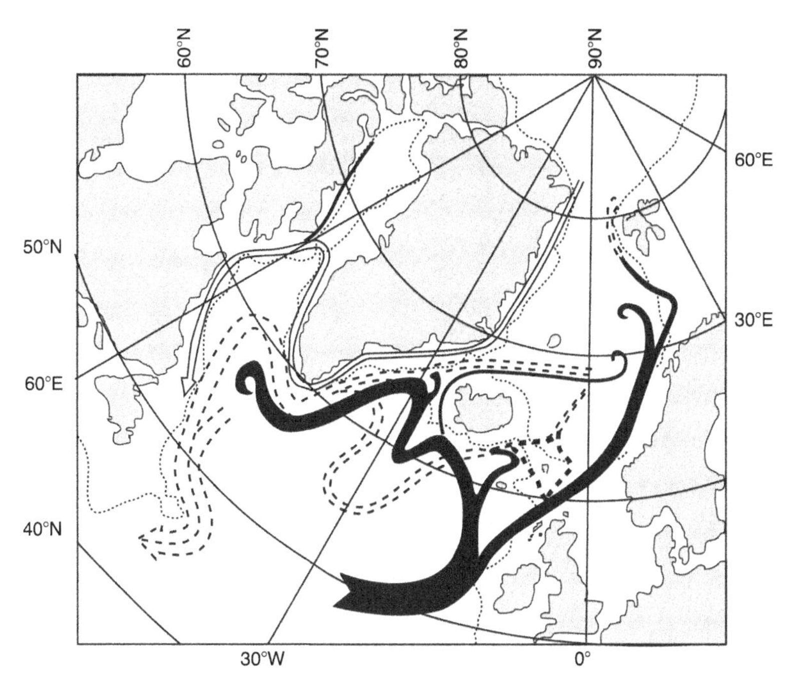

Figure 5.9 Warm water transports in the North Atlantic toward sites of deep convection shown in black, the curly ends indicating sinking and conversion of warm to cold water. Full lines ending in arrows show boundary currents; dotted lines show deep currents. From McCartney and Talley (1984).

The Labrador Sea Water (LSW) is one of the cold water masses formed in the North Atlantic. Others come from chimneys in the Norwegian Sea and around Iceland. According to Schmitz and McCartney (1993), the total rate of warm water conversion into these deep and cold waters is 13 Sv (Sv = sverdrup = 10^6 m^3 s^{-1}), out of which 7 Sv ends up as LSW, 3 Sv comes from the Norwegian Sea, the rest from around Iceland. Although off Ireland the warm water supply has a temperature of 11.5°C, it all originates in surface waters of the subtropical gyre, at 18°C and up. Thus, the total heat loss of the WWS associated with the conversion of warm to cold water is of the order of 200 K Sv ("degree-sverdrup"), which translates into 0.8 PW, or most of the northward oceanic heat transport.

5.3.1 Cold to Warm Water Conversion

The loss of 13 Sv through warm to cold water conversion would soon deplete the WWS of its waters – on a long-term average, it has to be replaced at the same rate. Moreover, the replacement process must start with waters colder than 3.5°C, the end product of the warm to cold water conversion, for otherwise this cold water mass would continuously accumulate in the world ocean. Indeed, because the distribution of water masses remains essentially the same over long period (the distribution seen in Figure 5.1), each temperature class must recover as much mass as it loses.

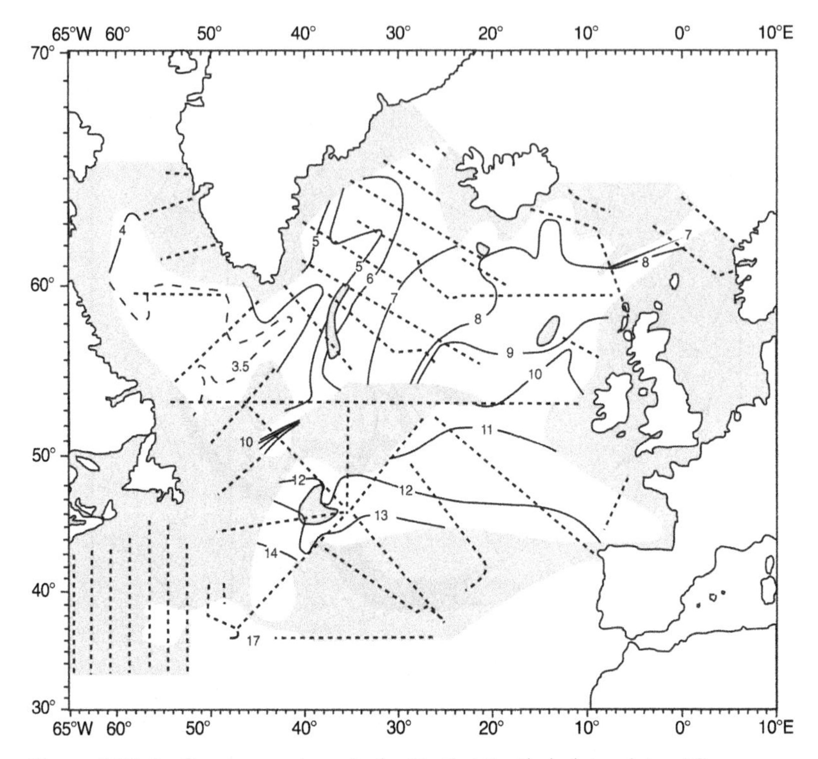

Figure 5.10 Surface temperatures in the North Atlantic in late winter, °C, contoured only where the mixed layer depth exceeds 200 m. Dropping temperatures follow the path of the cyclonic circulation in this region. From McCartney and Talley (1982).

Ocean waters can only be heated at significant rates while on the sea surface. One key question is, where can the temperature of, say, 2°C water be raised to 7°C, the latter being the low temperature limit of what oceanographers call "thermocline" waters, or of the waters of the WWS, from our point of view here. To accomplish this feat on 13 Sv of cold water, a heat gain of 65 K Sv $=$ 0.27 PW is needed. Scrutinizing the heat gain maps of Figures 5.2, 5.4, 5.5, and 5.6, together with the surface temperatures in Figure 5.3, we soon discover that the only conjunction of cold surface temperature and significant net heat gain is in the South Atlantic, where Bunker's map shows heat gain of order 25 W m^{-2}. In discussion of this above, we have attributed the gain to equatorward Ekman transport across the strong temperature gradient, the Ekman transport driven by the strong westerlies of the Southern Ocean.

An order of magnitude estimate of the total heat gain in the South Atlantic, from Bunker's map, is 0.10 PW, rather less than the total needed. The heat gain maps of the other oceans do not extend to the high southern latitudes to show similar heat gain over cold waters. There is good reason to suppose, however, that the factors influencing oceanic heat gain are much the same over the high latitude Indian and Pacific Oceans as over the Atlantic. The temperature distributions in the top 1000 m of the three oceans, between latitudes 40°S and 60°S (seen e.g., in meridional cross

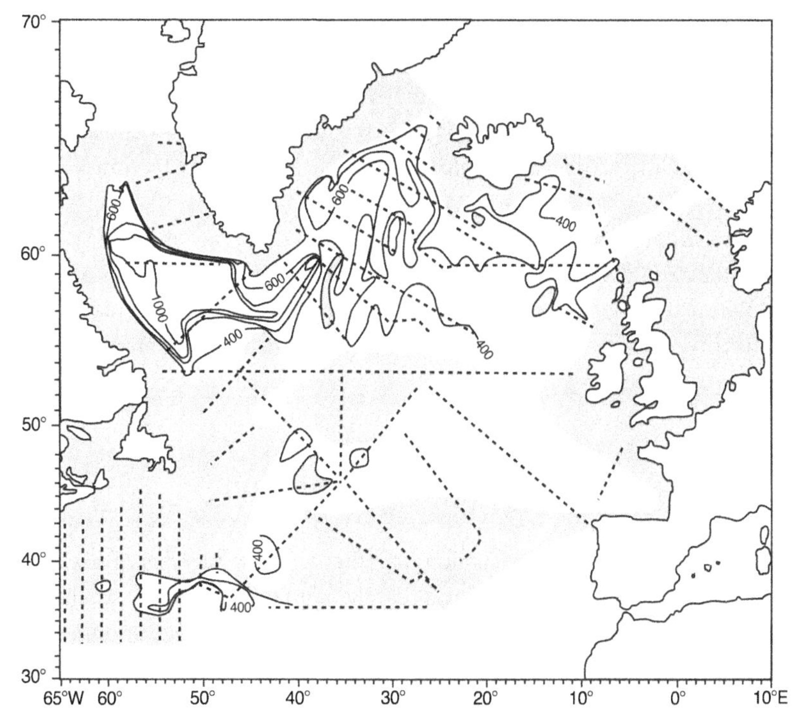

Figure 5.11 Mixed layer depth in the region of the previous figure, dropping suddenly to 1000 m in the Labrador Sea, at the end of the warm water path. From McCartney and Talley (1982).

sections of the GEOSECS Atlases, 1976, 1981, 1982), are very similar, as may be already inferred from the surface temperatures in Figure 5.3. This, together with the similar distribution of zonal mean winds in this region (mean westerly winds of speeds higher than $6\,\mathrm{m\,s^{-1}}$) imply equatorward Ekman transports higher than $1\,\mathrm{m^2\,s^{-1}}$. The strong winds extend from about 40°S to about 60°S in all three oceans. Our calculation above showed that this combination of winds and surface temperatures supports net oceanic heat gain to the tune of about $20\,\mathrm{W\,m^{-2}}$. Presumably then, all three sections of the Southern Ocean – Atlantic, Indian, and Pacific – contribute more or less equally to the conversion of cold to warm water.

Westerlies are at their peak strength at about 50°S. Southward of this latitude circle Ekman transport is divergent and induces upwelling; northward the convergence of Ekman transport causes subduction into the thermocline. The further fate of these subducted waters, by this time firmly in the WWS, depends on the complex currents of the thermocline, in all three oceans.

A final important point concerns the mass balance of the ocean's cold waters. The equatorward Ekman transport in the Southern Ocean originates in cold surface waters of between 2 and 3°C, at the surface outcropping of massive deep layers that further north are about 1000 m thick, with their centers at about 2 km depth, and extend over the entire South and North Atlantic, Indian, and Pacific Oceans. New batches of these waters are "formed" (in oceanographic jargon) from warmer waters in the North Atlantic at sites of deep convection. Only in the Southern Ocean are these waters exposed again to the atmosphere so that they may be "unformed," or converted into thermocline water. It is reasonable to conclude that the excess mass deposited in deep, cold layers in the North Atlantic is removed from the same layers by upwelling in the Southern Ocean.

This is not to imply that the Southern Ocean reheats the same "marked" water mass deposited in the deep waters of the North Atlantic. The cold deep waters of the world ocean circulate, exchanging mass between the different basins at rates greatly in excess of what is involved in the warm to cold water conversion and the corresponding cold to warm water reconversion.

5.4 The Ocean's Overturning Circulation

The evidence surveyed in the last section, together with the physics of oceanic deep convection discussed in Chapter 4, shows that the high latitude North Atlantic WWS imports warm water from lower latitudes, to lose it through surface cooling, sinking, and spreading out at depth. Tracing of water mass types and some direct observations of deep western boundary currents then show that the North Atlantic exports a mixture of cold water types, collectively known as NADW, North Atlantic Deep Water, to the Southern Hemisphere. Mass transports of surface inflow and deep outflow must balance on a yearly average, at a rate of some 13 Sv, according to Schmitz and McCartney's (1993) careful recent summary. This inter-hemispheric exchange amounts to an "overturning" circulation of the ocean, in Toggweiler's (1994) graphic phrase, albeit again with the caveat that not the same parcels of water go up and down, south and north. Similar overturning occurs near Antarctica where Weddell Sea deepwater forms, but on a smaller scale, involving only 2–3 Sv, and localized to the cold waters of the Antarctic Seas. The large-scale overturning circulation of the North Atlantic exchanges cold bottom waters for waters of the WWS, and is thus responsible for northward heat transport in the petawatt range.

No deep convection occurs in the North Pacific. This is usually attributed to the relatively low salinity of the mixed layer, compared to the underlying strata, that makes the mixed layer too light to sink when cooled. It is also true, however, that only in the Bering Sea are surface temperatures cold enough to initiate deep convection in waters of the same high salinity as prevails in the North Atlantic. Yet the temperatures in the eastern North Pacific are cold enough to ensure that the ocean gains heat here, similarly to the South Atlantic. Specifically along the 7°C or 8°C isotherm, the eastern North Pacific gains heat at a modest rate, while the North Atlantic loses it at a very high rate. Cold and dry winds from Greenland and Labrador make the difference, versus winds advecting moist maritime air from one region of the Pacific to another.

The overturning circulation in the Atlantic is closed at the south end by upward entrainment, followed by heating of cold waters to thermocline temperatures in the

Southern Ocean. This presumably takes place in all three of the Atlantic, Indian, and Pacific Ocean segments of that ocean. How the upwelled water continues from there on, and what precisely drives this worldwide circulation, are two questions that recently spawned a spirited debate in the oceanographic literature.

Gordon (1986b) tackled the question of pathways of the overturning circulation, beginning with the southward trip. He accepted the concensus on Deep Western Boundary Currents of the North Atlantic taking NADW across the equator and down to the Southern Ocean. Once there, according to Gordon: "The NADW upwells within the world ocean, returning water to the upper layer within the Antarctic region." Schmitz and McCartney (1993) emphasize this: "NADW must be upwelled in the vicinity of the Antarctic Circumpolar Current, where the relevant property surfaces do outcrop." They also point out that nowhere else in the world ocean are waters colder than 3.5°C exposed to surface heating. We have noted this before, but it bears emphasizing that surface waters are this cold only in the Antarctic region, and that they are cold all around Antarctica, over 360 degrees of longitude. Westerly winds over the Circumpolar Current similarly girdle the globe, so that upwelling, surface heating and warm water formation should occur in all sectors of the Southern Ocean, Indian, Pacific, and Atlantic.

Once converted into "thermocline water," that is water warmer than 7°C, Gordon (1986a) identified two alternative pathways of northward mass transport, involving also further heating. One pathway is via the Circumpolar Current, bringing thermocline water from the Pacific past Drake Passage into the South Atlantic. We may suppose this water further heated within the net heating zone in Bunker's map (Figure 5.2), before flowing northward in the Benguela Current. Gordon's pathway two has the Circumpolar Current transfer thermocline water to the Pacific, where it "upwells" and eventually flows westward via the Indonesian Throughflow into the Indian Ocean, then across that ocean into the Agulhas Current, then past the southern tip of Africa into the Benguela Current, and northward. One point in favor of this scenario is that it accounts for the observed increase of salinity through the addition of Indian Ocean water.

The worldwide range of scenario two, which Gordon (1986b) suggested was the more likely solution, inspired Broecker to name this the Great Ocean Conveyor, and to have a cartoon of it constructed by Monnier (see Broecker, 1991; Figure 5.12). This has attracted a great deal of attention. Broecker also suggested a mechanism for it: "The ocean's conveyor appears to be driven by the salt left behind as the result of water-vapor transport through the atmosphere from the Atlantic to the Pacific basin." Quite a flight of imagination. We will discuss the influence of salinity in a later section. A major weakness of the Great Ocean Conveyor illustration is the absence of round-the-globe upwelling in the Southern Ocean.

Recent work supports pathway one, with a few details filled in, not to the exclusion of pathway two, however. As we have already seen, Rintoul (1991) carefully analyzed available hydrographic data and other information on the heat and mass balance of the South Atlantic, between a zonal section at 32°S, a section separating the Weddell Sea region, and two sections from South Africa to Antarctica. His transport rates

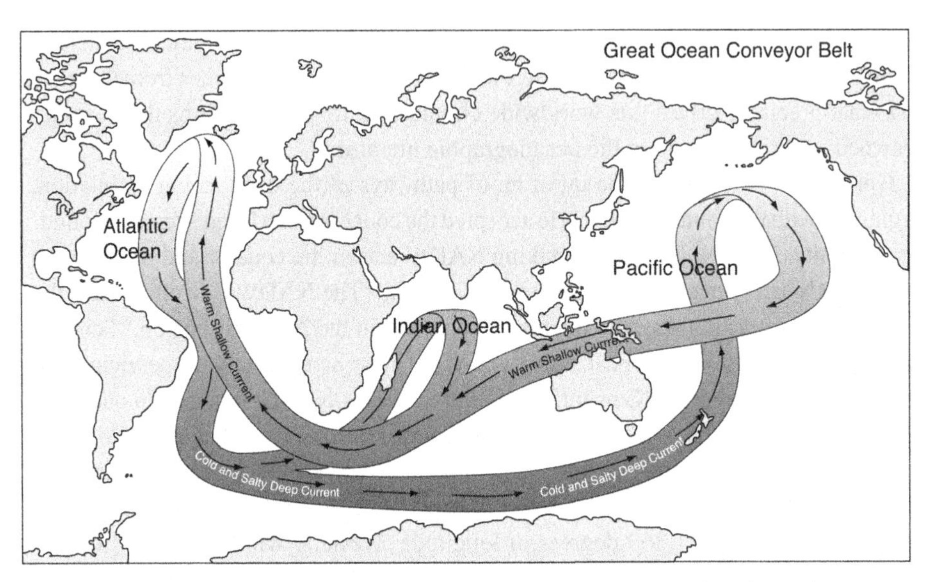

Figure 5.12 Broecker's celebrated cartoon of the ocean's overturning circulation.

in a surface layer, containing waters of the WWS or thermocline water, show 3 Sv coming from the Pacific through the Drake Passage. This increases to 8 Sv northward transport across 32°S, while another 3 Sv leaves eastward south of Africa. Most of the thermocline layer transport across 32°S should thus originate from warming in the South Atlantic, the positive heat gain region of Bunker's map (Figure 5.2), with about 40% coming from the South Pacific via the Circumpolar Currrent. Bunker's map shows only about half as much heat gain as this mass balance would imply. Higher northward heat transport across 32°S, suggested by the direct determination method of Saunders and King (1995), would be even harder to reconcile with either the known heat import via the Circumpolar Current, or with Bunker's map.

Gordon et al. (1992a) have discussed further the question of Indian Ocean contribution to the overturning circulation. Their cartoon of the circulation of waters warmer than 9°C shows waters of the South Atlantic subtropical gyre flowing eastward toward South Africa at the rate of 12 Sv. Half of this transport turns northward into the Benguela Current along the African coast, the other half intrudes into the Indian Ocean. 4 Sv returns after a sojourn there, augmented by Indian Ocean water transports of 3 Sv. Thus, most of the Indian-Atlantic Ocean circulation is mass exchange, with no net transfer. An important byproduct is, however, the transfer of salt from the Indian to the Atlantic Ocean, the main result of the substantial mass exchange. This leaves the origin of the northward heat transport in the South Atlantic unresolved: How does water upwelled in the Indian Ocean sector of the Southern Ocean get back to the South Atlantic?

As to the fate of water upwelled along the long stretch of the Pacific sector of the Southern Ocean, thermocline water transport through the Drake Passage into the Atlantic accounts perhaps for half, the other half presumably following Gordon's pathway two. Much uncertainty remains about the details of these worldwide balances.

The only reasonably solid point is that thermocline water must return through the South Atlantic and flow northward in the Benguela Current to replace the cold waters flowing southward.

On their pathway north in the South Atlantic, surface and thermocline waters of the Benguela Current gain heat as coastal upwelling exposes thermocline waters to the surface. These waters then cross in the South Equatorial Current westward to the coast of South America, where a fraction of them enters the equatorial circulation, the rest returning south via the Brazil Current.

5.4.1 The Role of the Tropical Atlantic

The surface and thermocline currents of the equatorial Atlantic play what is perhaps the decisive role in the ocean's overturning circulation. Figure 5.13 shows a schema of the main surface and thermocline currents, from Bourles et al. (1999). Stramma (1991) showed details of the broad and slow westward flow in the thermocline of the South Equatorial Current. Much of this flow ends up in the thermocline of the North Brazil Current, the major boundary current of the equatorial Atlantic. The NBC transports a total of some 24 Sv northwestward across the equator, 11 Sv of it in the surface layer, the rest in the thermocline (Schott et al., 1993).

The surface waters of the North Brazil Current retroflect well north of the equator, to form the North Equatorial Counter Current, but only in the months of northern summer, June to October inclusively. The seasonally varying surface currents of the entire equatorial Atlantic region, plotted by Richardson and Walsh (1986) from ship-drift

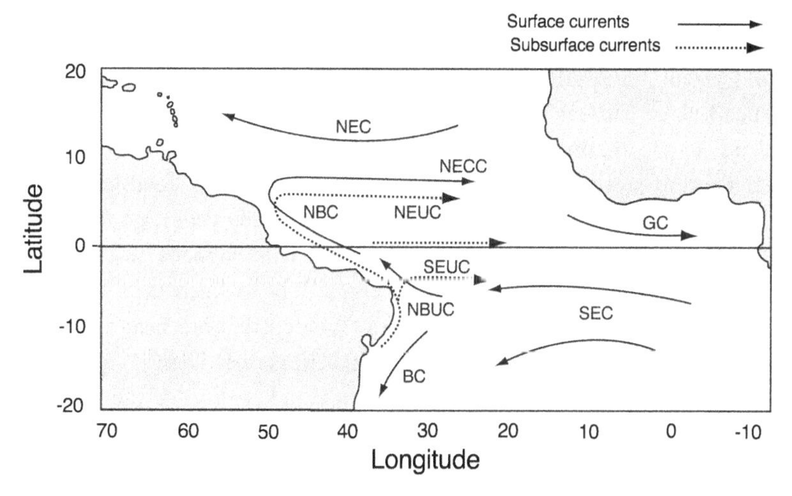

Figure 5.13 The complex currents of the equatorial Atlantic: full lines indicate surface currents. Mentioned in the text are the North Equatorial Current (NEC), North Equatorial CounterCurrent (NECC), North Brazil Current (NBC), South Equatorial Current (SEC), and the Brazil Current (BC). Dotted lines show thermocline currents; of those only the Equatorial UnderCurrent (EUC) plays a major role in cross-equatorial heat transport. Adapted from Bourles et al. (1999).

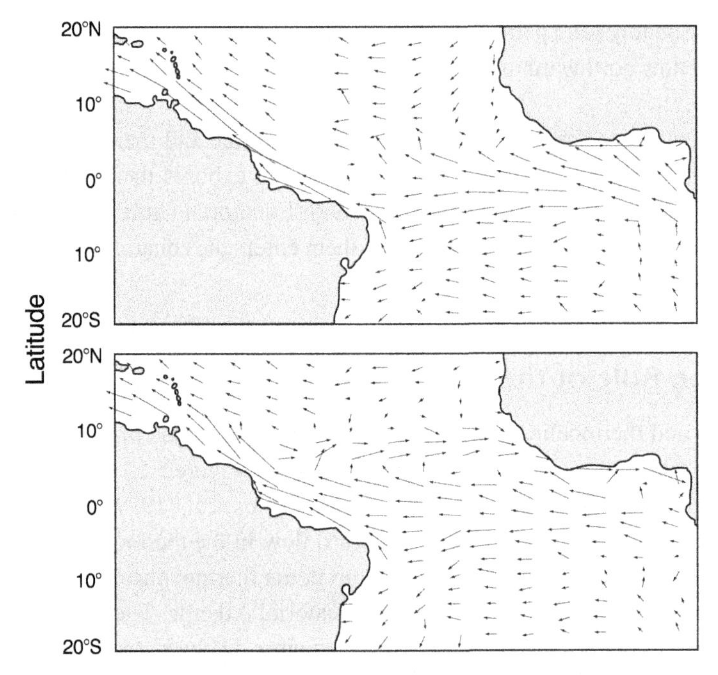

Figure 5.14 Winter (November to May) and summer (June to October) surface currents in the equatorial Atlantic. The longest arrows indicate speeds of about 1 m s^{-1}. From Richardson and Walsh (1994).

reports show this (Figure 5.14). Over the equator surface waters drift westward, driven by the easterly winds, at speeds of up to 0.5 m s^{-1}. In northern summer, together with the NBC and NECC, they form a clockwise gyre. North of that gyre a quiescent region separates the equatorial circulation from the broad sluggish flow of the North Equatorial Current, the southern leg of the subtropical gyre in the North Atlantic.

Underneath these surface currents the major drama of the equatorial circulation is played out in the thermocline waters: arriving via the Benguela Current-South Equatorial Current-North Brazil Current route, these waters feed the Equatorial UnderCurrent. Metcalf and Stalcup (1967) discovered many years ago that the waters of the EUC come from the Southern Hemisphere. This happens as the subsurface portion of the NBC retroflects eastward at about 3°N to form the Equatorial UnderCurrent (Flagg et al., 1986). Speeds in the Undercurrent are fast: a typical cross section of the current shows a peak velocity in excess of 0.8 m s^{-1} (Gouriou and Reverdin, 1992). The same authors also show a zonal section of the Undercurrent (Figure 5.15). At the eastern end of the section, the mixed layer shallows, allowing entrainment from below at the typical rate of $w_e = 2 \times 10^{-5}$ m s^{-1}. Taking place over a fairly large area, the total upward transfer of thermocline waters to the surface mixed layer is a key component of the equatorial circulation. Broecker et al. (1986) estimated the total upwelling rate at 17 Sv, from the distribution of a tracer (radiocarbon derived from nuclear bomb tests). Wunsch (1984) argued that this rate

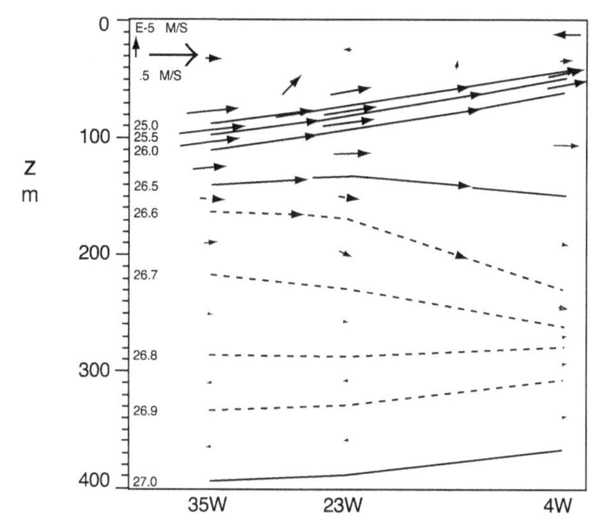

Figure 5.15 Mean circulation in the water column at the equator in a zonal (east-west) section. The thermocline tilt is exaggerated by more than a factor of 10^5. The Equatorial UnderCurrent is the dominant feature. From Gouriou and Reverdin (1992).

was too high to be reconciled with known features of the equatorial circulation. A later calculation by Gouriou and Reverdin (1992), based on a detailed observational study of the currents in the equatorial Atlantic, put the rate of equatorial upwelling at 10.6–14.7 Sv.

What makes equatorial upwelling in the Atlantic particularly important is that waters participating in it come from the Southern Hemisphere thermocline, and once in the surface layer, do not return. The absence of a southward escape route for the surface water is clear from the circulation maps in Figure 5.14. The implied northward mass transport via the thermocline-upwelling-surface current route must then be balanced by southward transport of other water types. As it happens, the southward transport of NADW at the known rate of about 13 Sv would approximately balance the mass budget. The northward transport of tropical surface water, at this rate and a temperature of up to 27°C, in exchange for the cold NADW is equivalent to a northward heat transport of some 1.3 PW. There is thus broad agreement on both the long term average heat and mass transports northward through the equatorial Atlantic, but the details remain sketchy. Further information comes from the heat budget and its changes with the seasons.

5.4.2 Heat Export from the Equatorial Atlantic

In two remarkable studies, Hastenrath and Merle (1986, 1987) documented the "annual march of heat storage and export," as well as the "annual cycle of subsurface thermal

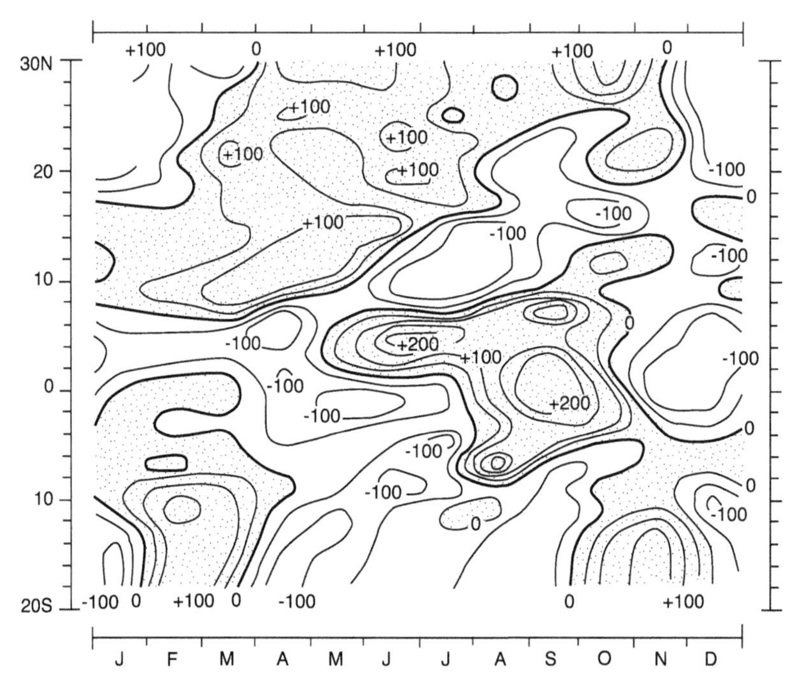

Figure 5.16 Heat storage in the equatorial Atlantic in the top 500 m of the water column, integrated between the coasts, month by month, in 2° latitude bins, expressed as heat stored in W m^{-2}. Stippling indicates heat gain, clear regions experience heat loss. From Hastenrath and Merle (1986).

structure" in the equatorial Atlantic. They have taken the net heat gain of the ocean, (A_w from Equation 5.2 above, with the mixed layer balance in Equation 5.1 extended to a 500 m deep layer) to consist of "storage" Q_t, or rate of change of heat content, and divergence of heat transport Q_v:

$$Q_t = \frac{\partial}{\partial t}\int_z^0 \rho c_p \theta \, dz = A_w - Q_v \tag{5.7}$$

where $z = 500$ m is a sufficiently deep level to encompass all significant change of heat content in the equatorial Atlantic. The net heat gain A_w is known from the balance of surface fluxes (Equation 5.2). Figure 5.16 shows the latitudinal and seasonal variation of the storage term in this heat balance. Conspicuous is the high heat storage in the equatorial belt in northern summer, May to November. In the same belt in the same months the transport divergence (determined by difference, from Equation 5.7) is high and negative, meaning advective import of heat.

High heat storage and transport divergence affect mainly the latitude belt 4°S to 6°N that acts as a seasonal warm water reservoir. Hastenrath and Merle (1986, 1987) listed the transport divergence Q_v integrated over 2° latitude and over the width of the equatorial Atlantic. Kapolnai and Csanady (1993) summed these over the warm water reservoir, to yield heat export from the reservoir in PW, with results shown in Figure 5.17. Between November and May, there is export of heat at a nearly

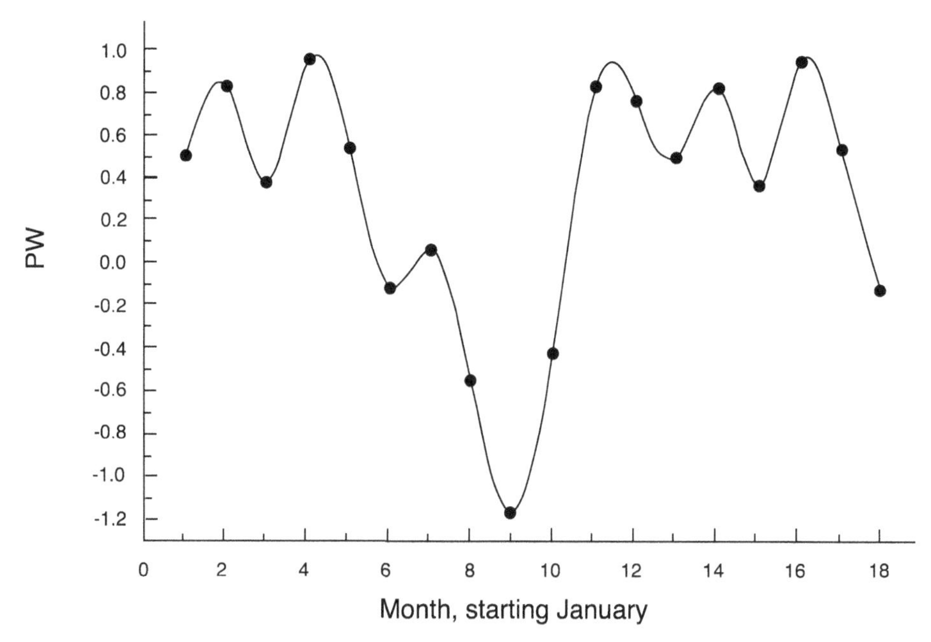

Figure 5.17 Heat export from the equatorial Atlantic "warm water reservoir" between 4°S and 6°N month by month, in PW. Calculated from data of Hastenrath and Merle (1986).

constant rate of 0.6 PW. In the remainder of the year imports scatter, reaching a peak of 1.2 PW.

Hastenrath and Merle (1986, 1987) show that the large storage and transport divergence changes follow the subsurface thermal structure, and that they are also a direct response to winds: "The basinwide subsurface thermal structure is dominated by the annual cycle of the surface wind field with extrema around April and August. As a result, two systems of annual cycle variation of mixed layer depth stand out. (i) Along the equator, the mixed layer depth increases from around April to about August, but (ii) In the North Equatorial Atlantic, a northward migration of a band of shallowest mixed layer is apparent from April to August, that follows the seasonal migration of the confluence zone between the northeast trades and the cross-equatorial airstreams from the Southern Hemisphere."

The "seasonal migration of the confluence zone" in the above quote refers to ITCZ movements. The easterlies are weakest just under the ITCZ, so that the northward oceanic Ekman transport is weak there, but it increases northward. There is then a divergent Ekman transport belt north of the ITCZ. The annual march of the ITCZ seen in Figure 5.18 (after Molinari et al., 1986), shows it to migrate roughly between the equator and 10°N. The divergent Ekman transport belt migrates with it. The ITCZ crosses the 6°N latitude circle in May on its northward trip and in November on the return trip. These months are also when heat export changes into import and back to export, when mixed layers start and stop deepening, and when heat storage begins and ends. Furthermore, the North Equatorial CounterCurrent develops and fades out at roughly

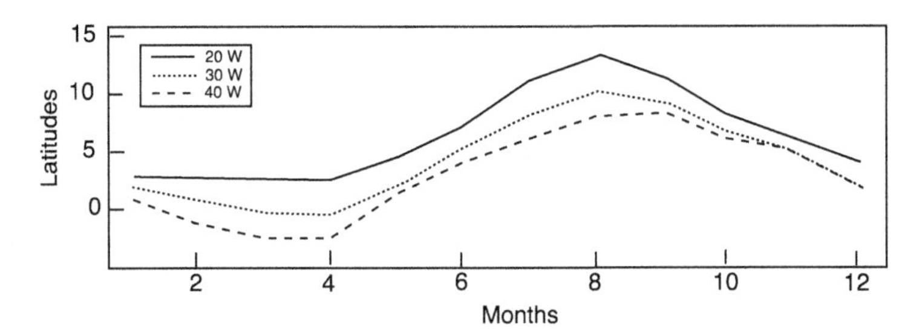

Figure 5.18 Annual north-south march of the ITCZ at different longitudes in the Atlantic. From Molinari et al. (1986).

the same time. Apparently, when the ITCZ is north of 6°N the clockwise gyre north of the equator is able to develop, and with it a deep warm pool, the warm water reservoir. As the ITCZ crosses this latitude circle southward, this circulation system collapses and the warm water reservoir is depleted under the divergent Ekman transport belt.

The net effect is the incorporation of equatorial surface water in the subtropical gyre of the North Atlantic. One may think of the ITCZ as a piston in a positive displacement pump, with a port in the south being uncovered when the piston moves north, allowing the cylinder to fill with warm water. As the piston moves back, it lays bare a northern port and closes the southern one, allowing the warm water to escape northward.

All in all, the peculiar seasonal changes of the wind field over the equatorial Atlantic effect a transfer of warm water northward, starting as thermocline water of the South Atlantic, picking up heat upon upwelling, and ending as tropical surface water in the North Equatorial Current or the Guyana Current. Both of these eventually end up in the Gulf Stream: This is the end of the return or heating leg of the overturning circulation. Once they are past the Straits of Florida, cooling starts in earnest along the North American coast.

5.5 What Drives the Overturning Circulation?

An alternative term for "overturning" circulation, favored by many authors, is "thermohaline" circulation. This at least suggests that temperature and salinity differences are somehow responsible for the phenomenon. There are more explicit statements: Gordon et al. (1992b) write: "There is a general appreciation today that the thermohaline circulation of the world ocean is to a large degree driven from the North Atlantic through the production of North Atlantic Deep Water (NADW). Sinking of surface water at a number of sites in the northern North Atlantic initiates an overturning cell on the meridional plane in which northward transport of upper ocean warm water is balanced by deep return flow of cold waters, imposing strong northward heat flux in both the North and South Atlantic." Also, the same authors: "The NADW-driven circulation is intimately linked to the large-scale hydrological cycle," a generalized and somewhat toned down version of Broecker's idea that the salt left behind by evaporation drives the overturning circulation.

These remarks imply that surface cooling in the North Atlantic is a source of mechanical energy powerful enough to sustain a meridional circulation extending almost pole to pole, crossing many latitude circles, and involving mass transports of order 10 Sv, as well as heat transports of the order of 1 PW from a cold to a warm ocean. Is this a realistic scenario? Suppose that ocean waters participating in the overturning circulation went through a thermodynamic cycle akin to the atmospheric hot tower cycle, and that this cycle delivered sufficient mechanical energy to counter bottom friction and other parasitic effects encountered along a return trip on the Great Ocean Conveyor. Such would be a heat engine cycle, requiring transfer of heat from a warm to a cold body, not the other way around, according to the second law of thermodynamics. The observed large northward heat transfer in the Atlantic conflicts with this scenario. One either has a heat engine thermodynamic cycle that produces mechanical energy, or a heat pump cycle that uses mechanical energy.

A major difficulty in identifying a thermodynamic cycle in the ocean, analogous to the atmospheric hot tower cycle, is that both heating and cooling in the ocean take place at the surface, at essentially constant (atmospheric) pressure. What surface pressure changes there are come from the deepening of a mixed layer, and are quite limited in magnitude, owing to the low thermal expansion coefficient of water, and a modest temperature range. Therefore, changes of entropy with temperature during heat gain are practically identical with changes during heat loss. There is then little mechanical energy gain or loss in any hypothesized cycle, as the area included in the *TS* diagram shrinks to the thickness of a line. We have noted this before in Chapter 4.

To effect the transfer of heat from the cold South Atlantic to the warm North Atlantic, another thermodynamic cycle would be required, in which cooling takes place at temperatures higher than heating. The same constraints on the temperature-entropy changes in the *TS* diagram as in the heat engine cycle would now imply little mechanical energy requirement for a lot of heat transfer. The wind is a known energy source so that the northward heat transport in the Atlantic is in principle easily reconciled with thermodynamics, as long as one blames the wind for it, rather than thermohaline driving. There is, however, little profit in pursuing the idea of an underlying thermodynamic cycle of either the heat pump or the heat engine kind. A more straightforward and satisfactory view of thermohaline energy supply to the deepwater leg of the overturning circulation emerges from analyzing the production of Convective Available Potential Energy, CAPE, by deep convection.

5.5.1 CAPE Produced by Deep Convection

The proximate cause of NADW formation, which supposedly inititates the overturning circulation, is surface heat flux and evaporation that make the surface layer heavier than layers below. The resulting deep convection mixes the fluid down to great depth, entraining fluid from below and creating a homogeneous mass denser than its surroundings. We discussed the details of the process in Chapter 4. The density excess of the convecting water column, multiplied by the acceleration of gravity and integrated over the depth of convection, is the potential energy available for conversion into kinetic energy, known as total CAPE.

The density of a convecting column increases as it cools and becomes more saline owing to evaporation. As Schmitt et al. (1989) have pointed out, surface fluxes of heat and water substance are conveniently combined into a density flux F_ρ:

$$F_\rho = -\rho(\alpha F_T - \beta F_S) \tag{5.8}$$

where α and β are thermal expansion and haline contraction coefficients defined by:

$$\alpha = -\rho^{-1}\frac{\partial \rho}{\partial T}$$

$$\beta = \rho^{-1}\frac{\partial \rho}{\partial S}$$

the derivatives to be taken at constant pressure and salinity for α, at constant pressure and temperature for β. The upward temperature flux on the water side of the air-sea interface is $F_T = -A_w/\rho c_p$, with A_w the net oceanic surface heat gain, and the salt flux is $F_S = S(E - P)/(1 - S)$, with E evaporation, P precipitation, and S salt concentration. The surface buoyancy flux B_0, the quantity that we related to mixed layer entrainment in Chapter 3, is proportional to the density flux, $B_0 = -(g/\rho)F_\rho$.

Surface density flux sustained for a period t changes the density of a convective column of depth h by an amount $\Delta\rho$:

$$\Delta\rho = -\frac{F_\rho t}{h} = \frac{\rho B_0 t}{gh} \tag{5.9}$$

CAPE per unit mass, generated by continuous surface density flux, is then $gh(\Delta\rho/\rho) = B_0 t$. Integrated over the mass of the convecting column, total CAPE, is $-ghtF_\rho$, the rate of change of total CAPE $-ghF_\rho$. This has the units of energy flux, W m^{-2}. When the density flux is due to heat flux alone, the ratio of total CAPE generation rate to surface heat loss is $ghF_\rho/A_w = \alpha gh/c_p$, a quantity typically of order 10^{-4}, that might be thought of as the "efficiency" of potential energy production.

Deep convection entrains fluid from below, further cooling the convecting column. We discussed the details in Chapter 3, and found that the negative buoyancy flux because of entrainment is a constant fraction (about a quarter) of the surface buoyancy flux, a relationship we called Carson's Law. The net result on a deep convecting column in the ocean is an increase in the cooling rate by 25%, over what surface cooling alone is responsible for. Entrainment thus increases the total CAPE generation rate to $-1.25ghF_\rho$. This still amounts to much the same very low efficiency, ratio of mechanical energy gain to surface heat flux, as from the surface density flux alone. The energy gain per unit mass of the convecting column, in a ten day long cooling episode, $(\Delta\rho/\rho)gh$ is typically between 0.1–1.0 J/kg. In the "capping" process of the deep convection much of this potential energy becomes the kinetic energy of the rim current, to be eventually dissipated. Some of it remains, however, in the form of potential energy in pycnostads, as discussed before.

5.5.2 Density Flux and Pycnostads in the North Atlantic

The distribution of density flux into the ocean, $-F_\rho$, reveals where, and how intensely, convection operates in the North Atlantic. Schmitt et al. (1989) constructed maps of this distribution, from data on sensible and latent heat flux, evaporation, and precipitation. Figure 5.19 shows their yearly mean map. High positive winter density fluxes over the Gulf Stream and vicinity dominate this map, with weak positive (convection generating) fluxes in the northern Seas, weak negative (stability enhancing) fluxes over the tropics and subtropics. The individual influence of temperature versus salt flux is compared in Figures 5.20 and 5.21, showing that the salt flux adds only a small perturbation to the temperature effect (contrary to Broecker's idea: "the salt left behind" is a feeble contributor to thermohaline forcing in the North Atlantic). Seasonal variations are large, however: in winter, the maximum density flux over the Gulf Stream reaches $34 \times 10^{-6}\,\mathrm{kg\,m^{-2}\,s^{-1}}$, versus $-4 \times 10^{-6}\,\mathrm{kg\,m^{-2}\,s^{-1}}$ in summer.

The large winter density flux in the Gulf Stream region generates the eighteen degree water pycnostad mentioned in Chapter 4, stretching from the Gulf Stream southward and eastward over some 15 degrees of latitude and 25 degrees of longitude or well beyond its generation area. Figure 5.22, from Warren (1972), shows the temperature and salinity distribution here in late winter (note that salinity decreases potential energy

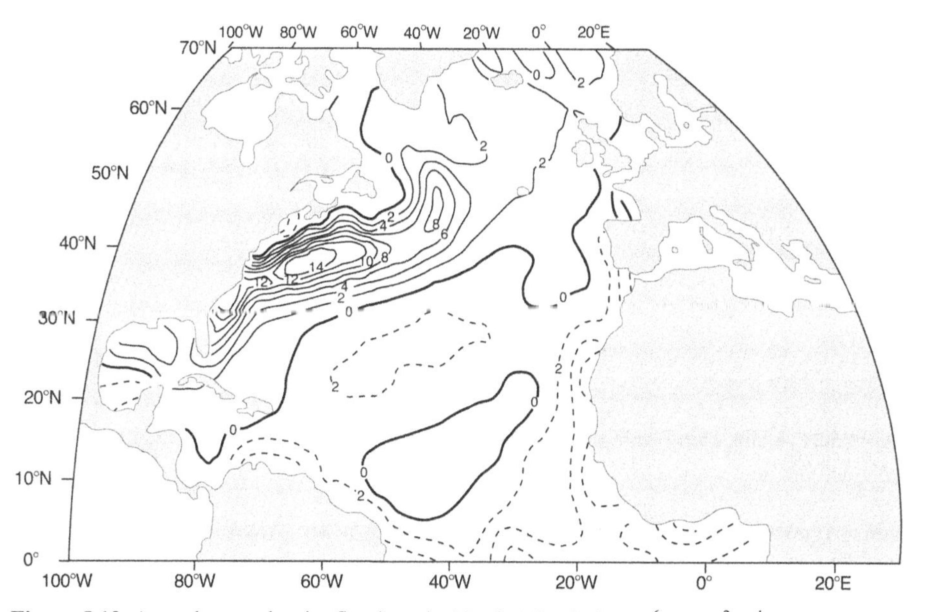

Figure 5.19 Annual mean density flux into the North Atlantic in $10^{-6}\,\mathrm{kg\,m^{-2}\,s^{-1}}$. Full line contours indicate density increase in the water column; broken lines indicate buoyancy gain. From Schmitt et al. (1989).

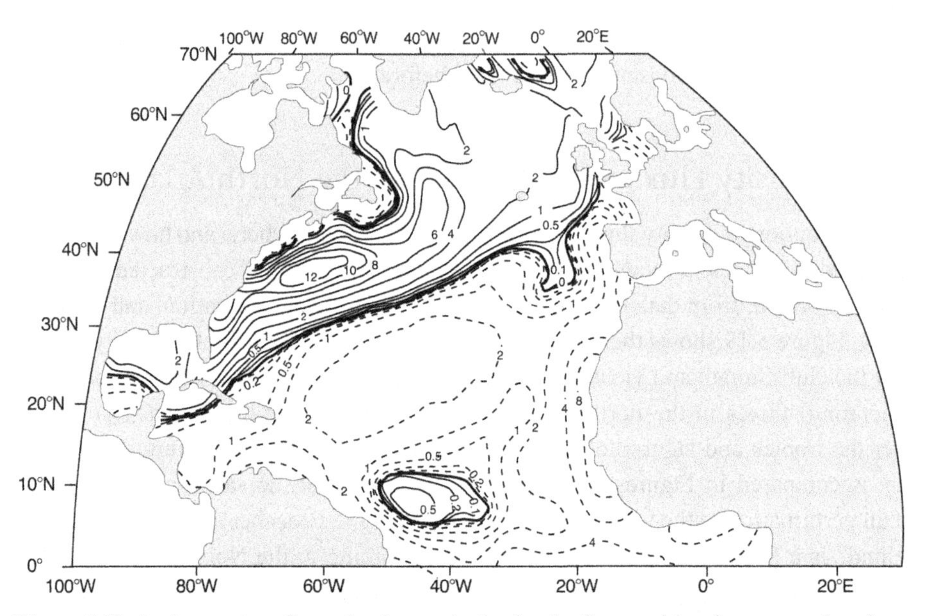

Figure 5.20 As the previous figure, but here only the density flux resulting from upward surface heat flux. From Schmitt et al. (1989).

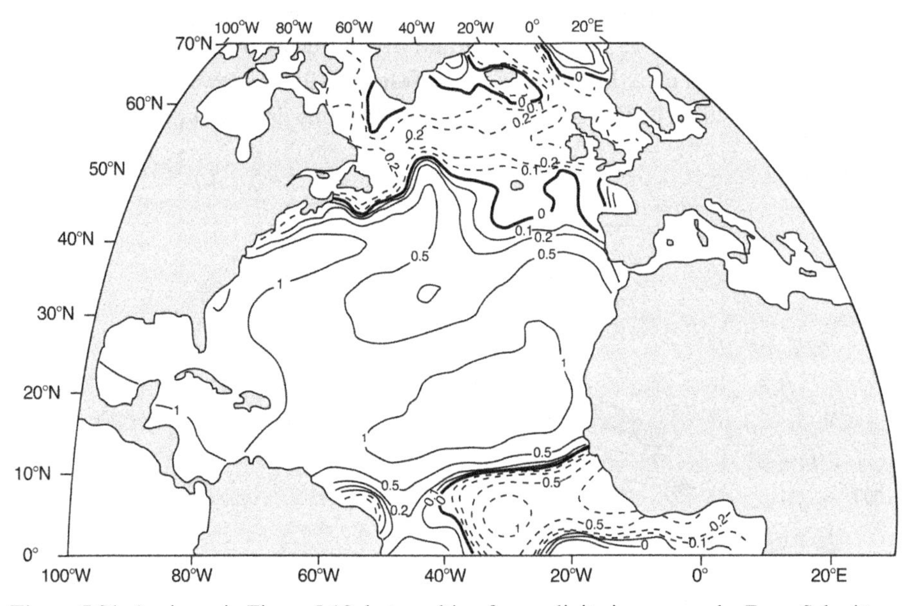

Figure 5.21 Again, as in Figure 5.19, but resulting from salinity increase only. From Schmitt et al. (1989).

in this pycnostad). The depth of the pycnostad is about 500 m. At the location of the peak winter density flux, the total CAPE generation rate approaches 0.2 W m^{-2}. In a brilliant insight, much ahead of his time and making many heads shake, Worthington (1972, 1977) recognized that the energy of this pycnostad sustains thermohaline circulation, in the form of an "anticyclone" as Worthington called it, that enhances Gulf Stream

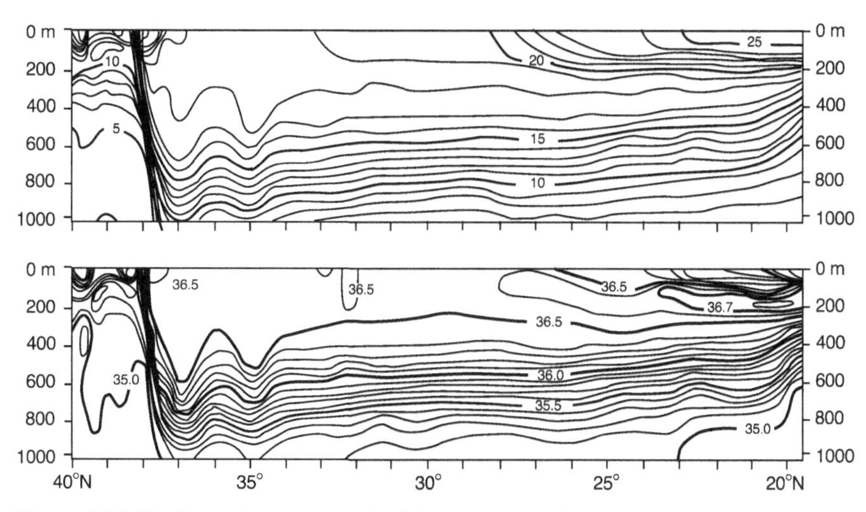

Figure 5.22 The late winter pycnostad of the North Atlantic, containing eighteen degree water, of salinity 36.5 parts per thousand. From Warren (1972).

transport on its north side, with its return leg in the eastern and southern parts of the eighteen degree pycnostad. The Gulf Stream's mass transport in winter increases on this account by some 10 to 15 Sv according to Worthington's estimate, or by one third to one half of the mass transport coming into the North Atlantic through the Florida Straits.

The lion's share of the potential energy gain from surface cooling in the North Atlantic no doubt goes into the eighteen degree pycnostad. There is, however, also significant thermohaline energy input to seas further north in winter. In the Labrador Sea, in particular, winter density flux is about one fifth of the peak flux in the Gulf Stream region, but the depth of convection is double, so that the LSW convecting column gains potential energy per unit mass of order 0.3 J kg^{-1}. The winter density flux here sustains a total CAPE generation rate of about 0.06 W m^{-2}. About one sixth of this, or 0.01 W m^{-2}, should be stored in the deep pycnostad remaining after convection, according to the calculation at the end of Chapter 4. By the same calculation, the eighteen degree pycnostad should store energy at the rate of some 0.03 W m^{-2}.

To put these numbers in perspective, consider the gain of mechanical energy by the ocean from the wind. The energy gain from wind stress is shear stress times surface velocity, the latter of order $U_s = 10u^*$ with u^* the friction velocity. In bands of strong westerlies or easterlies u^{*2} is typically $2 \times 10^{-4}\ \mathrm{m^2\,s^{-2}}$. Multiplied by density, this makes the wind energy input of order 0.03 W m^{-2}, of the same order as typical total CAPE generation rates in the northern seas. Again, some of this is dissipated in the turbulent mixed layer where kinematic dissipation rates are of order $\rho\varepsilon = 10^{-4}\ \mathrm{W\,m^{-3}}$ (Shay and Gregg, 1986), down to a depth of a few tens of meters, for a depth-integrated dissipation rate of perhaps 0.01 W m^{-2}, a third of the wind input.

Another way to look at this is that the part of the wind energy input that is available to support circulation is potential energy gain through the "uphill" Ekman transport of

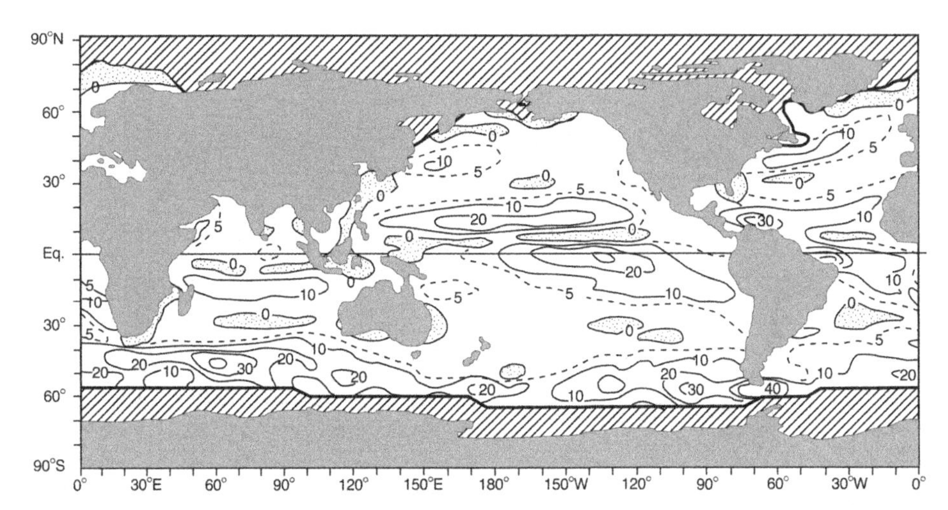

Figure 5.23 Annual mean energy input from the wind into the world ocean, in 10^{-3} W m^{-2}. Stippling indicates energy loss, wind blowing against the current. From Oort et al. (1994).

surface water toward higher surface pressure, which also equals the scalar product of wind stress and geostrophic velocity, $\tau \, . \, v_g$. Surface elevation gradients in the North Atlantic are of the order of one meter over a thousand kilometers or $\nabla\zeta = 10^{-6}$, equivalent to a potential energy gradient of $\rho g \nabla \zeta = 10^{-2}$ m s^{-2}. The order of the Ekman transport u^{*2}/f, with f the Coriolis parameter, is 2 m^2 s^{-1} at midlatitude. The product of energy gradient and Ekman transport gives a gain of potential energy of order 0.02 W m^{-2}, two thirds of the total wind input.

Given the crudeness of the order of magnitude estimates, the agreement of the sum of dissipation and potential energy gain with the "typical" total wind input is fortuitous. More definite are calculations of wind energy input based on observed surface currents and winds. Oort et al. (1994) have made such calculations and showed the results in a worldwide map (Figure 5.23). The observed velocities include the frictional departure from the geostrophic velocity, so that the energy inputs are the totals. They range from zero to 0.04 W m^{-2}, with the highest input rates in the Circumpolar Current, near the equator, and over the Gulf Stream and the Kuroshio. Oort et al. (1994) remark that in their opinion the wind inputs over the Circumpolar Current are "much underestimated".

Wunsch (1998) has recently determined the global distribution of the net energy input to the circulation, stress times geostrophic velocity, from a different database, altimeter data on surface pressure distribution and wind stress estimates. His results show similar concentrations over the Circumpolar Current and near the equator, somewhat less over the boundary currents. Absolute values range up to 0.02 W m^{-2} in the region of the Circumpolar Current, about half that in equatorial regions, even though the calculations were omitted within $\pm 3°$ of the equator. Wind energy input was found small over most of the world ocean, resulting in a worldwide average value about 0.003 W m^{-2}.

Potential energy inputs via surface density flux (total CAPE generation rate) and wind energy inputs are thus of broadly similar orders of magnitude, and share the property of being significant only over limited areas of the ocean surface (although the area-integrated wind energy input over the Circumpolar Current dwarfs other energy inputs, wind, or thermohaline).

To answer the question in the title of this section, thermohaline energy input in the North Atlantic certainly "initiates" the overturning circulation. Judging from the distribution of the LSW pycnostad depth seen in Figure 4.41, however, this energy supply runs out by the time the Deep Western Boundary Current reaches the tropics. On the rest of the way south, and on the northward return trip, the predominant energy source is wind input over the Circumpolar Current. Toggweiler (1994) puts the case this way: "A full understanding of the ocean's overturning may require some sort of hybridization of two theoretical frameworks. One framework contends that the overturning is driven by buoyancy forces, that is, the creation of dense and not-so-dense water masses by the addition or removal of heat and fresh water. The other sees the overturning as a largely mechanical process driven by the wind stress in the region of the Antarctic circumpolar current" (p. 45). Our consideration of thermodynamic constraints, of observed density fluxes and wind energy inputs, reveals that there is in fact no conflict between "two theoretical frameworks." Both are needed. An important addition to those frameworks is, however, the role of the equatorial wind field in the Atlantic. Southern Ocean upwelling may exert the "suction" that helps the NADW reach the high southern latitudes, but it is hard to see how it could force WWS waters across the equator.

To sum up this argument, there is significant thermohaline energy input in the North Atlantic into the overturning circulation. This constitutes, however, only a fraction of the total North Atlantic thermohaline energy, most of which drives the circulation of eighteen degree water. The bulk of the limited energy supply that goes into driving the Deep Water Boundary Current is exhausted by the time the DWBC reaches the tropics. From then on, and all along the northward return path, winds drive the overturning circulation. The strong wind energy input into the Southern Ocean, in particular, is key to raising deep waters to the surface and driving them equatorward. Similarly, strong energy input in the equatorial Atlantic, with a fortunate seasonal and geographic distribution, is another essential factor in sustaining the overturning circulation, and hence the ocean's role in keeping Western Europe warm.

References

Agrawal, Y.C. et al. Enhanced dissipation of kinetic energy beneath surface waves. *Nature* 359, 219–220, 1992.

Amorocho, J. and J.J. De Vries. A new evaluation of the wind stress coefficient over water surfaces. *J. Geophys. Res.* 85, 433–442, 1980.

Augstein, E., H. Schmidt, and F. Ostapoff. The vertical structure of the atmospheric planetary boundary layer in undisturbed trade winds over the Atlantic Ocean. *Bound. Layer Meteorol.* 6, 129–150, 1974.

Bacon, S. Circulation and fluxes in the North Atlantic between Greenland and Ireland. *J. Phys. Oceanogr.* 27, 1420–1435, 1997.

Ball, F.K. Control of inversion height by surface heating. *Quart. J. Roy. Meteorol. Soc.* 86, 483–494, 1960.

Banner, M.L. and W.K. Melville. On the separation of air flow over water waves. *J. Fluid Mech.* 77, 825–842, 1976.

Banner, M.L. and O.M. Phillips. On the incipient breaking of small scale waves. *J. Fluid Mech.* 65, 647–656, 1974.

Banner, M.L. and W.L. Peirson. Tangential stress beneath wind-driven air-water interfaces. Tangential stress beneath wind-driven air-water interfaces. *J. Fluid Mech.* 364, 115–145, 1997.

Barger, W.R., W.D. Garrett, E.L. Mollö-Christensen, and K.W. Ruggles. Effect of an artificial sea slick upon the atmosphere and the ocean. *J. Appl. Meteorol.* 9, 396–400, 1970.

Barnett, T.P. and A.J. Sutherland. A note on an overshoot effect in wind-generated waves. *J. Geophys. Res.* 73, 6879–6885, 1968.

Battjes, J.A. and T. Sakai. Velocity field in a steady breaker. *J. Fluid Mech.* 111, 421–437, 1981.

Betts, A.K. and B.A. Albrecht. Conserved variable analysis of the convective boundary layer thermodynamic structure over the tropical oceans. *J. Atmos. Sci.* 44, 83–99, 1987.

Betts, A.K. and W. Ridgway. Coupling of the radiative, convective and surface fluxes over the equatorial Pacific. *J. Atmos. Sci.* 45, 522–536, 1988.

Bigelow, H.B. and W.T. Edmondson. Wind waves at sea breakers and surf. Hydrographic Office, Washington, D.C. publication No. 602, 177 pp., 1947.

Bjerknes, J. A possible response of the atmospheric Hadley circulation to equatorial anomalies of ocean temperature. *Tellus* 18, 820–828, 1966.

Blanc, T.V. Variation of bulk-derived surface flux, stability and roughness results due to the use of different transfer coefficient schemes. *J. Phys. Oceanogr.* 15, 650–669, 1985.

Bolton, D. The computation of equivalent potential temperature. *Mon. Wea. Rev.* 108, 1046–1053, 1980.

Bourles, B., R.L. Molinari, E. Johns, and W.D. Wilson. Upper layer currents in the western tropical North Atlantic. *J. Geophys. Res.* 104, 1361–1375, 1999.

Bradley, E.F., P.A. Coppin, and J.S. Godfrey. Measurements of sensible and latent heat flux in the western equatorial Pacific Ocean. *J. Geophys. Res.* 96, Suppl. 3375–3389, 1991.

Brainerd, K.E. and M.C. Gregg. Diurnal restratification and turbulence in the oceanic surface mixed layer 1. Observations. *J. Geophys. Res.* 98, 22,645–22,656, 1993.

Bretherton, C.S., P. Austin, and S.T. Siems. Cloudiness and marine boundary layer dynamics in the ASTEX Lagrangian experiments. Part II: Cloudiness, drizzle, surface fluxes and entrainment. *J. Atmos. Sci.* 52, 2724–2735, 1995.

Broecker, W.S. The Great Ocean Conveyor. *Oceanography* 4, 79–89, 1991.

Broecker, W.S., J.R. Ledwell, T. Takahashi, R. Weiss, L. Merlivat, L. Memery, T-H. Peng, B. Jähne, and K.O. Münnich. Isotopic versus micrometorologic ocean CO_2 fluxes: a serious conflict. *J. Geophys. Res.* 91, 10,517–27, 1986.

Broecker, W.S. and T.-H. Peng. *Tracers in the Sea.* Eldigo Press, Palisades, N.Y., 1982.

Brooke-Benjamin, T. Shearing flow over a wavy boundary. *J. Fluid Mech.* 6, 161–205, 1959.

Brooke-Benjamin, T. Effects of a flexible boundary on hydrodynamic stability. *J. Fluid Mech.* 9, 513–532, 1960.

Bryden, H.L. Poleward heat flux and conversion of available potential energy in Drake Passage. *J. Marine Res.* 37, 1–22, 1979.

Bunker, A.F. Computations of surface energy flux and annual air-sea interaction cycles of the North Atlantic Ocean. *Mon. Wea. Rev.* 104, 1122–1139, 1976.

Bunker, A.F. Surface energy fluxes of the South Atlantic Ocean. *Mon. Wea. Rev.* 116, 809–823, 1988.

Bunker, A.F. and L.V. Worthington. Energy exchange charts of the North Atlantic Ocean. *Bull. Amer. Meteor. Soc.* 57, 670–678, 1976.

Businger, J.A. Equations and concepts. In: *Atmospheric Turbulence and Air Pollution Modeling*, Eds. Nieuwstadt and van Dop, Reidel, pp. 1–36, 1982.

Byers, H.R. and R.R. Braham. Thunderstorm structure and circulation. *J. Meteorol.* 5, 71–86, 1948.

Carslaw, H.S. and J.C. Jaeger. *Conduction of Heat in Solids*. Oxford Univ. Press, 1959.

Carson, D.J. The development of a dry inversion capped convectively unstable boundary layer. *Quart. J. Roy. Meteorol. Soc.* 99, 450–467, 1973.

Carissimo, B.C., A.H. Oort, and T.H. Vonder Haar. Estimating the meridional energy transports in the atmosphere and ocean. *J. Phys. Oceanogr.* 15, 82–91, 1985.

Caughey, S.J. Observed characteristics of the atmospheric boundary layer. In: *Atmospheric Turbulence and Air Pollution Modeling*, Eds. Nieuwstadt and van Dop, Reidel, pp. 107–158, 1982.

Caughey, S.J. and S.G. Palmer. Some aspects of turbulence structure through the depth of the convective boundary layer. *Quart. J. Roy. Meteorol. Soc.* 105, 811–827, 1979.

Charnock, H. Wind stress on a water surface. *Quart. J. Roy. Meteorol. Soc.* 639–640, 1955.

Churchill, J.H. and G.T. Csanady. Near-surface measurements of quasi-Lagrangian velocities in open water. *J. Phys. Oceanogr.* 13, 1669–1680, 1983.

Clarke, R.A. and J.-C. Gascard. The formation of Labrador Sea Water. Part I: Large scale processes. *J. Phys. Oceanogr.* 13, 1764–1778, 1983.

Cox, C.S. Measurements of slopes of high-frequency wind waves. *J. Marine Res.* 16, 199–225, 1958.

Cox, C.S. and W.H. Munk. Statistics of the sea surface derived from sun glitter. *J. Marine Res.* 13, 198–227, 1954.

Csanady, G.T. The 'roughness' of the sea surface in light winds. *J. Geophys. Res.* 79, 2747–2751, 1974.

Csanady, G.T. Turbulent interface layers. *J. Geophys. Res.* 83, 2329–2342, 1978.

Csanady, G.T. Momentum flux in breaking wavelets. *J. Geophys. Res.* 95, 13,289–13,299, 1990a.

Csanady, G.T. The role of breaking wavelets in air-sea gas transfer. *J. Geophys. Res.* 95, 749–759, 1990b.

Danckwerts, P.V. Significance of liquid-film coefficients in gas absorption. *Ind. Eng. Chem.* 43, 1460–1467, 1951.

Deacon, E.L. Gas transfer to and across an air-water interface. *Tellus* 29, 363–374, 1977.

Deacon, E.L. Sea-air gas transfer: The wind speed dependence. *Bound. Layer Meteorol.* 21, 31–37, 1981.

Deardorff, J.W. Dependence of air-sea transfer coefficients on bulk stability. *J. Geophys. Res.* 73, 2549–57, 1968.

Deardorff, J.W. Convective velocity and temperature scales for the unstable planetary boundary layer and for Rayleigh convection. *J. Atmos. Sci.* 27, 1211–1213, 1970.

Deardorff, J.W. On the distribution of mean radiative cooling at the top of a stratocumulus-capped mixed layer. *Quart. J. Roy. Meteorol. Soc.* 107, 191–202, 1981.

DeCosmo, J., K.B. Katsaros, S.D. Smith, R.J. Anderson, W.A. Oost, K. Bumke, and H. Chadwick. Air-sea exchange of water vapor and sensible heat: The Humidity Exchange Over the Sea (HEXOS) results. *J. Geophys. Res.* 101, 12,001–12,016, 1996.

De Groot, S.R. *Thermodynamics of Irreversible Processes.* North Holland Publishing Co., 1963.

De Groot, S.R. and P. Mazur. *Non-Equilibrium Thermodynamics.* Dover Publications, 1984.

Dickson, R.R. and J. Brown. The production of North Atlantic Deep Water: sources, rates and pathways. *J. Geophys. Res.* 99, 12,319–12,341, 1994.

Dobson, F.W., S.D. Smith and R.J. Anderson. Measuring the relationship betwen wind stress and sea state in the open ocean in the presence of swell. *Atmos.-Ocean*, 32, 237–256, 1994.

Donelan, M.A. Air-Sea interaction. In: *The Sea*, Vol. 9, J. Wiley and Sons, pp. 239–292, 1990.

Donelan, M.A., F.W. Dobson, S.D. Smith, and R.J. Anderson. On the dependence of sea surface roughness on wave development. *J. Phys. Oceanogr.* 23, 2143–2149, 1993.

Drennan, W.M., M.A. Donelan, E.A. Terray, and K.B. Katsaros. Oceanic turbulence dissipation measurements in SWADE. *J. Phys. Oceanogr.* 26, 808–814, 1996.

Duncan, J.H. An experimental investigation of breaking waves produced by a towed hydrofoil. *Proc. Roy. Soc. London A* 377, 331–348, 1981.

Duncan, J.H. The breaking and non-breaking resistance of a two-dimensional aerofoil. *J. Fluid Mech.* 126, 507–520, 1983.

Duynkerke, P.G., H.Q. Shang, and P.J. Jonker. Microphysical and turbulent structure of nocturnal stratocumulus as observed during ASTEX. *J. Atmos. Sci.* 52, 2763–2777, 1995.

Dvorak, V.F. Tropical cyclone intensity analyses using satellite data. *NOAA Tech. Rep.* NESDIS 11, 1984.

Ebuchi, N., H. Kawamura, and Y. Toba. Bursting phenomena in the turbulent boundary layer beneath the laboratory wind-wave surface. In: *Natural Physical Sources of Underwater Sound*, Ed. B.R. Kerman, Kluwer Acad. Publ., 1993.

Emanuel, K.A. Sensitivity of tropical cyclones to surface exchange coefficients and a revised steady-state model incorporating eye dynamics. *J. Atmos. Sci.* 52, 3969–3976, 1985.

Emanuel, K.A. An air-sea interaction theory for tropical cyclones. Part I: steady-state maintenance. *J. Atmos. Sci.* 43, 585–604, 1986.

Emanuel, K.A. The theory of hurricanes. *Ann. Rev. Fluid Mech.* 23, 179–196, 1991.

Emanuel, K.A. *Atmospheric Convection.* Oxford Univ. Press, 1994.

Flagg, C.N., R.L. Gordon, and S. McDowell. Hydrographic and current observations on the continental slope and shelf of the western equatorial Atlantic. *J. Phys. Oceanogr.* 16, 1412–1429, 1986.

Fillenbaum, E.R., T.N. Lee, W.E. Johns, and R.J. Zantopp. Meridional heat transport variability at 26.5°N in the North Atlantic. *J. Phys. Oceanogr.* 27, 153–174, 1997.

Fortescue, G.E. and J.R.A. Pearson. On gas absorption into a turbulent liquid. *Chem. Eng. Sci.* 22, 1163–1176, 1967.

Frank, W.M. and J.L. McBride. The vertical distribution of heating in AMEX and GATE cloud clusters. *J. Atmos. Sci.* 46, 3464–3478, 1989.

Garratt, J.R. Review of drag coefficients over oceans and continents. *Mon. Wea. Rev.* 105, 915–929, 1977.

Garratt, J.R. *The Atmospheric Boundary Layer.* Cambridge Univ. Press, 1992.

Gascard, J.-C. and R.A. Clarke. The formation of Labrador Sea Water. Part II: mesoscale and smaller-scale processes. *J. Phys. Oceanogr.* 13, 1779–1797, 1983.

Geernaert, G.L., S.E. Larsen, and F. Hansen. Measurements of the wind stress, heat flux, and turbulence intensity during storm conditions over the North Sea. *J. Geophys. Res.*, 92, 13,127–39, 1987.

Georgi, D.T. and J.M Toole. The Antarctic Circumpolar Current and the oceanic heat and freshwater budgets. *J. Marine Res.* 40 Supplement, 183–197, 1982.

Gill, A.E. *Atmosphere-Ocean Dynamics*. Academic Press, 1982.

Gordon, A.L. Interocean exchange of thermocline water. *J. Geophys Res.* 91, 5037–5046, 1986a.

Gordon, A.L. Is there a global scale ocean circulation? *EOS*, March 1986, 1986b.

Gordon, A.L., R.F. Weiss, W.M. Smethie, and M.J. Warner. Thermocline and intermediate water communication between the South Atlantic and Indian Oceans. *J. Geophys. Res.* 97, 7223–7240, 1992a.

Gordon, A.L., S.E. Zebiak, and K. Bryan. Climate variability and the Atlantic Ocean. *EOS*, April 1992, 1992b.

Gouriou, Y. and G. Reverdin. Isopycnal and diapycnal circulation of the upper equatorial Atlantic Ocean in 1983–1984. *J. Geophys. Res.* 97, 3543–3572, 1992.

Häkkinen, S. and D.J. Cavalieri. A study of oceanic surface heat fluxes in the Greenland, Norwegian, and Barents Seas. *J. Geophys. Res.* 94, 6145–6157, 1989.

Hall, M.M. and H.L. Bryden. Direct estimates and mechanisms of ocean heat transport. *Deep Sea Res.* 29, 339–359, 1982.

Hasselman, K. On the non-linear energy transfer in a gravity wave spectrum. Part 1, *J. Fluid Mech.* 12, 481–500, 1962.

Hasselman, K. On the non-linear energy transfer in a gravity wave spectrum. Part 2, *J. Fluid Mech.* 15, 273–281; Part 3, *ibid.* 15, 385–398, 1963a,b.

Hasselman, K. et al. Measurements of wind-wave growth and swell decay during the Joint North Sea Wave Project (JONSWAP). *Ergänzungsheft zur Deutsch. Hydrogr. Z.* A(8°), Nr. 12, 1973.

Hastenrath, S. Heat budget of tropical ocean and atmosphere. *J. Phys. Oceanogr.* 10, 159–170, 1980.

Hastenrath, S. On meridional heat transports in the World Ocean. *J. Phys. Oceanogr.* 12, 922–927, 1982.

Hastenrath, S. and P.J. Lamb. On the heat budget of hydrosphere and atmosphere in the Indian Ocean. *J. Phys. Oceanogr.* 10, 694–708, 1980.

Hastenrath, S. and J. Merle. The annual march of heat storage and export in the tropical Atlantic Ocean. *J. Phys. Oceanogr.* 16, 694–708, 1986.

Hastenrath, S. and J. Merle. Annual cycle of subsurface thermal structure in the tropical Atlantic Ocean. *J. Phys. Oceanogr.* 17, 1518–1538, 1987.

Higbie, R. The rate of absorption of a pure gas into a still liquid during short periods of exposure. *Trans. Am. Inst. Chem. Eng.* 31, 365–388, 1935.

Holland, G.J. The maximum potential intensity of tropical cyclones. *J. Atmos. Sci.* 54, 2519–2541, 1997.

Houze, R.A., S.A. Rutledge, M.I. Biggerstaff, and B.F. Smull. Interpretation of Doppler weather radar displays of mesoscale convective systems. *Bull. Amer. Meteorol. Soc.* 70, 608–619, 1989.

Hsiung, J. Estimates of global oceanic meridional heat transport. *J. Phys. Oceanogr.* 15, 1405–1413, 1985.

Hsiung, J. Mean surface energy fluxes over the global ocean. *J. Geophys. Res.* 91, 10,585–10,606, 1986.

Hsiung, J., R.E. Newell, and T. Houghtby. The annual cycle of oceanic heat storage and oceanic meridional transport. *Q. J. Roy. Meteorol. Soc.* 115, 1–18, 1989.

Jähne, B. New experimental results on the parameters influencing air-sea gas exchange. In: *Air-Sea Mass Transfer*, Eds. S.C. Wilhelms and J.S. Gulliver, *Am. Soc. Civil Eng.*, pp. 582–591, 1991.

Jähne, B, K.O. Münnich, R. Bösinger, A. Dutzi, W. Huber, and P. Libner. On the parameters influencing air-water gas exchange. *J. Geophys. Res.* 92, 1937–1949, 1987.

Jähne, B., T. Wais, L. Memery, G. Gaulliez, L. Merlivat, K.O. Münnich, and M. Coantic. He and Rn gas exchange experiments in the large wind-wave facility of IMST. *J. Geophys. Res.* 90, 11,989–11,998, 1985.

Johnson, D.H. The role of the tropics in the global circulation. In: *The Global Circulation of the Atmosphere*, pp. 113–136, Royal Meteorol. Soc., London, 1969.

Johnson, D.R. On the distribution of heat sources and sinks and their relation to mass and energy transport. Presented at the FGGE Workshop Committee Meeting WMO FGGE Seminar, Tallahassee, FL, 1984.

Jones, H. and J. Marshall. Convection with rotation in a neutral ocean: a study of open-ocean deep convection. *J. Phys. Oceanogr.* 23, 1009–1039, 1993.

Jones, H. and J. Marshall. Restratification after deep convection. *J. Phys. Oceanogr.* 27, 2276–2287, 1997.

Jorgensen, D.P. and M.A. LeMone. Vertical velocity characteristics of oceanic convection. *J. Atmos. Sci.* 46, 621–640, 1989.

Jorgensen, D.P., M.A. LeMone, and S.B. Trier. Structure and evolution of he 22 February 1993 TOGA COARE squall line: aircraft observations of precipitation, circulation and surface energy fluxes. *J. Atmos. Sci.* 54, 1961–1985, 1997.

Jorgensen, D.P., E.J. Zipser, and M.A. LeMone. Vertical motions in intense hurricanes. *J. Atmos. Sci.* 42, 839–856, 1985.

Kapolnai, A. and G.T. Csanady. Heat export from the equatorial Atlantic. Old Dominion University, unpublished manuscript, 1993.

Kawai, S. Generation of initial wavelets by instability of a coupled shear flow and their evolution to wind waves. *J. Fluid Mech.* 93, 661–703, 1979.

Kawai, S. Visualization of airflow separation over wind-wave crests under moderate wind. *Boundary Layer Meteorology* 21, 93–104, 1981.

Kawai, S. Structure of air flow separation over wind wave crests. *Bound. Layer Meteorol.* 23, 503–521, 1982.

Kawamura, H. and Y. Toba. Ordered motion in the turbulent boundary layer over wind waves. *J. Fluid Mech.* 197, 105–138, 1988.

Keulegan, G.H. Interfacial instability and mixing in stratified flows. *J. Res. Nat. Bur. Stand.* 43, 487–500, 1949.

Kinsman, B. Surface waves at short fetches and low wind speed—a field study. Chesapeake Bay Inst. Johns Hopkins University, Tech Report No. 19, 1960.

Kitagorodskii, S.A. *The Physics of Air-Sea Interaction.* Israel Program of Scientific Translations and U.S. Dept. of Commerce, 1973.

Kleinschmidt, E. Grundlagen einer Theorie der tropischen Zyklonen. *Arch. Meteorol., Geophys. and Bioclimatol.* 4, 53–72, 1951.

Kolmogorov, A.N. The local structure of turbulence in incompressible viscous fluid for very large Reynolds numbers. Doklady ANSSSR 30, 301, 1941.

Lamb, H. *Hydrodynamics.* Cambridge Univ. Press, 1957.

Large, W.G. and S. Pond. Sensible and latent heat flux measurements over the ocean. *J. Phys. Oceanogr.* 12, 464–482, 1982.

Larson, T.R. and J.W. Wright. Wind-generated gravity-capillary waves: laboratory measurements of temporal growth rates using microwave backscatter. *J. Fluid Mech.* 70, 417–436, 1975.

Leaman, K.D and F.A. Schott. Hydrographic structure of the convection regime in the Gulf of Lions: winter 1987. *J. Phys. Oceanogr.* 21, 575–598, 1991.

Levitus, S. *Climatological Atlas of the World Ocean.* NOAA Professional Paper 13, U.S. Dept of Commerce, 1982.

Lilly, D.K. Models of cloud-topped mixed layers under a strong inversion. *Quart. J. Roy. Meteorol.* Soc. 94, 292–309, 1968.

Lin, X. and R.H. Johnson. Heating, moistening and rainfall over the western Pacific warm pool during TOGA COARE. *J. Atmos. Sci.* 53, 3367–3383, 1996.

Lock, R.C. Hydrodynamic instability of the flow in the laminar boundary layer between parallel streams. *Proc. Cambridge Phil. Soc.* 50, 105–124, 1953.

Lofquist, K. Flow and stress near an interface between stratified liquids. *Phys. Fluids* 3, 158–175, 1960.

Lombardo, C.P. and M.C. Gregg. Similarity scaling of viscous and thermal dissipation in a convecting surface boundary layer. *J. Geophys. Res.* 94, 6273–6284, 1989.

Longuet-Higgins, M.S. On the statistical distribution of the heights of sea waves. *J. Marine Res.* 11, 245–266, 1952.

Lorenz, E.N. The nature and theory of the general circulation of the atmosphere. WMO, 1967.

Lucas, C., E.J. Zipser, and M.A. Lemone. Vertical velocity in oceanic convection off tropical Australia. *J. Atmos. Sci.* 51, 3183–3193, 1994.

Lumley, J.L. and H.A. Panofsky. The structure of atmospheric turbulence. Interscience Publishers, 1964.

Maat, N., C. Kraan, and W.A. Oost. The roughness of wind waves. *Bound. Layer Meteorol.* 54, 89–103, 1991.

Malkus, J.S. Some results of a trade cumulus cloud investigation. *J. Meteorol.* 11, 220–237, 1954.

Malkus, J.S. Large-scale interactions. In: *The Sea, Section II: Interchange of Properties between Sea and Air*, pp. 43–294, Interscience Publishers, 1962.

Mapes, B.E. Gregarious tropical convection. *J. Atmos. Sci.* 50, 2026–2037, 1993.

Marks, F.D. and R.A. Houze. Inner core structure of Hurricane Alicia from airborne Doppler radar observations. *J. Atmos. Sci.* 44, 1296–1317, 1987.

Marshall, J., and the "Lab Sea Group." The Labrador Sea deep convection experiment. *Bull. Am. Meteorol. Soc.* 79, 2019–2058, 1998.

Martin, G.M., D.W. Johnson, D.P. Rogers, P.R. Jonas, P. Minnis, and D.A. Hegg. Observations of the interaction between cumulus clouds and warm stratocumulus clouds in the marine boundary layer during ASTEX. *J. Atmos. Sci.* 52, 2902–2922, 1995.

Maxworthy, T. and S. Narimousa. Unsteady, turbulent convection into a homogeneous, rotating fluid, with oceanographic applications. *J. Phys. Oceanogr.* 24, 865–887, 1994.

McCartney, M.S. The subtropical recirculation of Mode Waters. *J. Marine Res.* 40 Supplement, 427–464, 1982.

McCartney, M.S. and L.D. Talley. The subpolar mode water of the North Atlantic Ocean. *J. Phys. Oceanogr.* 12, 1169–1188, 1982.

McCartney, M.S. and L.D. Talley. Warm-to-cold water conversion in the northern North Atlantic Ocean. *J. Phys. Oceanogr.* 14, 922–935, 1984.

MEDOC Group. Observation and formation of deep water in the Mediterranean. *Nature* 227, 1037–1040, 1970.

Merrill, R.T. Environmental influences on hurricane intensification. *J. Atmos. Sci.* 1678–1687, 1988.

Metcalf, W.G. and M.C. Stalcup. Origin of the Atlantic Equatorial Undercurrent. *J. Geophys. Res.* 72, 4959–4975, 1967.

Miles, J.W. On the generation of surface waves by shear flows. Part 1. *J. Fluid Mech.* 3, 185–204, 1957.

Miles, J.W. On the generation of surface waves by shear flows. Part 2, *J. Fluid Mech.* 6, 568–582; Part 3, *ibid* 583–598, 1959.

Miles, J.W. On the generation of surface waves by shear flows. Part 4. *J. Fluid Mech.* 13, 433–448, 1962.

Miller, M.A. and B.A. Albrecht. Surface-based observations of mesoscale cumulus-stratocumulus interaction during ASTEX. *J. Atmos. Sci.* 52, 2809–2826, 1995.

Moisan, J.R. and P.P. Niiler. The seasonal heat budget of the North Pacific: net heat flux and heat storage rates (1950–1990). *J. Phys. Oceanogr.* 28, 401–421, 1998.

Molinari, R.L., S.L. Garzoli, E.J. Katz, D.E. Harrison, P.L. Richardson, and G. Reverdin. A synthesis of the First GARP Global Experiment (FGGE) in the equatorial Atlantic Ocean. *Prog. Oceanogr.* 16, 91–112, 1986.

Monin, A.S. and A.M. Yaglom. Statistical Fluid Mechanics, MIT Press, Cambridge, Mass. 769 pp., 1971.

Morton, B.R., G.I. Taylor, and J.S. Turner. Turbulent gravitational convection from maintained and instantaneous sources. *Proc. Roy. Soc. A* 234, 1–23, 1956.

Nichols, S. and J. Leighton. An observational study of the structure of stratiform cloud sheets. Part I: Structure. *Quart. J. Roy. Meteorol. Soc.* 112, 431–460, 1986.

Obukhov, A.M. Turbulence in an atmosphere with a non-uniform temperature. Trudy Akad. Nauk. USSR, English translation in *Boundary Layer Meteorol.* 2, 7–29, 1971, 1946.

Ohlmann, J.C., D.A. Siegel, and L. Washburn. Radiant heating of the western equatorial Pacific during TOGA-COARE. *J. Geophys. Res.* 103, 5379–5395, 1998.

Okuda, K., S. Kawai, and Y. Toba. Measurement of skin friction distribution along the surface of wind waves. *J. Oceanog. Soc. Japan* 33, 190–198, 1977.

Oort, A.H. and P.H. Vonder Haar. On the observed annual cycle in the ocean atmosphere heat balance over the Northern hemisphere. *J. of Phys. Oceanogr.* 6, 781–800, 1976.

Oort, A.H., L.A. Anderson, and J.P. Peixoto. Estimates of the energy cycle of the oceans. *J. Geophys. Res.* 99, 7665–7688, 1994.

Oost, W.A. The KNMI HEXMAX stress data – A revisit. *Bound. Layer Meteorol.*, 1997.

Palmén, E. and C.W. Newton. *Atmospheric Circulation Systems*. Academic Press, 1969.

Pandya, R.E. and D.R. Durran. The influence of convectively generated thermal forcing on the mesoscale circulation around squall lines. *J. Atmos. Sci.* 53, 2924–2951, 1996.

Paulson, C.A. and J.J. Simpson. Irradiance measurements in the upper ocean. *J. Phys. Oceanogr.* 7, 952–956, 1977.

Peng, T.-H., W.S. Broecker, G.G. Mathieu, Y.H. Li, and A.E. Bainbridge. Radon evasion rates in the Atlantic and Pacific oceans as determined during the GEOSECS Program. *J. Geophys. Res.* 84, 2471–2486, 1979.

Peters, H., M.C. Gregg, and J.M. Toole. On the parameterization of equatorial turbulence. *J. Geophys. Res.* 93, 1199–1218, 1988.

Phillips, O.M. *The Dynamics of the Upper Ocean*. Cambridge Univ. Press, 1977.

Pierson, W.J. and L. Moskowitz. A proposed spectral form for fully developed wind seas based on the similarity theory of S.A. Kitagorodskii. *J. Geophys. Res.* 69, 5181–90, 1964.

Portman, D.J. *An Improved Technique for Measuring Wind and Temperature Profiles over Water and Some Results Obtained for Light Winds*. Publ. 4, pp. 77–84, Great Lakes Res. Div. Univ. Michigan, 1960.

Ramamonjiarisoa, A. Contribution à l'étude de la structure statistique et des méchanisms de génération des vagues de vent. Thèse de Doctorat d'Etat, Université de Provence, 1974.

Rennó, N.O. and A.P. Ingersoll. Natural convection as a heat engine: a theory for CAPE. *J. Atmos. Sci.* 53, 572–585, 1996.

Richardson, P.L. and D. Walsh. Mapping climatological seasonal variations of surface currents in the tropical Atlantic using ship drifts. *J. Geophys. Res.* 91, 10,537–10,550, 1986.

Riehl, H. *Tropical Meteorology*. McGraw Hill, 1954.

Riehl, H. and J.S. Malkus. On the heat balance in the equatorial trough zone. *Geophysica* 6, 503–538, 1958.

Rintoul, S.R. South Atlantic interbasin exchange. *J. Geophys. Res.* 96, 2675–2692, 1991.

Roll, H.U. *Physics of the Marine Atmosphere*. Academic Press, 1965.

Saunders, P.M. and B.A. King. Oceanic fluxes on the WOCE A11 section. *J. Phys. Oceanogr.* 25, 1942–1958, 1995.

Schlichting, H. *Boundary Layer Theory*. Transl. J. Kestin, McGraw Hill Book Co., New York, 1960.

Schmitt, R.W., P.S. Bogden, and C.E. Dorman. Evaporation minus precipitation and density fluxes for the North Atlantic. *J. Phys. Oceanogr.* 19, 1208–1221, 1989.

Schmitz, W.J. and M.S. McCartney. On the North Atlantic circulation. *Rev. Geophys.* 31, 29–49, 1993.

Schott, F. and K.D. Leaman. Observations with moored acoustic Doppler current profilers in the convection regime in the Golfe de Lion. *J. Phys. Oceanogr.* 21, 558–574, 1991.

Schott, F., M. Visbeck, U. Send, J. Fischer, L. Stramma, and Y. Desaubies. Observations of deep convection in the Gulf of Lions, Northern Mediterranean, during the winter of 1991/92. *J. Phys. Oceanogr.* 26, 505–524, 1996.

Schubauer, G.B. and H.K. Skramstad. Laminar boundary layer oscillations and stability of laminar flow. *J. Aero. Sci.* 14, 69–78, 1947.

Schubert, W.H., P.E. Cieselski, C. Lu, and R.H. Johnson. Dynamical adjustment of the Trade Wind Inversion layer. *J. Atmos. Sci.* 52, 2941–2952, 1995.

Shay, T.J. and M.C. Gregg. Convectively driven turbulent mixing in the upper ocean. *J. Phys. Oceanogr.* 16, 1777–1798, 1986.

Sheppard, P.A., D.T. Tribble, and J.R. Garratt. Studies of turbulence in the surface layer over water. *Q. J. Roy. Meteorol. Soc.* 98, 627–641, 1972.

Simpson, J.J. and T.D. Dickey. The relationship between downward irradiance and upper ocean structure. *J. Phys. Oceanogr.* 11, 309–323, 1981.

Slingo, A., R. Brown, and C.L. Wrench. A field study of nocturnal stratocumulus: III High resolution radiative and microphysical observations. *Quart. J. Roy. Meteorol. Soc.* 108, 145–165, 1982.

Smith, S.D. Eddy fluxes of momentum and heat measured over the Atlantic Ocean in gale force winds. In: *Turbulent Fluxes Through the Sea Surface, Wave Dynamics, and Prediction*. pp. 35–48, Eds. Favre and Hasselman, Plenum Press, 1978.

Smith, S.D. Wind stress and heat flux over the ocean in gale force winds. *J. Phys. Oceanogr.* 10, 709–726, 1980.

Smith, S.D. Coefficients for sea surface wind stress, heat flux, and wind profiles as a function of wind speed and temperature. *J. Geophys. Res.* 93, 15,467–15,472, 1988.

Smith, S.D. et al. Sea surface wind stress and drag coefficients: the HEXOS results. *Boundary Layer Meteorol.* 60, 109–142, 1992.

Smith, S.D. and E.P. Jones. Evidence for wind-pumping of air-sea gas exchange based on direct measurements of CO_2 fluxes. *J. Geophys. Res.* 90, 869–875, 1985.

Smith, S.D. and E.P. Jones. Isotopic and micrometeorological ocean CO_2 fluxes: different time and space scales. *J. Geophys. Res.* 91, 10,529–10,532, 1986.

Stewart, R.W. Mechanics of the Air-Sea interface. *Physics of Fluids* 10, Suppl. S47–55, 1967.

Stramma, L. Geostrophic transport of the South Equatorial Current in the Atlantic. *J. Mar. Res.* 49, 281–294, 1991.

Talley, L.D. Meridional heat transport in the Pacific Ocean. *J. Phys. Oceanogr.* 14, 231–241, 1984.

Talley, L.D. and M.S. McCartney. Distribution and circulation of Labrador Sea water. *J. Phys. Oceanogr.* 12, 1189–1205, 1982.

Tennekes, H. Free convection in the turbulent Ekman layer of the atmosphere. *J. Atmos. Sci.* 27, 1027–1034, 1970.

Terray, E.A., M.A. Donelan, Y.C. Agrawal, W.M. Drennan, K.K. Kahma, A.J. Williams III, P.A. Hwang, and S.I. Kitagorodskii. *J. Phys. Oceanogr.* 26, 792–807, 1996.

Thompson, S.M. and J.S. Turner. Mixing across an interface due to turbulence generated by an oscillating grid. *J. Fluid Mech.* 67, 349–368, 1975.

Toba, Y., M. Tokuda, K. Okuda, and S. Kawai. Forced convection accompanying wind waves. *J. Oceanogr. Soc. Japan* 31, 192–198, 1975.

Toba, Y. Local balance in the air-sea boundary processes, I: On the growth process of wind waves. *J. Oceanogr. Soc. Japan* 28, 109–120, 1972.

Toba, Y. Wind waves and turbulence. In: *Recent Studies of Turbulent Phenomena*, Eds. T. Tastumi, H. Maruo and H. Takami Assoc. for Sci. Doc. Inform., Tokyo, 1985.

Toba, Y. Similarity laws of the wind-wave, and the coupling process of the air and water turbulent boundary layers. *Fluid Dyn. Res.* 2, 263–279, 1988.

Toba, Y., N. Iida, H. Kawamura, N. Ebuchi, and I.S.F. Jones. Wave dependence of sea-surface wind stress. *J. Phys. Oceanogr.* 20, 705–721, 1990.

Toba, Y. and H. Kawamura. Wind-wave coupled downward-bursting boundary layer (DBBL) beneath the sea surface. *J. of Oceanography*, 52, 409–419, 1996.

Toggweiler, J.R. The ocean's overturning circulation. *Physics Today* November, 45–50, 1994.

Townsend, A.A. *The Structure of Turbulent Shear Flow*. Cambridge Univ. Press, 1956.

Turner, J.S. *Buoyancy Effect in Fluids*. Cambridge Univ. Press, 1973.

Ursell, F. Wave generation by wind. In: *Surveys in Mechanics*, Ed. G.K. Batchelor, pp. 216–249, Cambridge Univ. Press. 1956.

Valenzuela, G.R. The growth of gravity-capillary waves in a coupled shear flow. *J. Fluid Mech.* 76, 229–250, 1976.

von Ficker, H. Die Passat-Inversion. *Veröff. Meteorol. Inst. Univ. Berlin* 1, Heft 4, 1936.

Warren, B.A. Insensitivity of subtropical mode water characteristics to meteorological fluctuations. *Deep Sea Res.* 19, 1–19, 1972.

Wesely, M.L. Response to "Isotopic versus micrometeorologic ocean CO_2 fluxes: a serious conflict" by Broecker et al. *J. Geophys. Res.* 91, 10,533–10,535, 1986.

Wheless, G.H. and G.T. Csanady. Instability waves on the air-sea interface. *J. Fluid Mech.* 248, 363–381, 1993.

Williams, A.G., J.M. Hacker, and H. Kraus. Transport processes in the tropical warm pool boundary layer. Part II: Vertical structure and variability. *J. Atmos. Sci.* 54, 2060–2082, 1997.

Willis, G.E. and J.W. Deardorff. A laboratory model of the unstable planetary boundary layer. *J. Atmos. Sci.* 31, 1297–1307, 1974.

Willoughby, H.E., F.D. Marks, and R.J. Feinberg. Stationary and moving convective bands in hurricanes. *J. Atmos. Sci.* 41, 3189–3211, 1984.

Worthington, L.V. The 18° water in the Sargasso Sea. *Deep Sea Res.* 5, 297–305, 1959.

Worthington, L.V. Anticyclogenesis in the oceans as a result of outbreaks of continental polar air. A Tribute to Georg Wüst on his 80th Birthday, pp. 169–178, Ed. A.L. Gordon, Gordon and Breach, New York, 1972.

Worthington, L.V. Intensification of the Gulf Stream after the winter of 1976–77, *Nature* 270, 415–417, 1977.

Wuest, W. Beitrag zur Enstehung von Wasserwellen durch Wind. *Z. Angew. Math. Mech.* 29, 239–252, 1949.

Wüst, G. *The Stratosphere of the Atlantic Ocean*. Scientific results of the German Atlantic expedition of the research vessel "Meteor" 1925–27, Berlin and Leipzig, 1935.

Wunsch, C. An estimate of the upwelling rate in the equatorial Atlantic based on the distribution of bomb radiocarbon and quasi-geostrophic dynamics. *J. Geophys. Res.* 89, 7971–7978, 1984.

Wunsch, C. The work done by the wind on the oceanic general circulation. *J. Phys. Oceanogr.* 28, 2332–2340, 1998.

Wyrtki, K. An estimate of equatorial upwelling in the Pacific. *J. Phys. Oceanogr.* 11, 1205–1214, 1981.

Yaglom, A.M. Comments on wind and temperature flux-profile relationships. *Bound. Layer Meteorol.* 11, 89–102, 1977.

Yelland, M. and P.K. Taylor. Wind stress measurements from the open ocean. *J. Phys. Oceanogr.* 26, 541–558, 1996.

Yelland, M.J., B.I. Moat, P.K. Taylor, R.W. Pascal, J. Hutchings, and V.C. Cornell. Wind stress measurements from the open ocean corrected for airflow distortion by the ship. *J. Phys. Oceanogr.* 28, 1511–1526, 1998.

Yoshikawa, I., H. Kawamura, K. Okuda, and Y. Toba. Turbulent structure in water under laboratory wind waves. *J. Oceanogr. Soc. Japan* 44, 143–156, 1988.

Yutter, S.E. and R.A. Houze. The natural variability of precipitating clouds over the western Pacific warm pool. *Quart. J. Roy. Meteorol. Soc.* 124, 53–99, 1998.

Index

CAPE (Convective Available Potential Energy)
  and deep convection, 217
  definition of, 158
  generation of, in North Atlantic, 219–21
capillary waves, 56, *see also* celerity, cat's paws
Carnot cycle
  representation of, 155
  thermodynamic efficiency of, 153
Carson's law, 107
  and cloud top cooling, 108–10
  and oceanic convection, 110
cat's paws, 59, 62, *see also* capillary waves
celerity, 54
  of classical inviscid wave, 55, *see also* capillary waves, gravity waves
  role in wave breaking, 82
characteristic wave, 12, 65, 68, *see also* windsea
  properties of, 69–70, *see also* wave age
  tail of, 71–2
Charnock's law, 14
  corrected for buoyancy, 20
  evidence for, 22–5
  limitations of, 25–8
chimney(s), 99–100, 147
  and plumes, 179
  and pycnostads, 180
  in the North Atlantic, 204–5
Circumpolar Current
  heat transport by, 200
Clausius-Clapeyron equation, 149
clouds
  cumulonimbus, 146
  isolated, 101–2
  liquid content of, 122
  processes in, 101–3, *see also* cloud top cooling
  stratiform, 100–1, 120
  trade cumuli, 116
cloud top cooling, 108–10, 124
conservation laws
  with open boundaries, 133–6
Coriolis parameter
  definition of, 196

Dalton's law, 149
deep convection, 99
  CAPE produced by, 217–9
  observations of, 181–2
  oceanic, properties of, 178–81
  preconditioning phase of, 182
dewpoint depression, 139
discrete propagation, 164, 165–6
dissipation method, 9, 22

Ekman transport, 129, 132, 173, 190
  and heat gain, 196–7
  definition of, 195–6
  in the Southern Ocean, 207–8
entrainment
  and shear flow, 110–3
  breaker related, 113–5
  definition of, 98, 105
  laws of, 104, 107–10, *see also* Carson's law, Turner-Lofquist law
entrainment velocity
  caveats, 114–5
  definition of, 105

entropy production, 3, 5, 34–5, 41–3, 104
  in hurricanes, 172–5
equatorial Atlantic
  heat export from, 213–5
equatorial upwelling, 129–32
  role in Atlantic circulation, 212–3, *see also* EUC
equivalent potential temperature, 115
EUC (Equatorial UnderCurrent), 130, 132, 211
  and overturning circulation, 212–3, *see also* equatorial upwelling

fetch, 12, 66, *see also* windsea
  and wave age, 69
friction velocity, 9, 10, 23, 24
  water-side value, 47

gas constant
  of dry air, 149
  of water vapor, 149
gas transfer
  mechanisms of, 87–90
  surface divergence, 88–9
  surface renewal, 87
gravity waves, 56, 61, *see also* celerity
Great Ocean Conveyor, 210

heat and vapor transfer
  mechanisms of, 90–2
  surface divergence, 91
Henry's law, 44
hot tower(s)
  ascent of air in, 158–60
  clusters of, 160–4
  drying out process in, 150–2
  drying rate in, 163–4
  origin of term, 146–7
  updrafts and downdrafts in, 159–64
hurricanes
  eyewalls in, 1–2, 167–8
  mechanical energy gain in, 178
  MSLP (Minimum Sea Level Pressure) in, 171–2
  MSW (Maximum Sustained Wind) in, 171–2
  rainbands in, 169
  structure of, 1–2, 167–9
  thermodynamic cycle of, 175–8
  updrafts and downdrafts in, 170

instability waves, 51, *see also* Orr-Sommerfeld equation
  on the air-water interface, 52–3
  properties of, 56–9
irradiance, 99
  absorption of, 125
ITCZ (InterTropical Convergence Zone), 1, 98
  annual march of, 215–6
  hot towers in, 116–7, 120
  role in overturning circulation, 216

Keulegan number, 25, 40, 91

laboratory wind waves, 77–81
  roller on breaking wave, 81, *see also* roller, wave breaking
  shear stress distribution, 78
latent heat, values of, 149
LCL (Lifting Condensation Level), 98, 101
  calculation of, 150

mixed layer budgets
  atmospheric, 136
  combined, 137–9
  in various locations, 140–5
  oceanic, 136–7
mixing ratio
  definition of, 149
  relationship to specific humidity, 149
monsoon, source of water vapor for, 194, 197

NADW (North Atlantic Deep Water), 208
  upwelling of, 209
nonequilibrium thermodynamics, 2, *see also* Onsager's theorem, entropy production, laws of general form
  and buoyancy flux, 41–3
  and Charnock's law, 15–6

Obukhov length, 18, 35–6
oceanic heat gain
  distribution of, 189–94
  mechanisms of, 195–7
oceanic heat transport, 197–203
  calculation from heat gain, 199–200
  direct estimation of, 198–9
  from satellite data, 202
  global distribution of, 202–3
  in the Atlantic, 202
oceanic mixed layer
  compensation depth in, 125
  diurnal thermocline in, 125–7
  structure of, 125–8
Onsager's theorem, 3, 5–6, 15–6, 34–5, *see also* nonequilibrium thermodynamics
Orr-Sommerfeld equation, 51, 54–5, *see also* instability waves
overturning circulation
  mechanism of, 216–23
  of the ocean, 208
  pathways of, 209–11
  role of tropical Atlantic in, 211–3

Pathways of air-sea momentum transfer, 92–6
  long-wave route, 93–5
  mechanism of, 216–23
  shear flow route, 95–6
  shortwave route, 95
peak downward buoyancy flux, *see also* entrainment, laws of
  at cloud top, 108–9
  in the atmospheric mixed layer, 106
  in the oceanic mixed layer, 110–4

perfect gas law, 149
potential temperature, 43, 97
pseudoadiabatic process
  energy balance of, 148
  representation of, 158–9
pycnostads, 179
  energy of, 180–1
  in eighteen degree water, 183, 219–21
  in Labrador Sea Water, 184–5

radiant energy flux, 102–3
  at cloud top, 108–9, 124
  in the oceanic mixed layer, 125
roller
  and capillary waves, 86
  interaction with wave, 80
  properties of, 79–81

saturation pressure, 148
  variation with temperature, 149, *see also* Clausius-Clapeyron equation
Schmidt number, 46
sea and swell, 60, *see also* windsea
significant wave height
  definition of, 62
  similarity law for, 65
specific humidity, 149, *see also* mixing ratio
squall lines, 164–7
  flow pattern in, 165–6
  interaction with troposphere, 166–7
Stokes drift, 70
  and growth laws, 70
stratocumulus, 120
  structure of mixed layer under, 120–4
subsidence, 99–100
surface slope
  mean square of, 76
  spectrum of, 75

thermocline waters
  and the Circumpolar Current, 209
  definition of, 206, *see also* WWS
thermodynamic cycle
  efficiency of, 157
  of hurricanes, 175–8
  of the atmospheric overturning circulation, 152–7
  of the oceanic overturning circulation, 216–7
  representation of, 155
THV (Turbulent Humidity Variance) equation, 104
TKE (Turbulent Kinetic Energy) equation, 16
  with buoyancy, 18, 42
  with changing wind direction, 101
Toba's law, 68
Trade Inversion, 97, 116
  mixed layer structure under, 117–20
TTV (Turbulent Temperature Variance) equation, 34
turbulence, 7
  and buoyancy, 17–20, *see also* Obukhov length
  convective, similarity theory of, 106–7
  TKE (Turbulent Kinetic Energy), 7
Turner-Lofquist law, 111–2
  and the oceanic mixed layer, 112–3

upwelling, 99–100
  coastal, 132
  equatorial, 129–32

virtual potential temperature, 97
  and CAPE, 158–9

wave age, 12, 26, *see also* windsea
  and wave properties, 69–70
  effect of, on Charnock's law 27–8
wave breaking
  and fishingline effect, 85–6, *see also* roller and capillary waves
  criterion of, 82
  dynamics of, 83–6
windsea, 10
  definition of, 61
  momentum transport by, 70
  properties of, 11–2
WWS (WarmWaterSphere)
  definition of, 187
  heat and mass loss from, 204